Cognition and Motor Processes

Edited by

W. Prinz and A. F. Sanders

With Contributions by

D. A. Allport P. Bach-y-Rita R. B. Freeman, Jr. D. Gopher
L. Hay H. Heuer B. G. Hughes H. H. Kornhuber
D. M. MacKay G. W. McConkie D. J. K. Mewhort
O. Neumann R. W. Pew H. L. Pick W. Prinz
D. A. Rosenbaum E. Saltzman A. F. Sanders E. Scheerer
W. L. Shebilske G. E. Stelmach C. Trevarthen P. Wolff
D. Zola

With 34 Figures

Springer-Verlag
Berlin Heidelberg New York Tokyo 1984

Professor Dr. Wolfgang Prinz
Abteilung für Experimentelle und Angewandte Psychologie, Universität Bielefeld, Postfach 8640, 4800 Bielefeld 1, Federal Republic of Germany

Professor Dr. Andries F. Sanders
Institute for Perception TNO, Postbus 23, Kampweg 5, 3769 DE Soesterberg, The Netherlands

ISBN-13: 978-3-642-69384-7 e-ISBN-13: 978-3-642-69382-3
DOI: 10.1007/978-3-642-69382-3

Library of Congress Cataloging in Publication Data. Main entry under title: Cognition and motor processes. Bibliography: p. Includes index. 1. Cognition. 2. Motor ability. I. Prinz, Wolfgang. II. Sanders, A. F. (Andries Frans), 1933-. BF311.C548 1984 153 83-20173

2126/3130-543210

Preface

The issue of the relationship between cognition and motor processes can be – and has been – raised at different levels of analysis. At the neurophysiological level it refers to the interactions between afferent and efferent information. At the neurological and neuropsychological level it relates to the mutual dependencies between the sensory and the motor part of the brain, or, more precisely, between sensory and motor functions of various parts of the brain. In psychology, the issue under debate concerns, at a molecular level, the relationship between perception and movement or, at a more molar level, the relations between cognition and action.

For the title of this book we deliberately decided to combine two terms that are taken from two of these levels in order to emphasize both the multilevel structure of the issues involved and the multidisciplinary nature of the following contributions. Although the term "cognition" has been tremendously misused in recent years (at least in psychology), it is still the only term available to serve as a convenient collective name for all sorts of cognitive processes and functions.

The book is divided into five sections each of which is devoted to a particular sub-issue of the field. The first two sections are on cognitive contributions to motor control and action planning (I) and on motor contributions to perception and cognition (II). Whereas the first of these two topics has always been included in the traditional scope of the disciplines concerned, the second sub-issue, which traces the less familiar path from efferent to afferent information and thereby reverses the usual direction of inquiry, has often been neglected. This is probably due to the bottom-up oriented approach, which has come to dominate our understanding of the structure of organisms as a consequence of the general prevalence of empiristic and, particularly, sensualistic doctrines in scientific thinking.

The subsequent sections deal with structures and processes that mediate between cognitive and motor processes in general (III) and with the relative merits of various types of models of attention and skilled performance in particular (IV). In the last section interactions between cognitive and motor functions are considered in development (V). As it is sometimes assumed that sensory and motor functions are more closely related to and less dissociated from each other in earlier than in later stages of development, the developmental perspective should play a crucial role for building theories about the links between cognition and motor processes.

In each section the reader will find different types of chapters (e.g., with respect to level of analysis, disciplinary background, tutorial vs. single-issue coverage of the field, or empirical vs. theoretical emphasis). Tutorial papers with broad coverage will precede single-issue papers that discuss in more detail one or the other theoretical or empirical problem.

The chapters of this book have developed from papers presented at a conference at the Center for Interdisciplinary Research (ZiF) at the University of Bielefeld, which was held in July 1982 under the same title. We are indebted to the authors of the individual chapters and to all who helped edit the book by reviewing the contributions.

Bielefeld and Soesterberg
January 1984

W. Prinz
A.F. Sanders

Contents

I Motor Control and Action Planning

II Motor Contributions to Perception and Cognition

III Mediating Structures and Operations Between Cognition and Action

IV Attention, Cognition, and Skilled Performance

V Interactions Between Cognition and Action in Development

Contributors

D.A. Allport, Department of Experimental Psychology, University of Oxford, South Parks Road, Oxford OX1 3UD, Great Britain

P. Bach-y-Rita, Department of Rehabilitation Medicine, School of Medicine, University of Wisconsin, 600 Highland Avenue, Madison, WI 53792, USA

R.B. Freeman, Jr., Fachbereich Psychologie und Soziologie, Universität Konstanz, Postfach 733, 7750 Konstanz, Federal Republic of Germany

D. Gopher, Technion – Israel Institute of Technology, Faculty of Industrial Engineering and Management, Technion City, Haifa 32000, Israel

L. Hay, CNRS – Institut de Neurophysiologie et Psychophysiologie, B.P. 71, 31, Chemin Joseph Aiguier, 13277 Marseille, Cédex 9, France

H. Heuer, Abteilung für Experimentelle und Angewandte Psychologie, Universität Bielefeld, Postfach 8640, 4800 Bielefeld 1, Federal Republic of Germany

B.G. Hughes, Motor Behavior Laboratory, The School of Education, Department of Physical Education and Dance, University of Wisconsin-Madison, 2000 Observatory Drive, Madison, WI 53706, USA

H.H. Kornhuber, Abteilung Neurologie, Universität Ulm, Steinhövelstraße 9, 7900 Ulm, Federal Republic of Germany

D.M. MacKay, Department of Communication and Neuroscience, University of Keele, Keele, Staffordshire ST5 5BG, Great Britain

G.W. McConkie, Center for the Study of Reading, University of Illinois, 51 Gerty Drive, Champaign, IL 61820, USA

D.J.K. Mewhort, Department of Psychology, Queen's University, Kingston, Ontario K7L 3N6, Canada

O. Neumann, Abteilung für Experimentelle und Angewandte Psychologie, Universität Bielefeld, Postfach 8640, 4800 Bielefeld 1, Federal Republic of Germany

R.W. Pew, Bolt Benarek and Newman Inc., 10 Moulton Street, Cambridge, MA 02238, USA

H.L. Pick, Institute of Child Development, University of Minnesota, 51 East River Road, Minneapolis, MN 55455, USA

W. Prinz, Abteilung für Experimentelle und Angewandte Psychologie, Universität Bielefeld, Postfach 8640, 4800 Bielefeld 1, Federal Republic of Germany

D.A. Rosenbaum, School of Communications and Cognitive Science, Hampshire College, Amherst, MA 01002, USA

E. Saltzman, School of Communications and Cognitive Science, Hampshire College, Amherst, MA 01002, USA

A.F. Sanders, Institute for Perception TNO, Postbus 23, Kampweg 5, 3769 DE Soesterberg, The Netherlands

E. Scheerer, Psychologisches Institut, Universität Oldenburg, Birkenweg 3, 2900 Oldenburg, Federal Republic of Germany

W.L. Shebilske, Department of Psychology, University of Virginia, Charlottesville, VA 22901, USA

G.E. Stelmach, Motor Behavior Laboratory, The School of Education, Department of Physical Education and Dance, University of Wisconsin-Madison, 2000 Observatory Drive, Madison, WI 53706, USA

C. Trevarthen, Department of Psychology, University of Edinburgh, 7 George Square, Edinburgh EH8 9TA, Great Britain

P. Wolff, Zentrum für Interdisziplinäre Forschung, Universität Bielefeld, Postfach 8640, 4800 Bielefeld 1, Federal Republic of Germany

D. Zola, Center for the Study of Reading, University of Illinois, 51 Gerty Drive, Champaign, IL 61820, USA

I Motor Control and Action Planning

1 Cognitivism and Future Theories of Action: Some Basic Issues

GEORGE E. STELMACH and BARRY G. HUGHES

Contents

1 Introduction

The information processing revolution that came to psychology in the wake of behaviorism has by now established two cornerstones upon which motor behavior theories as well as methodology are based. One is the *conceptual distinction* between the nature of various cognitive processes (attention, memory, intentions, reasoning, etc.) and overt behavior itself, and the other is the notion that motor acts are *centrally represented,* that like other memorial representations they are stored, modified, and retrieved through the distinct cognitive processing. To the extent that theorists in motor behavior have conceptualized action from an information processing approach, they have found it useful, if not necessary, to incorporate these two views.

The information processing paradigm has in recent years undergone a series of changes: e.g., the idea that there exists a concatenation of distinct processing stages has come under increasing fire; conceptions of memory storage, retrieval, and depth of processing have all changed; definitions of what comprises an elementary mental process are rare. With respect to motor behavior, the nature of programming, especially, has become an issue of increasing theoretical consequence. Further, a relationship has rapidly developed between cognitive psychologists and computer scientists, particularly in the area known as computer simulation, where cognitive psychology and artificial intelligence (AI) overlap.

Cognition and Motor Processes
Ed. by W. Prinz and A.F. Sanders

In recognition of the widespread acceptance that cognitive processes and overt behavior are intricately related, this paper is a brief exploration of the implications of attempting to fit common cognitive science, or cognitivist approaches, to issues of motor behavior. Cognitivism, according to Haugeland (1978), is the scientific position that "intelligent behavior can (only) be explained by appeal to internal cognitive processes," and although it has many individual orientations, the approach seems to have one common goal: the production of a rigorous *formal* analysis of cognition (Shaw & McIntyre, 1974), an algorithmic and/or heuristic basis for cognition set apart from any *immediate* commitment to a description of the physical structures through which such processes would occur (Block & Fodor, 1972; Pylyshyn, 1978). As Haugeland (1982) has described it, "the guiding light of cognitive science is that, at a suitable level of abstraction, a theory of natural intelligence should have the same basic form as the theories that explain sophisticated computer systems" (p. 2). In this way, artificial intelligence and psychology may be regarded as having basically the same pure form.

We are proposing to introduce briefly some of the more intriguing issues that accompany such an approach to cognition, especially in light of what they may hold for future theories of motor behavior. The paper is divided into two major parts. First, we seek to establish a theoretical foothold for cognitivism by introducing its *modus operandi* as it is seen in cognitive science, and by presenting some of the current theoretical orientations in motor behavior, specifically those that might eventually be couched in cognitivist terms. The second section presents some basic developments associated with cognitivism and its present and possible utility in developing psychological theories about human motor performance. Can a cognitivist information processing orientation, in other words, provide viable and coherent psychological theories of action? We provide no clear answers, nor closed issues. In fact, we hope that this paper will be read not so much as definitive arguments for or against cognitivism, but rather as a series of questions and issues that may well face us more directly in the next few years. Rather than treat the issues in great depth or with any large measure of conclusiveness, we hope to show their relevance to contemporary and future theories of motor behavior. Certainly, they seem to us to be among cognitive psychology's constitutive issues.

2 Action, Cognition, and Computation

In this section, we would like to introduce some of the basic ideas that have supported and encouraged the development of cognitivism (as well as cognitive science, AI, and computer simulation). The cognitivist approach differs from the more traditional information processing approach in a number of respects. While it still conceives of humans as information processing systems, its formulation of the style of processing is very different. It is, for one thing, a more theoretical enterprise because it often works from general conceptions of a problem domain (e.g., playing chess, tying knots) *down to* specific details about how such a problem might be solved by one or more systems (human or otherwise), and because it is very mindful of philosophical and metatheoretical dilemmas (mind interacting with matter; explaining the relevance of meanings without recourse to homunculi, the empirical testability of mentalistic explanations; Haugeland, 1982).

Another characteristic difference between cognitivism and more traditional information processing accounts is that the former often uses computer terminology more frequently and with a more literal flavor. According to Block (1980), the cognitivist enterprise aims to decompose mental processes into combinations of processes, each of which has "no internal goings on" (p. 426) that can be further decomposed into other operations. It permits "abstract information handling principles" (Pylyshyn, 1978, p. 94) irrespective of functional entities to be carefully established. To paraphrase Haugeland (1982) cognitivists hold that, at a suitably abstract level, *all* intelligent beings are formal automatic systems. People and computers are "merely different manifestations of the same underlying phenomenon" (p. 31). "Formal" essentially means systems that are self-contained, perfectly definite, and finitely checkable, and thereby digital; "automatic" means to device that, by itself, manipulates the symbols of a formal system according to rules. It is a relative of the famous universal Turing machine that can, in principle, imitate *any* formal system. Thus, any automatic system can in principle be formally imitated by a computer if it has the appropriate program. In this way, cognitivists can view AI as a particularly pure and abstract form of psychology, without even having to argue about computers "not really being like humans," or creating homunculi to do our mind's perceiving, acting, or controlling for us.

A third difference that may be apparent from the above discussion is the growing view that the traditional information processing decomposition of intelligence along such lines as sensation, perception, memory, attention, and action may be of more theoretical harm than use. As Pylyshyn (1982) has remarked, "this is not to imply that we cannot study problems in these areas, but rather to suggest that categorizing them in such terms may not reveal (indeed, may cover up) the way in which they individually and collectively contribute to producing intelligent behavior" (p. 70). The cognitivist approach carves up the information processing system in a different way, something along the lines sketched above, and as we note below, not very differently from the style of control theorists and systems analysts.

Clearly, however, the cognitivist view is not universally accepted in all respects, even by those *in* AI or cognitive science: there are many pilosophical, theoretical, and methodological grounds for dispute. What motivates more our discussion here is the potential methodologies and conceptions that such an approach introduces (either by design or less wittingly) to the established information processing approach to motor behavior. It is apparent that the traditional division of the information processing system (see above) conceptually distinguished motor processes from perceptual and cognitive ones (they traditionally follow stimulus encoding and cognition, for example). Yet the cognitivist approach advocates a radical reorganization of theoretical approaches to cognition. For example, Pylyshyn (1981) has argued that it might be more appropriate to distinguish, for *all* traditional cognitive functions, between two distinct kinds of processes: those that are knowledge-based (which "explains certain aspects of behavior as a rational consequence of the organism's processing specific beliefs and goals", p. 267) and those that require a "mechanical-cause explanation" (p. 268), or one based on physiological or physical properties. In a sense, motor behavior research has long been guided by both cognitive (psychological) and physiological orientations; but rarely has this coupled relationship been examined. Cognitivism threatens to press this relationship beyond anything that has been previously attempted.

It seems appropriate, therefore, to ponder the potential influence of the cognitivist approach to action: if cognition is to be considered a formal symbol operation and programmatic, can action (particularly its preparation or programming) be likewise considered organized and represented in the same formal notation? A major issue confronting students of motor behavior working within the information processing approach is whether it is, in principle, possible and/or desirable to have a theory of action that makes similar appeals to representational, formal, cognitive processes such as are made in other areas of cognition (language acquisition, mental imagery, knowledge representation, etc.). To our knowledge such an issue has yet to be raised publicly, although there is evidence that it is being actively attempted. In all likelihood, there are probably few researchers in motor behavior who, if asked, would admit to being cognitivists or even overtly allied with cognitivism. But the common use of "motor programs" and similar terminology by so many suggests that if the affiliation is not explicitly obvious, research trends imply that an indirect alliance is recognizable.

Consider Fodor's (1968, p. 627) familiar description of one particular skill:

> Here is the way we tie our shoes:
>
> There is a little man who lives in one's head. The little man keeps a library. When one acts upon the intention to tie one's shoes, the little man fetches down a volume entitled *Tying One's Shoes.* The volume says such things as: "Take the left free end of the shoelace in the left hand. Cross the left free end of the shoelace over the right free end of the shoelace. . . , etc."
>
> When the little man reads the instruction 'take the left free end of the shoelace in the left hand', he pushes a button on a control panel. The button is marked 'take the left free end of a shoelace in the left hand'. When depressed, it activates a series of wheels, cogs, levers and hydraulic mechanisms, one's left hand comes to seize the appropriate end of the shoelace. Similarly, *mutatis mutandis,* for the rest of the instruction.
>
> The instructions end with the word 'end'. When the little man reads the word 'end', he returns the book of instructions to his library.
>
> That is the way we tie our shoes.

We offer this passage as a vehicle for introducing here several basic concepts (caricatured, we realize) about the ways in which action may be reviewed from a cognitivist perspective. First, it reveals the important role that central representation is held to have in psychological theories of perception, cognition, and action. Second, Fodor's account limits the distinction made between an information representative (the book that Fodor's little man retrieves from the library) and an execution representation (what happens when the man pushes the button) that lies at the heart of the motor program notion, for several years a dominant account of motor behavior. Third, it suggests the cognitivist view that behavior, and not just overt behavior, can be formalized (into rules) and that, further, this formalization could in principle be used to reproduce that behavior on some other device. Fourth, it serves to raise the issue of whether motor programs should be regarded as metaphors or as literal entities. All these points are discussed in more detail in Section 3.

Perhaps among the reasons the motor program notion has not been explicitly defined are the strong ties that research in this field has had with the neurosciences and the perceived strength of the bottom-up research strategy. The historical association with the neurosciences is probably not surprising and is certainly well justified. In fact, the original conception of a motor program was mainly the result of researchers of motor skills extracting data and conceptions almost simultaneously from neurophysiology and information processing psychology; for example, by evidence that "sensory factors play a minor part in regulating the intensity and duration and nervous discharge" (Lashley, 1951, p. 122). Such data led directly to (1) the idea of central representations of actions (Keele, 1968); (2) the problems of serial order and memory (i.e., how such preplanned sequences of behavior could be generated in an appropriate sequence); (3) the constituent components of this program (to the contents of the volumes titled *tying one's shoes, riding a bicycle, writing one's signature,* or practically any other volitional motor act).

It seems more likely, however, that the almost universal use of "programs", "schema", "subroutines", and similar terms relates to what has been termed the "wait-and-see strategy" or the "principle of least commitment" (see Pylyshyn, 1979). Although we are not aware of any explicitly outlined acceptance of this approach, we feel that many researchers in the field are nonetheless applying it by avoiding, in the longer term interests of theoretical orientation, any definitive statements regarding the nature of programs that subsequent data may prove erroneous. This tendency has some ramifications that we will discuss below; essentially, it introduces the notion of the use of metaphors, and it relates to the power of appealing to programs in explaining behavior.

The present state of representational theorizing on motor behavior is not explicitly aligned with the cognitivist approach although there appears to be at least a trend toward such an orientation (e.g., Shaffer, 1981; Monsell & Sternberg, Note 1). It remains to be seen, of course, whether cognitivist trends will develop similarly and to the same extent that they have in cognitive psychology. The history of psychology seems to indicate that there is a rather consistent, if delayed, adoption of theoretical conceptions by theorists of action – perhaps the trend will continue. Much of the following section is speculation upon the issues that may well accompany cognitivist accounts into the realm of action.

3 Issues in the Cognitivist Debate

While there are many issues that have surfaced with the emergence of cognitivism, there are several that appear to be sufficiently general as to exert a significant influence on theorizing and research in motor behavior as well as on cognitive psychology in general. We introduce and highlight a change in research strategy that has become popular in a variety of disciplines in which psychologists hold methodological or theoretical interest. We also discuss the many theoretical gains and controversial issues that are the result of research approaches, both top-down and bottom-up. Among the issues that arise from a top-down strategy is that of the use of metaphors and the explanatory nature of appeals to programs, schema, and other apparent metaphors.

3.1 Cognitivism and a Top-Down Research Strategy: An Introduction

It is a decade since Newell wrote a paper entitled "You can't play 20 questions with nature and win" (1973) in which he railed against a tendency of psychological research toward an obsession with the noncumulative gathering of data and primarily focused on the resolution of binary oppositions (e.g., peripheral vs. central control, single vs. dual memory, serial vs. parallel processing, analog vs. digital processing). This approach, commonly called bottom-up, typically results in either (a) an inference as to a mechanism that accounts for certain experimental results or (b) an operational definition of the parameters useful in predicting further experimental data. Examples from the motor behavior literature might include the "memory drum" of Henry and Rogers (1960) (an example of a mechanism created to account for robust movement latency data) and Rosenbaum's (1980) arm-direction-extent parameters for the prediction of other data in precuing paradigms. Regardless of whether (a) or (b) results, the bottom-up strategy leaves temporarily unaddressed the issues of how these mechanisms might fit into a general conception of the theoretical domain, be it cognition or action (Pylyshyn, 1979).

There is clearly a very good reason for theories to be based on well-established experimental data: to do so is to locate scientific rigor in observable replicable evidence. This was one of the major tenets of behaviorism: to have objective observable data upon which psychological theories would firmly rest. Unfortunately, behaviorist approaches ignored an equally important locus of rigor, that which rests in consistent and coherent theoretical ideas, and some other mechanisms proposed to be part of the motor system may be subject to similar criticism. The cognitivist approach places importance on both the experimental and theoretical loci of rigor; it is currently advocating the potential benefit, *at certain stages* in the development of a particular theory, of what has been termed a top-down research strategy. The latter approach often attempts to establish a theoretical coherence *prior to* the complete collection of relevant experimental data. It is viewed by many as a valuable, flexible combination of both theoretical and empirical science (Allport, 1980; Dennett, 1978; Marr, 1977; Pylyshyn, 1979).

There are two scientific domains in which the top-down approach seems to be particularly commonplace, and both are relevant in contemporary motor behavior: it is very typical in much of the work on systems design and control theory and is also a key aspect of functional analysis of systems.

3.2 Systems Design and Control Theory

With respect to control theory and the design of systems, very different questions are asked with the top-down approach than are asked in traditional psychological approaches. The motivation for asking these questions often has little if anything to do with the development of a particular theory of cognition or motor control. The top-down approach in control theory and in the design of systems (e.g., robotic devices for vision or motor skill) is essentially an engineering search for a solution to a particular problem domain (e.g., Stark, 1968); it changes the question from being directed uniquely at any one particular system (e.g., the information processing system, the perceptual system, or the motor system) to one directed at any system with similar features. As Dennett (1978) put it, the question is changed from "How does the nervous system accomplish X, Y, or Z?" to the (relatively) easier question "How can *any* system possibly accomplish X?" An earlier paper (Stelmach and Diggles, 1982) was devoted to exploring the particular characteristics of contemporary motor control theory, and it contained reference to the rapidly increasing impact of such a top-down strategy. Many neuroscientists and brain theorists, long accustomed to studying smaller "theoretical atoms" (muscle spindles, particular afferent pathways, etc.), and perhaps becoming frustrated at the slow construction of larger theories from them, have begun to turn to control theory and bioengineering for a catalog of *possible* solutions. For example, Arbib (1975, p. 239) has argued the case

> for a top-down brain theory, which will use, modify and contribute to AI techniques so as better to delimit the functions of brain regions in a form suitable for bottom-up . . . analysis.

In a similar vein, Greene (1972) has remarked on the value of a strategy that takes "a handful of biological experiments and engineering techniques" and "distills a few essential guiding principles of operation of the nervous system" (p. 305). Top-down approaches and computer modelling in particular, while far from addressing how the nervous system works, may guide our subsequent experimental work, suggesting the important empirical questions to ask. This can be contrasted with the traditional bottom-up approach upon which Greene (1972, p. 306) also comments:

> Neurophysiologists have learned much about the operation of individual neurons and their interactions in small groups, but little about the meaning of the circuits they compose . . . . Knowledge of functional units and what they accomplish, is almost nil, despite the continual flow of physiological findings.

Thus, in systems design (as practiced in AI and robotics) the primary objective is to capture some method or methods that solve a particular problem, but that also delay certain other questions. Greene (1982) again remarks on such work in studying robotic arm design:

> One technique for learning how to ask significant questions about how brains control movement is to distill out some style of operation we think is being exploited, and try and identify the problems faced by artificial systems that exploit this style . . . Perpaps this practice will help us discern relevant units of information and problems of control when brains do it their way, and perhaps it will not (p. 262).

The strategy may provide a means for solution of one problem but it makes no concurrent attempt to compare this method with human mechanisms. Such crucial questions are temporarily postponed (Pylyshyn, 1979). The approach "tolerates incompleteness in specifiability of detailed correspondence with experiments in favor of accounting for the possibility of certain performance skills" (p. 429). Thus, the top-down approach in systems design seeks to "discover informational pathways and possibilities which then must be tested to learn whether they are (in fact) used by nervous systems" (Greene, 1982, p. 276).

Have advances already been made in theoretical motor behavior that might be attributed to a top-down research strategy? The answer to this question seems to be a definite "Yes". Even though the strategy has not been adopted for any great period of time, there are several notions about motor control especially that can be attributed to the sorts of discoveries and questions a systems design, top-down enterprise advocates.

Perhaps the most obvious example relates to the distribution of control. Just one of several aspects of the organization of complex systems (along with others such as the allocation of organization and of constraints; see Pylyshyn, 1982), the control issue has been especially central in motor behavior: the idea that motor control could be gained and improved through more and more feedback loops has gradually, and for a variety of reasons, given way to more centralist-mentalist ideas about knowledge and movement representation, to complicated hierarchies with their modularity and quasi-independence. Further work has improved even hierarchical designs: now heterarchies and coalitions with their emphasis on multiple control loci (Greene, 1972; Shaw & Turvey, 1981; Turvey, Shaw, & Mace, 1978) have come to represent more sophisticated understanding of complex systems in general, not just human motor systems (see Stelmach & Diggles, 1982, for a review). In cognitivist approaches too, the value of distributed control (rather than executive control, with its specter of homunculi) has long been recognized. However, although the practical utilization of this concept has been valuable in many AI projects, it has yet to find a securely established place in much of the work on motor programming, preparation, and representation issues. In 1981 Arbib introduced one such distributed control computational approach that also exemplifies the top-down style.

The top-down strategy, as utilized in systems design and control theory, is not without potential theoretical dangers. Dennett (1978) has offered the following as being among the chief dangers: (a) devising a system whose component "capacities are *miraculous* given the constraints one is accepting" (p. 112), e.g., assigning homunculi-like qualities to a part of the system; (b) confusing one particular solution with the constraints offered by general human mechanisms (see the above quotation from Greene, 1982, p. 262); (c) having a solution for one restricted subsystem that cannot be blended into a more general system or offering no plausible way for the particular system to be expanded to include a wider "natural" task domain. Dennett (1978), however, also notes that one strength of the top-down systems design approach is that "when one of its products succumbs to any of these dangers, this can usually be quite conclusively demonstrated" (p. 113).

3.3 System Reduction and Functional Analysis

It is with respect to the functional analysis of systems that the top-down approach utilized by cognitivists has most relevance to familiar *psychological* theories of motor behavior. Control theory has generally been concerned with the *execution* aspects of action (Stelmach & Diggles, 1982) rather than with cognitive influences on behavior: preparation, attention, goal-setting, intention, etc. (Greene, 1982, offers several reasons why it is desirable to have a different design approach to the preparation for action and why traditional control theory does not deal with it). This approach is also closely connected with the conventional information processing approach in motor behavior. Generally, functional analysis attempts to account for the behavior of a system (e.g., the motor system) as a "black box", and it does so "in terms of the interaction of smaller black boxes, which carry out subfunctions, and each of which is also explained in a similar way" (Pylyshyn, 1979, p. 427). In other words, the functional analytic strategy proceeds by separating a system (or its disposition to act in a particular way) into a series of less problematic dispositions, with particular focus on not only the functional components but also their interactive abilities (Cummins, 1977); indeed this is a prerequisite of the approach (Haugeland, 1978). Ideally, this reduction of each functional capacity is continued until, at the lowest level in the system hierarchy, a different – and to avoid an infinite regress, probably physiological – style of explanation is required.

Thus, we can contrast the top-down and bottom-up approaches by recognizing that the former involves taking the whole system and separating its functional (rather than structural) components, whereas the latter, in its extreme, contrives mechanisms in a more ad hoc manner to account for certain experimental data. A perennial problem for the bottom-up strategy, therefore, is that the categories (established on the basis of empirical data) may have to be changed or additional ones added to account for further empirical discoveries. For example, one might add a certain memory buffer mechanism to one's model to account for certain linear characteristics of RT data only perhaps to "require" an additional buffer mechanism to account for certain other, subsequently collected data. In addition to risking a proliferation of task domain-specific mechanisms that can be viewed as accounting for the data, such a bottom-up approach remains mute on the issues of *how* these mechanisms operate in a more general cognitive-motor system (they risk being overly narrow in domain utility) and how they might be instantiated in a physical system, human or otherwise.

While these problems have done much to dissuade a number of theorists from advocating or pursuing a traditional bottom-up approach (Allport, 1980; Newell, 1973; Simon, 1980), the top-down approach in functional analysis is not without several controversial aspects. One of the more obvious problems of a top-down approach lies in the nature of the systematic reduction that characterizes the cognitivist approach and in the nature of the "style of operation" the control theorist thinks is being exploited in the performance of some (motor) skill (Greene, 1982). As mentioned above, the top-down strategy seeks to avoid creating ad hoc devices to account for certain data (because a mechanism that accounts for a domain of data A may bear no resemblance to one that accounts for domains A *and* B; see Pylyshyn, 1982). In the cognitivist top-down approach one seeks to specify the crucial modes of interaction among a few functional components; preferably as few as possible. This is similar (with a different explanatory style) to the nature of Newtonian mechanics where a few parameters and equational laws cover everything relevant to the motions of moving bodies, leaving as irrelevant other characteristics such as color and texture (Haugeland, 1978). Finding the appropriate components and specifying which of their interactions fully encapsulate the behavior of interest is crucial to the whole long-term strength of the enterprise. However, as Haugeland (1978) has argued, "not every contiguous collection of components (e.g., neurons or neuronal networks) constitutes a single component of what's relevant into a few specific abilities and interactions – usually different in kind from those of any of the smaller components" (p. 221). Thus, there exists a certain circularity in the approach: the reconstruction of a system on the basis of (a) systematic division and (b) putative functional components fitting neatly back into the original division both requires and supports the other.

Perhaps a more pressing issue in discussions of information processing concepts and motor behavior is the nature of common terms such as "programs": are they to be regarded as metaphors or can they offer a literal description of motor- or cognitive-related functional organization? Consider, for example, this passage from Greene (1982) on how systems design and limb control might be envisioned:

> Before the overt behavior starts, the controller must decide how to partition variables into principal variables of the active functional subsystem, variables that make fine adjustments, and variables that must be held fixed, for these three types . . . require different subroutines. After the controller had declared the type of each variable, subroutines are called to implement this distinction. Finally, the subsystems and subroutines are activated that realize the intended action. These proceed relatively autonomously (p. 281).

Clearly, this description leaves totally unspecified, for now, how this might actually work in a human central nervous system, and for good reason: it is not immediately expected to. But one might also wonder whether the use of terms such as "decide", "declare", "subroutines" are similar examples of the theorist trying to "provide a way to introduce terminology for features of the world whose existence seems probable, but many of those fundamental properties have yet to be discovered" (Boyd, 1979, p. 364): i.e., whether they are being used metaphorically or whether they are to be regarded as literal functions and/or mechanisms. While we cannot speak for Greene (1982) or anybody else who uses this technique, for us a crucial issue is still how such a distinction can be made, whether the imprecision is to be removed at a later date, and what kinds of explanation such terminology provides. These are very substantial issues that we cannot hope to cover fully here. However, several authors have discussed the issues in depth (see Cummins, 1977; Haugeland, 1978; Pylyshyn, 1979) and the following interesting points are taken from their work.

The above passage from Greene (1982) introduces several terms that, while not giving a detailed explanation, nonetheless serve to constrain the manifestations of such a system (e.g., it is clearly not a traditional executive-dominated hierarchy; its subroutines are "relatively autonomous", and their parameters have differing qualities). Although there is no precise operational definition of the subroutines, "partitioning", the controller's "declarations", etc., the system is nevertheless tightly constrained for it must be able to account for certain motor actions (Greene offers the examples of certain gaits as well as goal-directed limb movements). In this case, the requirement that the system be coherent imposes tight constraints (Pylyshyn, 1979) on the potential properties of such "subroutines", "controllers" and similar exemplars.

Differentiation between terms that are offered as metaphors and those that are supposed to literally capture systematic qualities has also been made on the basis of their openness to explication; that is, whether their use "leaves one feeling that a phenomenon has been 'explained', even though only a superficial level of functional reduction or process explanation has been offered" (Pylyshyn, 1979, p. 431). It might be argued that expressions such as motor "programs" and "subroutines" deserve an assessment in terms of this explanatory openness: are they adequate as explanations of behavior? Could a program be a coherent theory of behavior? They certainly *have* been postulated as theories of all sorts of behaviors and appeal to programs in explaining behavior as a common strategy in the functional reduction of systems (Cummins, 1977; Haugeland, 1978). In some instances, "following a program" can be used to explain how someone makes a cake from a recipe or dismantles or rebuilds an automobile engine: the task as a whole can be decomposed cleanly into a series of simpler and antecedently comprehensible steps. From this *external* "program", almost everybody could, just by following the instructions in the correct order, perform some quite sophisticated task. But action is clearly not "programmed" this way: we have no externally represented set of instructions to follow. So it has seemed reasonable to *internalize* this set of instructions, "to suppose that the required program is 'there' all right, but 'internally represented' in the brain, and 'tacitly known' and followed" (Cummins, 1977). This imperative use of the term appears to be inappropriate, but what of a descriptive sense? Could we not, as Cummins suggests, approach the idea with an analytic strategy, by merely claiming that an actor's execution of a program can be analyzed into the execution of certain processes (Pylyshyn, 1980), capacities (Chomsky, 1980), or subroutines (Greene, 1982) he or she has? Certainly; in fact, if we have read the motor programming literature with a sufficiently accommodating eye, this would seem to be the purpose of much of the work focused on "units of motor programming" (Monsell & Sternberg, Note 1), "movement feature specification" (Goodman & Kelso, 1980; Rosenbaum, 1980), and "levels of representation" (Shaffer, 1981). A tolerance of metaphoric expression depends greatly on how the metaphor is subsequently used. The intuitive appeal of a certain metaphor may encourage a lot of experimental research, as is certainly the case with "programs". However, the very intuitiveness that it provides may also have a negative consequence: it may inhibit further functional reduction (Pylyshyn, 1979) because it appears to explain how a particular element of a performance takes place. For example, the idea of "motor programming" is mainly based on the very robust data showing, among other things, that reaction time increases when "task complexity" increases. But can we also say that it *explains* why this is so? Its intuitive appeal coupled with its ability to predict certain other performances only makes it *appear* to explain the data. It can be argued that it does not explain how programming occurs, even though we know quite a bit about how some other systems (a computer, for instance) are programmed. Similarly, the claim that since the program is internally represented in memory, execution can be explained by saying that the actor performs a certain act because his or her memory "directs" it, does the same thing with the metaphor "directs": it fails to detail *how* it directs, thereby forestalling further reduction.

However, philosophers of science and mind have rarely offered any explicit criteria by which the metaphoric might be distinguished from the literal. Pylyshyn (1979), for one, has suggested that those seeking such a distinction might consider rather pragmatic points such as general acceptance of terms (particularly their consistency in usage and referential specificity). Such considerations in terms of motor programming would, we think, produce no clear answer as to whether "program" is one or the other: it is a term that has almost universal usage in the field, yet it also is a term that has an extremely loose definition. In one sense, it conforms to the "principle of procrastination" alluded to earlier (i.e., it avoids restricting theorists to an operational definition that may later have to be restated) yet it also seems to bear all the hallmarks of a scientific metaphor that appears to explain certain behavioral characteristics but really does not. We can only hope that researchers who favor usage of this and similar terminology are also aware that differing consequences attend metaphoric and literal uses: if the former, then in the interests of clarity exactly what is metaphoric about the term should be spelled out and how the term may eventually be divested of its metaphoric content should be part of a longer term research agenda. If one already takes "program" or "schema", for example, to be literal terms, such imprecision as is tolerated with metaphors cannot be allowed; in fact, by taking this stance one is committed to explaining, in functional terms, *how* it is literal (Pylyshyn, 1979). No one suggests that this is an easy task, and it is one that few have ventured to attempt. Yet among cognitivists there seems to be an optimism that top-down strategies represent major improvement over bottom-up techniques, at least those that are extreme in their commitment to laboratory methodologies and mini-models (Abelson, Note 2).

4 Conclusions

In a recent paper, Johnson-Laird (1981) noted that two particular questions ought to haunt cognitive psychologists: (1) Can there ever be a scientific understanding of the mind? (2) Are there deep invariances or uniformities in the ways in which the mind works? This paper has attempted to raise some of the issues surrounding the latter question: are there certain regularities between cognitive processes and the organization and planning of action, regularities that effectively transcend many underlying cognitive and motor processes? Cognitivism, and cognitive science as a discipline, appears to indicate that the answer to the second question is "Yes".

We introduced the cognitivist concept because it seems to be a means, but not necessarily the only one, by which the search for such principles can be established. This means, on a broad sense, is termed a top-down strategy, not because its ultimate goals (theory construction and explanation) differ from a bottom-up approach, but because it goes about attaining them by first seeking some general principles and workable solutions, and then subjecting such solutions to an empirical test. Cognitivism, according to its advocates, is already a detailed and worked-out account of how all cognitive processing in principle can be understood as formal symbol computation and representation. This paper briefly discussed the extent to which such a view might be applied to issues in motor behavior, including (particularly) those issues such as intention, goal-setting, attention, etc., that seem especially to beg for psychological (rather than just physiological) theorizing.

We noted that the motor programming literature has its roots in the information processing paradigm that has long dominated cognitive psychology. We also believe that, of the various accounts of motor behavior, programming is the most likely to absorb the techniques of the cognitivist paradigm. It already has a substantial pretheoretical orientation. Additionally, the rapid recent developments in brain theory (of which Arbib, 1975, 1981, is the most eloquent advocate) seem explicitly focused on forming a convergence between the long separated theoretical axes of the neurosciences and computer simulation of cognition, and especially in areas such as motor behavior, speech control, and linguistics. Much of the literature on motor behavior seems to reveal a significant pretheoretical orientation to cognitivist ideas.

A major portion of this paper is devoted to several key issues that arise via a cognitivist approach. These issues were raised more in the hope that researchers in the areas of cognition and motor processing would recognize and contemplate their significance for ongoing theoretical and experimental endeavors than in the hope of resolving them. There are many additional issues – some of them entirely philosophical –

that we have not addressed. We did not discuss the major philosophical strategies for arguing that cognitivism is misconceived (see Haugeland, 1982), nor those that argue that cognitivism merely reflects technological advances (rather than theoretical improvements) or that its actual successes are too narrow to justify theoretical upheavals.

These issues, however, are important. Many working scientists have tended to dismiss them as *purely* philosophical arguments that are largely irrelevant to ordinary science, or that such science effectively skirts them. Such light dismissal of issues crucial to the advancement of cognitive psychology and motor behavior are, we would argue, dangerous: philosophical problems should be faced up to if, as we believe to be the case, the goals include understanding and explaining cognitive behavior.

In any event, there have been major advances in one field using techniques common in cognitive science. Brain theory, barely a recognizably distinct scientific field a decade ago (Arbib, 1975), has made major advances with the aid of com puter experts, electrical engineers, and control theorists. The motor control area already bears the influence of such work (e.g., Greene, 1982) and we suspect that the psychology of motor behavior will, in the near future, face the reality of theoretical solutions to perennial problems coming not so often from behavioral research as from some combination of research in brain theory and cognitive science. We should welcome the opportunity to share the labor but realize that along with the benefits we will also acquire the intriguing puzzles that cognitive psychology currently faces.

Reference Notes

1. Monsell, S., & Sternberg, S. *Speech programming: a critical review, a new experimental approach, and a model of the timing of rapid utterances.* Part 1. Unpublished manuscript. (Available from Bell Laboratories, Murray Hill, N.J.)
2. Abelson, R.P. *Constraint, construal, and cognitive science.* Paper presented at the Third Annual Conference of the Cognitive Science Society, Berkeley, CA, August 1981

References

Allport, D.A. Patterns and actions: cognitive mechanisms are content-specific. In G.L. Claxton (Ed.), *New directions in cognitive psychology.* London: Routledge, 1980.

Arbib, M.A. Perceptual structures and distributed control. In V.B. Brooks (Ed.), *Section on neurophysiology, Handbook of physiology,* (Vol. 3). Bethesda, MD: American Psychological Society, 1981.

Arbib, M.A. Artificial intelligence and brain theory: Unities and diversities. *Annals of Biomedical Engineering,* 1975, *3,* 238–274.

Block, N. What intuitions about homunculi don't show. *Behavioral and Brain Sciences,* 1980, *3,* 425–426.

Block, N., & Fodor, J. What psychological states are not. *Philosophical Review,* 1972, *81,* 159–181.

Boyd, R. Metaphor and theory change: what is 'metaphor' a metaphor for? In A. Ortony (Ed.), *Metaphor and tought*. Cambridge: Cambridge University Press, 1979.

Chomsky, N. *Rules and representations.* New York: Columbia University Press, 1980.

Cummins, R. Progams in the explanation of behavior. *Philosophy of Science,* 1977, *44,* 269–287.

Dennett, D. *Brainstorms: Philosophical essays on mind and psychology.* Montgomery, VT: Bradford Brooks, 1978.

Fodor, J.A. *Psychological explanations.* New York: Random House, 1968.

Goodman, D., & Kelso, J.A.S. Are movements prepared in parts? Not under compatible (naturalized) conditions. *Journal of Experimental Psychology: General,* 1980, *109,* 574–495.

Greene, P.H. Problems of organization of motor systems. In R. Rosen and F. Snell (eds.), *Progress in theoretical biology.* New York: Academic Press, 1972.

Greene, P.H. Why is it easy to control your arms? *Journal of Motor Behavior,* 1982, *14,* 260–286.

Haugeland, J. The nature and plausibility of cognitivism. *Behavioral and Brain Science,* 1978, *2,* 215–260.

Haugeland, J. Semantic engines: an introduction to mind design. In J. Haugeland (Ed.), *Mind design.* Cambridge, MA: MIT Press, 1982.

Henry, F.M., & Rogers, D.E. Increases response latency for complicated movements and a "memory drum" theory of neuro-motor reaction. *Research Quarterly,* 1960, *31,* 448–458.

Johnson-Laird, P.N. Cognition, computers, and mental models. *Cognition,* 1981, *10,* 139–143.

Keele, S.W. Movement control in skilled motor performance. *Psychological Bulletin,* 1968, *70,* 378–403.

Lashley, K.S. The problem of serial order in behavior. In L.A. Jeffress (Ed.), *Cerebral mechanisms in behavior.* New York: John Wiley & Sons, 1951.

Marr, D. Artificial intelligence – a personal view. *Artificial Intelligence,* 1977, *9,* 37–48.

Newell, A. You can't play 20 questions with nature and win: projective comments on the papers of this symposium. In W.G. Chase (Ed.), *Visual information processing.* New York: Academic, 1973.

Pylyshyn, Z.W. Computational models and empirical constraints. *Behavioral and Brain Sciences,* 1978, *1,* 93–127.

Pylyshyn, Z.W. Metaphorical imprecision and the 'top-down' research strategy. In A. Ortony (Ed.), *Metaphor and thought.* Cambridge: Cambridge University Press, 1979.

Pylyshyn, Z.W. Computation and cognition: issues in the foundations of cognitive science. *Behavioral and Brain Sciences,* 1980, *3,* 111–169.

Pylyshyn, Z.W. Psychological explanations and knowledge-dependent processes. *Cognition,* 1981, *10,* 267–274.

Pylyshyn, Z.W. Complexity and the study of artificial and human intelligence. In J. Haugeland (Ed.), *Mind design.* Cambridge, MA: MIT Press, 1982.

Rosenbaum, D.A. Human movement initiation: specification of arm, direction and extent. *Journal of Experimental Psychology: General,* 1980, *109,* 444–474.

Shaffer, L.H. Performances of Chopin, Bach and Bartok: studies in motor programming. *Cognitive Psychology,* 1981, *13,* 326–376.

Shaw, R., & McIntyre, M. Algoristic foundations to cognitive psychology. In W.B. Weimer and D.S. Palermo (Eds.), *Cognition and the symbolic processes.* Hillsdale, N.J.: Erlbaum, 1974.

Shaw, R., & Turvey, M.T. Coalitions as models for ecosystems: a realist perspective on perceptual organization. In M. Kubovy and J.R. Pomerantz (Eds.), *Perceptual organization.* Hillsdale, N.J.: Erlbaum, 1981.

Simon, H.A. How to win at twenty questions with nature. In R.A. Cole (Ed.), *Perception and production of fluent speech.* Hillsdale, N.J.: Erlbaum, 1980.

Stark, L. *Neurological control systems: Studies in bioengineering.* New York: Plenum Press, 1968.

Stelmach, G.E., & Diggles, V.A. Control theories in motor behavior. *Acta Psychologica,* 1982, *50,* 83–105.

Turvey, M.T., Shaw, R.E., & Mace, W. Issues in the theory of action: Degrees of freedom, co-ordinative structures and coalitions. In J. Requin (Ed.), *Attention and performance VII.* Hillsdale, N.J.: Erlbaum, 1978.

Summary. One of the tenets of a current theoretical account of human information processing, that termed cognitivism, is that cognitive psychology and the study of formal symbol systems may, at an appropriate level of abstraction, be concerned with very similar, and perhaps the same, problem. Quite aside from the philosophical issues that such a claim raises, cognitivism introduces several issues that may have a far larger impact on motor control theorizing than is currently realized. While not all theoretical ideas from cognitivism will take root in motor behavior (for the obvious reasons of subject matter, for instance), there are several that may well be very influential. This paper is an introduction of and speculation upon these potentially influential issues.

In introducing the cognitivist approach we note that several major differences exist between this and more traditional information processing approaches from which substantial cognitive-motor theory derives, especially with respect to their approaches to theory construction and their respective divisions of the information processing system. Cognitivists often advocate a top-down research strategy, one that seeks a rigorous theoretical base upon which to undertake subsequent experimental work, and thus contrasting with the more familiar bottom-up, data-oriented approach to theory construction and explanation.

We discuss some of the repercussions of this approach as well as demonstrate the extent to which such an approach has already influenced (the less psychological) aspects of human control theory and systems design without always having been motivated to do so. We go on to discuss how functional analysis and a top-down approach blend in a cognitivist approach and what some of the difficulties are. Consideration is also given to the prevalent use of metaphoric terms in cognitive psychology and motor behavior; how they can be beneficial to theory development as well as how they are limited as explanations of certain psychological phenomena.

2 A Distributed Processing View of Human Motor Control

RICHARD W. PEW

Contents

1 Introduction

This chapter has three purposes: (1) to provide a selective review and overview of current issues in motor control; (2) to argue for use of the metaphor of a distributed processing system in connection with motor control; (3) to encourage the reader to consider what does *not* need to be explained by the integration of cognition and motor theory because it has an adequate representation already either in terms of known neuromotor control or biomechanics. In the course of developing these themes, the chapter will introduce a new perspective on the concept of a motor program.

In computer science there is a growing interest in the technology of distributed information processing systems. A distributed system is one in which each independent processing entity can act in response to a query from another system. Each system has its own data base and is capable of initiating programs utilizing this data base given only the request for information from an alterantive source. Such systems operate by passing messages among processors. When one system needs output from another, it broadcasts a request. Then, the unit with the capability to respond carries out the requested task and in turn transmits the result back to the relevant units.

The analogy with the motor system suggests that there are a variety of centers of activity – cortical, subcortical, spinal and in the muscles themselves – which

Cognition and Motor Processes
Ed. by W. Prinz and A.F. Sanders

contribute to the production of coordinated movement. Rather than thinking of them, as we have in the past, as a unitary integrated system, I am suggesting that the centers of activity may be thought of as relatively autonomous structures, which are coordinated by passing specifications to each other up and down the hierarchy.

I will begin by describing the series of processing activities in terms of both the outflow and feedback that may be undertaken by the motor system. The elaboration of these activities might be thought of as the genesis of movement and feedback.

2 The Genesis of Movement

2.1 Choice of Action

Perhaps the highest level of motor control, namely, the level at which the motor activity to be performed is selected, is the least well understood. Most of our representations gloss over this process by one of two dodges. Either we deny the admissibility of such questions because they cannot be addressed without retreat into mentalism (Gallistel, 1980) or else we draw a box and label it "Executive Routines", letting that substitute for an explanation. The issue remains that choice of an action among all the actions that might be initiated at a particular point in time is of fundamental interest and we have only general motivational theory, such as the law of effect, or perhaps the Hullian concept that goal-directed action leads to drive reduction, on which to draw.

Gallistel (1980) argues, probably appropriately, that drive or motivation for action can be interpreted in a hierarchical framework: Motivation serves to select a goal and action plan. The action plan in turn leads to the instantiation of a specific schema, and so forth. Thus, goal selection is just another level in the same hierarchy of action that is described in the remainder of this paper.

The next step after goal selection is the formulation of an action plan. A popular view of this next stage has been to express it in terms of schemata. A schema is an expression for what is laid down with practice and retrieved from memory at the time of action (Pew, 1974; Schmidt, 1975; Arbib, 1980; Rumelhart & Norman, 1982).

Although the concept of schemata can be traced at least to Bartlett (1932, 1958; see Schmidt, 1975, for a more complete history), an excellent defining experiment is that of Posner and Keele (1968). They created several dot patterns that were defined as standards. Then for each standard they created a set of distortions by moving any particular dot a radial distance from its origin according to a given probability distribution. The dots were moved systematically to control the uncertainty associated with each set of distortions (see Posner, Goldsmith, & Welton, 1967). The subjects were trained to classify the collection of distortions into sets, based on the standards from which they were derived. After achieving reliable classification, the subjects were shown two types of previously unseen patterns: (a) new random distortions and (b) the standards themselves. The subjects were able to classify the standards that they had never seen, reliably better than the new distortions. In addition, in a separate experiment, after 1 week delay, there was an increase in classification errors for the distortions that had been previously learned, but no such increase for the classification of the standards. What appears to have been retained with practice was a generalization from the aggregated exposure to distortions that corresponded to the standard.

The generalization of this idea to movement schemata suggests that each execution of a purposeful movement pattern represents an instance. Whereas the visual patterns were distributed in space, movement patterns require a representation for spatial layout as a function of time. Over repeated executions of a given movement, there are variations in the exact patterns produced because of changing conditions under which the pattern is initiated, as well as random perturbations. What is built up in memory with practice is a standard or generalization derived from these aggregated instances.

In the case of movements, then, a schema is a generalized representation of a class of movements that exists in memory. Presumably, with further practice, the schema becomes refined both in terms of the level of generality that it has accrued and in terms of the constants and variables that characterize it. There are at least three, and probably many more researchable questions about this kind of representation. In what terms is it encoded and what is unique and invariant about the representation? At the time of execution, what are the variable parameters that must be specified for each instance that is actually carried out? How large (or small) a unit of behavior can be represented by a schema?

The best evidence to date supports Bernstein's hypothesis, formulated as early as 1935, that it is the topology and not the metrics of a movement that are represented in memory. [See Bernstein (1967) for a translation.] By the topology of a movement, he meant its spatio-temporal geometry. By the metrics, he meant its measures of amplitude and time. Viviani and Terzuolo (1980) have provided compelling evidence that, in typing and handwriting, regardless of the speed (for typing) or speed and amplitude (for handwriting), there is remarkable invariance in the spatio-temporal organization within an individual on repeated productions. It is as if size and speed of production are free parameters that may be set at "run time," that is, at the time when the movements are executed. As for the code of the representation, Bernstein (1967) argued that the topology of the movement is specified in terms of its spatial path or targeted end point. While this has not been proven, it is certainly clear that the central representation is *not* in terms of commands to individual muscles. A person can generate substantially similar movement patterns under a variety of environmental conditions that require different specific muscle action. One does not put chalk to the blackboard in an identical orientation of the body, arm, and hand each time, yet the results are substantially identical.

Conceding that the memory representation of a movement cannot consist of specific muscle commands, it is also necessary to ask the contrasting question: How abstract can the representation be? If we think of memory as an ordered complex, with abstract ideas at one end and sensory primitives at the other, then a particular motor schema may be simply a sample point within such a complex. The schema may vary in its degree of abstraction from a specific movement primitive, such as extension or pronation, to a very general goal. I am told that it takes a player who knows both squash and racquetball at championship level about three playing hours to become fully adapted to the conditions of one or the other game. This anecdote is suggestive of some of the parameters that are *not* set at "run time" and therefore of the level of specificity associated with this kind of motor schema.

2.2 Action Programs and Neuromotor Execution

The stored representation of action plans must evolve into action programs for execution. In Pew (1974), this process was described as selecting a specific schema instance, particularizing it for the specific environmental conditions, and executing it. This process is now seen as a much more distributed activity among lower centers (Turvey, 1977). In order to understand this idea it is necessary to digress.

2.2.1 Simple Movement Execution

Consider, first, the level of simple movement execution. It appears possible to formulate a parameter set that defines desired final position, tension, and orientation and to have the muscles produce the required temporal sequence of muscle tensions and lengths to carry it out.

In an ingenious combination of modelling and experimentation with monkeys, Bizzi, Dev, Morasso, and Polit (1978) showed that the effect of introducing an unexpected mechanical load during a visually elicited head movement could be explained largely by examining the kinematic response to the tension and length specifications defined prior to the introduction of the disturbance. Central corrections did not contribute, and reflexive load compensation played but a minor role. This formulation has been confirmed by Polit and Bizzi (1979) and Kelso and Holt (1980).

The importance of this finding for the distributed processing argument is twofold: (1) simple movements can be interpreted in terms of a parameter set passed to the muscle system for execution at that level, and (2) the resultant movement pattern can be explained largely in terms of the kinematic response of the limb to a relatively simple motor command input sequence. Thus, is not necessary to postulate sophisticated feedback or central processing activity to account for what superficially appears to be a rather complicated response sequence.

2.2.2 Pattern Generators at the Spinal Level

Consider next a more complex pattern such as locomotion. There is clear evidence in both cats and in lower animals that "pattern" generators, controlling elemental locomotion processes, exist at the spinal level (Grillner, 1975). In the most relevant experiment, a cat with its spinal cord severed just below the brain stem, when placed on a treadmill, will execute primitive but coordinated walking movements and will even modify the stepping process in response to encountering an obstacle with the paw (Grillner, 1973). One can think of these patterns generators as receiving the message to walk from higher centers and executing walking movements under local control. The pattern generator for walking in turn generates and sends commands to the muscles to carry out a particular movement, perhaps by specifying the length-tension parameters for each simple positioning response.

2.2.3 Relation to Distributed Processing

The purpose for having elaborated these examples is only to illustrate how a distributed neural architecture with "lower centers" could be controlled with simpler but less complete commands from each higher level in the motor system. This formulation bears some relation to that expressed by Arbib (see Arbib, 1980, for examples). Presumably still higher structures involving vision and other sensors send "messages" that set the parameters of these structures in turn. I think of the process as the inverse of the visual pattern recognition process, discovered by Hubel and Wiesel (1962), in which visual primitives at the optic nerve become the building blocks for more and more integrated forms as the signals are traced to higher levels in the cortex. In the case of motor schema the integrated goals or action plans are systematically decomposed as they are relayed to lower levels in the nervous system.

The examples in this section can be exploited to highlight one further point: Feedback can be expected to operate at multiple levels. In the experiment of Bizzi et al. (1978), reflexive load compensation was shown to provide fine tuning to the simple movements. In the experiment of Grillner (1973), the obstacle hitting the cat's paw was found to signal the spinal pattern generator to call for a disruption in the normal gait. If either of these mechanisms failed to provide the needed corrections, then higher level centers would be brought into play.

Taken together, these concepts illustrate rather well what Turvey (1977) meant when he spoke of the control of coordinative structures, a term he admits to having borrowed from Easton (1972) but to using with greater latitude. A coordinative structure is a hierarchical control system from one perspective; however, it is also a two-way system because afferent sensory information may be utilized to modify the operation at each level, with longer latency, as higher level structures are involved.

3 The Genesis of Feedback

Thus far, the discussion has focused on the outflow of the motor system, that is, on the distributed processes involved in movement production. It is equally important to consider the genesis of feedback. Feedback is the information derived from the environment through neural signals that originate at the distal nodes, beginning with the peripheral sensors. These signals relay data about the success of the movement execution at each level to the succession of higher levels. It is an essential component of finely coordinated movement and a key contributor to virtually all skill learning.

Feedback is a multi-level activity just as is outflow. At the lowest level, the level of the individual muscles, the muscle spindles report whether each muscle "pulled its own weight". If the local load compensation response is not sufficient, then signals are relayed to higher levels to initiate new corrective responses.

At the next highest level the system must evaluate whether, within each component action, the desired terminal location and orientation was achieved. If not, remedial action components must be initiated. Thus, the obstacle hitting the cat's paw signalled the pattern generator to call for a disruption of the normal gait to support the higher level goal of walking.

Feedback at the next level contributes to the build-up and refinement of the schemata out of which the specific movement sequence was drawn and finally, at the highest level, the system must ascertain whether the goal of the action plan has been satisfied. Usually there are multiple criteria associated with achievement of a goal. Some of these criteria, such as whether the task was completed satisfactorily, are self evident. Others, such as the amount of energy used to accomplish the task, are not. In the latter case it is necessary that the system be provided with supporting instrumentation to make explicit whether the goal has been achieved.

These feedback tasks are accomplished through the use of a wide range of sensors including semicircular canals, otoliths, eyes, ears, mechano-receptors, Golgi tendon organs, and muscle spindles. They also involve different degrees of processing and central integration. Just as on the outflow side, information fed back within different levels of specificity may be utilized throughout the processing hierarchy. As these various sources of feedback are combined and interpreted by higher level structures, the most effective and timely evaluative information is made available at each moment in time.

While the structure I have proposed is basically hierarchical in the sense that one level of specificity follows another, I *do* want to suggest that these levels are capable of operating as autonomous processes on the basis of parameters passed from the next highest level. I also want to suggest that feedback from each level may be utilized, not only at the level from which it originates, but also at all higher levels. This type of interaction between levels takes the proposed structure out of the domain of a strict hierarchy and makes further features of the distributed processing metaphor relevant since now higher levels in the hierarchy receive input from multiple processes. I am not arguing, however, that signals are sent down to multiple processes for independent execution. I believe the outflow is highly constrained and coordinated.

4 What is a Motor Program?

With this background, the stage is now set for discussion of the concept of a motor program. As early as 1960, Henry introduced what he referred to as the "Memory Drum" theory of movement control. The idea was that the effect of practice was to "carve in" a pattern of response much like the grooves on a phonograph record. More recently computer programs have become the more popular analogy for motor programs. The program analogy has several convenient properties that make it attractive. A computer program exists as a stored representation, but it is also "run" and can produce outputs as a function of time. Typically, a computer program is designed as a collection of subroutines or subsidiary units that are organized hierarchically and are "called" in response to a higher level goal. Subroutines are frequently parameterized, that is, at the time they are to be executed they are supplied with initial conditions and with the specification of parameter settings appropriate for this particular execution. Sometimes a program is set up so that during execution it is prepared to accept a new input that may be contingent on what the program has produced thus far. All of these properties are appropriate to the concept of a motor program.

What are the implications of the perspectives presented here for the concept of a motor program? It is not that motor programs are in any way denied. Rather, the problem is the converse. It is possible for the concept of a motor program to have different manifestations at different levels in the distributed system.

The spinal control of the walking response of the cat can be explained as a local reflexive stimulus-response chain in which feedback from each small unit of behavior serves as the stimulus for the next unit. While this response may not be representative of what is achieved in intact higher mammals, it may be a suitable description of the apparent programmed behavior at this level in the motor system. In the intact animal, the autonomy of the chain is subjected to higher level supervision and control.

Similarly, analysis of the kinematic response of a biomechanical system may reveal that behavior which is sufficiently regular and repeatable to be classed as programmed may be nothing more than a product of the kinematic constraints to movement. At the level of physical response, it is clear that the biomechanical properties of the responding member play a role in explaining the exact spatio-temporal pattern produced. Thus one does not need to postulate a sophisticated central pro gram in order to explain that pattern. At this level the "motor program" has a very simple and concrete explanation.

It is in this sense that I argue that it is just as important to know what *does not* need to be explained by cognitive theories as what does need further theoretical understanding. At the spinal level, reflexive S-R chains may be the appropriate program representation. At the level of simple movements, a kinematic analysis of the biomechanical response may provide a suitable prediction. It is at the level of schemata at which further research is needed to understand the defining characteristics of a motor program. In short, we should not be searching for a unitary integrative concept of a motor program: It may have different representations at different levels in the motor system.

As a research strategy, it is especially important that the experiments be formulated in a way that activity at any level can be examined independently of the activities at other levels. The lower level responses should not contaminate the results of higher level analyses. It takes ingenuity and careful experiment design to accomplish this result. As a way to examine programming implications at a specific level in the system, Sternberg's analysis of speech and typing behavior in terms of the lengthening of preparatory reaction time (Sternberg, Monsell, Knoll, & Wright, 1978) strikes me as a cogent example. In these experiments, care was taken to use highly overlearned responses and to isolate the measurements of the retrieval of prestored "programs". Rosenbaum's (1980) experiment on programming choice of arm, direction, and extent is an example of an experiment that confounded these levels of analysis. Goodman and Kelso (1980) found a change in result with a change in the structure of the stimulus information. The fact that the impact of the programs being examined changed with a simple rearrangement of the stimulus and response conditions suggests that they were the product of interacting levels of processing.

Acknowledgments. Portions of this chapter are derived from material for a forthcoming chapter by this author to be published in a revision of Stevens, S.S., *Handbook of Experimental Psychology* (1951). I wish to thank Marilyn Jager Adams for many helpful suggestions.

References

Arbib, M.A. Interacting schemas for motor control. In G.E. Stelmach & J. Requin (Eds.), *Tutorials in motor behavior,* (pp. 71–82). Amsterdam: North-Holland, 1980.

Bartlett, F.C. *Remembering.* London: Cambridge University Press, 1932.

Bartlett, F.C. *Thinking: An experimental and social study.* London: Allen & Irwin, 1958.

Bernstein, N. *The co-ordination and regulation of movements.* Oxford: Pergamon Press, 1967.

Bizzi, E., Dev, P., Morasso, P., & Polit, A. Effect of load disturbances during centrally initiated movements. *Journal of Neurophysiology,* 1978, *41,* 542–556.

Easton, T.A. On the normal use of reflexes. *American Scientist,* 1972, *60,* 591–599.

Gallistel, C.R. *The organization of action: A new synthesis.* Hillsdale, N.J.: Erlbaum, 1980.

Goodman, D., & Kelso, J.A.S. Are movements prepared in parts? Not under compatible (naturalized) conditions. *Journal of Experimental Psychology: General,* 1980, *109,* 475–495.

Grillner, S. Locomotion in the spinal cat. In R.B. Stein, K.G. Pearson, R.S. Smith, & J.B. Redford (Eds.), *Control of posture and locomotion,* (pp. 515–535). New York: Plenum Press, 1973.

Grillner, S. Locomotion in vertebrates: Central mechanisms and reflex interaction. *Physiological Reviews,* 1975, *55,* 247–306.

Hubel, D.H., & Wiesel, T.N. Receptive fields, binocular interaction and functional architecture in the cat visual cortex. *Journal of Physiology,* 1962, *160,* 106–154.

Kelso, J.A.S., & Holt, K. Exploring a vibratory systems analysis of human movement production. *Journal of Neurophysiology*, 1980, *43*, 1183–1196.

Pew, R.W. Human perceptual-motor performance. In B.H. Kantowitz (Ed.), *Human information processing: Tutorials in performance and cognition.* New York: Erlbaum, 1974.

Polit, A., & Bizzi, E. Characteristics of motor programs underlying arm movements in monkeys. *Journal of Neurophysiology*, 1979, *42*, 183–194.

Posner, M.I., Goldsmith, R., & Walton, K.E. Perceived distance and the classification of distorted patterns. *Journal of Experimental Psychology*, 1967, *73*, 28–38.

Posner, M.I., & Keele, S.W. On the genesis of abstract ideas. *Journal of Experimental Psychology*, 1968, *77*, 353–363.

Rosenbaum, D.A. Human movement initiation: Specification of arm, direction and extent. *Journal of Experimental Psychology: General*, 1980, *109*, 444–474.

Rumelhart, D.E., & Norman, D.A. Simulating a skilled typist: A study of skilled cognitive-motor performance. *Cognitive Science*, 1982, *1*, 1–36.

Schmidt, R.A. A schema theory of discrete motor skill learning. *Psychological Review*, 1975, *82*, 225–260.

Sternberg, S., Monsell, S., Knoll, R.L., & Wright, C.E. The latency and duration of rapid movement sequences: Comparisons of speech and typewriting. In G.E. Stelmach (Ed.), *Information processing in motor control and learning*, (pp. 117–152). New York: Academic Press, 1978.

Turvey, M.T. Preliminaries to a theory of action with reference to vision. In R. Shaw & J. Bransford (Eds.) *Perceiving acting and knowing*. Hillsdale, N.J.: Erlbaum, 1977.

Viviani, P., & Terzuolo, C. Space-time invariance in learned motor skills. In G.E. Stelmach & J. Requin (Eds.), *Tutorials in motor behavior.* Amsterdam: North-Holland, 1980.

Summary. The human motor system should be viewed as a hierarchically distributed processing system. The hierarchical property has always been accepted, but here it is argued that the various levels in the hierarchy operate more autonomously than previously assumed. Higher level nodes provide parameter specifications to satisfy successively more restricted goals; peripheral sensors, in turn send feedback messages to successively higher levels of control that are utilized to evaluate the success in achieving the goal or subgoals at each autonomous level.

One of the implications of this perspective is that a motor program can no longer be considered a unitary explanatory concept. It has potentially different explanations at different levels. Thus, walking movements in the cat may be largely controlled at a spinal level on a relatively reflexive basis. No central motor program needs to be postulated for this activity.

This perspective also suggests aspects of motor performance that need not be a part of cognitive theories of motor control. Careful experimental design is required to partition the levels of analysis so that valid interpretation of results at a particular cognitive level can be achieved.

3 The Apraxias, Purposeful Motor Behavior, and Left-Hemisphere Function

ROBERT B. FREEMAN, Jr.

Contents

1 Introduction

The relationship between cognition and motor processes has been investigated perhaps no more extensively than in connection with the effects of natural and artificial lesions in the human brain. Together with studies of aphasia and agnosia, the analysis of the various clinical forms of apraxia and their cognitive and motor correlates has interested clinical neurologists at least since the beginning of this century (Liepmann, 1900). Although the neurologist and, more recently, the clinical neuropsychologist have been largely concerned with questions of diagnostic procedures and clinical classification of the apraxias, the theoretical problems associated with the identification of the anatomical substrates of the praxias and their cognitive concomitants offer the attractive possibility of arriving at a taxonomy of complex motor behavior via an analysis of the integrative mechanisms of the cortical and subcortical neural systems of the motor brain.

Until recently investigation of the apraxias has been conducted almost entirely outside the domain of theoretical and experimental psychology with the result that there has been little exchange of ideas between neurologists interested in the apraxias as a neuropsychological phenomenon and psychologists concerned with the relation-

Cognition and Motor Processes
Ed. by W. Prinz and A.F. Sanders

ship between motor processes and cognitive phenomena. For this reason it would be presumptuous to attempt an integration of the theories and methods of apraxia research with those of experimental psychology at this time. On the other hand, recent findings of neurophysiology and neuropsychology have had increasing influence on the development of general psychological theory, particularly with respect to perceptual learning and language processes. There is also good reason to suppose that the study of the pathology of motor behavior as a result of brain lesions may also come to influence the direction of thinking about normal complex motor behavior.

With this in mind, this chapter discusses the traditional classification of the motor, as opposed to constructional and other forms of apraxias, the assumed anatomical correlates of the apraxias, and the relationship of motor learning to the occurrence of apraxia as a result of brain pathology. Particular attention is paid to the role of disconnection syndromes in apraxia behavior as posited originally by Liepmann and developed further by Geschwind, Gazzaniga, Sperry, and others, and to the relationship of the apraxias to other forms of pathological behavior, in particular as they affect motor speech. Finally, the paradoxical lack of apraxic deficits in split-brain patients with no focal pathology of the left hemisphere and possible resolutions of this paradox are discussed in relation to current theories of the apraxias.

2 Definition of the Apraxias

As suggested by the use of the plural form of apraxia in the present paper and many others (e.g., Ajuriaguerra & Tissot, 1969; Hécaen, 1968), it has been the custom to classify disturbances of motor behavior consequent to brain lesions into a number of categories. The categories and the behavioral phenomena they describe have been determined on both theoretical and clinical grounds. The plethora of types of apraxia that have been suggested has led some investigators to argue that some are not apraxias at all (Poeck, 1975), or that there is only one form of apraxia, or indeed that the term should be dropped altogether. This section describes some of the issues associated with the definition of the apraxias that have persisted to the present day and that will have to be resolved before a clear picture may emerge of the mechanisms underlying disturbances of motor behavior.

There is fairly general agreement that a person is apraxic when he exhibits certain disturbances of motor behavior that cannot be attributed to muscular weakness, incoordination, or sensory loss, or to lack of attention or the incomprehen sibility of directions to act. There is also general agreement that the behavior disturbed by the apraxias is the result of protracted premorbid learning (e.g., Ajuriaguerra & Tissot, 1969; Geschwind, 1975), although opinions differ with respect to the importance of prior experience (Kimura & Archibald, 1974). Affected behaviors are generally thought to be "complex" in the sense that they are nonreflexive, are the result of intentional effort, and are functional or purposive in their effects on the physical and social environment.

Over the years, a number of different types of apraxia have been suggested, based largely on distinctions of different categories of motor behavior. Of these, constructional (occasionally called constructive) apraxia, or a disability to produce two- and three-dimensional constructions, has recently received much attention, partly because of its frequent occurrence in brain-damaged patients, but also because of the apparent involvement of right-hemisphere visuospatial function (cf. Warrington, 1969). Other forms, such as dressing apraxia, bucco-linguo-facial apraxia, and apraxia of gait have been suggested, each of which presents interesting clinico-anatomical problems. Useful reviews of the various types of apraxia have been made by Ajuriaguerra and Tissot (1969), Hécaen (1968), Kerschensteiner, Poeck, and Lehmkuhl (1975) and others. Because of some controversy concerning the legitimacy of so many different apraxic "syndromes" (Hécaen, 1968; Poeck, 1975), and for the sake of clarity of presentation, this paper will be limited to a discussion of the theoretical, experimental, and anatomical correlates of the „classical" motor apraxias.

2.1 Hugo Liepmann's Classification of the Apraxias

The revolutionary theoretical contributions of the commanding figure of Hugo Liepmann have been attested to by virtually every neuropsychologist concerned with the problem of the definition and classification of the apraxias (Geschwind, 1965, 1975; Wilson, 1908). Liepmann's investigations were conducted in a scientific climate that was greatly influenced by Finkelnburg (1870) and Liepmann's mentor, Carl Wernicke (1874). Finkelnburg had taken over the concept of aphasia and recommended its extension to include asymboly, or the loss of the ability to understand concepts on the basis of conventional signs as well as to make use of such signs (Liepmann, 1908, p. 18). Wernicke (1874; cf. Eggert, 1977) turned the Finkelnburg concept of asymboly around to mean solely the disturbance of recollection of objects as opposed to aphasia, which was a disturbance of the recognition of linguistic signs. The resulting confusion of terminology and diagnostic concepts was mitigated by Freud (1891/1953) in his classic monograph *On Aphasia: A Critical Study,* in which he suggested the now current term *agnosia* to refer to disturbances in the recognition of objects.

The final step on the way to Liepmann's motor apraxia was Meynert's (1890) distinction between sensory and motor asymboly. For Meynert, sensory asymboly was, like Wernicke's asymboly and Freud's agnosia, a disturbance in the recognition of objects, while motor asymboly characterizes a disturbance in object *use.* It is the concept of motor asymboly, due to Meynert, which represents the point of departure for the development of the concept of motor apraxia by Liepmann. The term is, indeed, included in the title of Liepmann's first publication on the case of the Imperial Counselor (*Regierungsrat*) M.T., *The symptomatology of apraxia* (*'motor asymboly'*) *based on a case of unilateral apraxia* (Liepmann, 1900). Liepmann required M.T. to perform such tasks as pointing out objects, copying simple figures, pointing out parts of his own body, putting on a hat, making a first, pointing out the necktie of the examiner, as well as more complicated tasks such as lighting a cigar, pouring water from a pitcher into a glass, putting on his trousers, or opening a lock with a key.

An unusual aspect of this case was the fact that the apraxia was much more marked in the right than in the left hand. Particularly remarkable was the fact that, when M.T. failed to perform a task correctly with the right hand, Liepmann need only say: "Do it with the left", and M.T. could generally perform the task correctly. In addition to ruling out the possibility of a severe sensory aphasia, this observation provided evidence that M.T. had fully understood the instructions, not only at the verbal level, but also with regard to their psychomotor meaning. Finally, there was a marked difference in the ability of M.T. to perform on command and spontaneously. For example, M.T. was generally unable to button or unbutton a button with his right hand on command, but if it was led to the button by the examiner, it would perform correctly. In writing, M.T. was able to generate many latters correctly, but he was unable to write "a" to dictation, or to copy it from a sample, or to write it voluntarily.

In his monograph, *On disturbances of action in the brain damaged,* Liepmann (1905a) distinguishes between motor or innervatory apraxia and ideatory or ideational (*ideatorische*) apraxia. Ideational apraxia was conceived to be a disturbance of what he called the *formula for movement* (*Bewegungsformel*), which in turn is the mechanism for converting the concept of the main goal (*Hauptzielvorstellung*) of action into concepts of intermediary goals (*Teilzielvorstellungen*). As such, ideational apraxia represented for Liepmann part of a more general disturbance of ideation, such as a disturbance of memory or attention. Motor (or innervatory) apraxia, on the other hand, was thought to be a disturbance in which motion becomes dissociated from the ideational process, such that the corticomuscular apparatus remains intact, but does not act in the service of the psychic process as a whole (Liepmann, 1905b, pp. 156–157)[1].

2.2 Conceptual Evolution of the Apraxias

As in the earliest stages of conceptual development of the Liepmann apraxias, the delineation of the various types has remained controversial to this day. In the early period, definition was based largely on assumed anatomical location of lesions re-

1 Pick (1905) introduced a further distinction between purely *motor* apraxia and *ideomotor* apraxia (Pick's term). Liepmann accepted Pick's concept of *ideomotor* apraxia, but called it ideational (*ideatorisch*) while retaining the concept of motor apraxia, which later came to be known as *ideomotor* apraxia (Kleist, 1912, p. 357).

sponsible for the different types of apraxia, whereas more recently many investigators have chosen to lay more emphasis on the operational definition of the apraxias. I shall return later to the question of neuroanatomical correlates.

In an influential monograph, Morlaas (1928) suggested a differentiation between object utilization and gesture as a basis for a classification of apraxia. The ideomotor apraxic subject reveals disturbances in gesture while the ideational apraxic subject is characterized by problems in handling actual objects. Furthermore, the disturbed gestures of the ideomotor apraxic subject need not be of symbolic form, or serve any purpose, but are defined by a spatial dyskinesia, while the problem of someone with ideational apraxia is not strictly speaking a motor disturbance at all, but rather an agnosia of object utilization (cf. Ajuriaguerra & Tissot, 1969). The ideomotor apraxic subject is quite capable of appreciating the symbolic form of the gesture that he is incapable of performing just as the ideational apraxic subject may be able to name and describe the use of the object that he is incapable of using.

More recently, ideomotor apraxia has been defined by DeRenzi, Pieczuro, and Vignolo (1968) to refer to the inability to *imitate* intransitive gestures, while Morlaas' definition of ideational apraxia, in the sense of the inability to demonstrate the use of actual objects, was retained. Heilman (1973) suggested yet another definition of ideational apraxia in terms of the inability to pantomime the use of objects on *verbal command* while retaining the ability to handle actual objects appropriately as well as to imitate gestures made by the examiner.

Still more recently, Kerschensteiner et al. (1975) returned essentially to the classic distinctions. Ideomotor apraxia is considered to be the incapacity to carry out a movement sequence, whether in response to verbal command or in imitation, whether the motion represents a transitive or intransitive act, whether it is merely a simple gesture or an actual act involving the use or pantomiming of objects. Ideational apraxia, on the other hand, is characterized by a disturbance in the sequence of multiple individual acts. These sequences are typified by functional behaviors such as preparing a cup of instant coffee, spreading butter and marmelade on a piece of bread, opening a can with a can-opener and pouring the contents into a cooking pot, and eating a soft-boiled egg from an egg-cup (Poeck & Lehmkuhl, 1980). In such tasks, it is the *sequence* in which the individual elements should be performed that is mixed up by the ideational apraxic subject, whereas the individual movements themselves are performed satisfactorily. By way of strengthening the central or planning nature of the motor process disrupted in ideational apraxia, Lehmkuhl and Poeck (1981) have demonstrated a „disturbance in the conceptual organization of actions" in patients with ideational apraxia by showing that such patients not only confuse the sequence of actions, e.g., to look up a telephone number in a telephone book, then pick up the receiver, dial the number and listen, but they are also incapable of arranging a series of *photographs* of the various stages of such a sequence in the correct order. This evidence suggests that the disturbance is essentially unrelated to the initiation of the behavioral components themselves but rather connected to the programming of the sequence per se.

Finally there are others (Geschwind, 1965; Kimura, 1977; Kimura & Archibald, 1974) who have argued that such distinctions are either not demonstrable, or are based merely on hypothetical stages in the initiation of motor behavior whose existence has yet to be proven. Statistical evidence suggesting that ideomotor and ideational apraxia may well result from the loss of highly similar functions has recently been obtained by DeRenzi et al. (1980), who showed that imitation scores (ideomotor apraxia) and demonstration-of-use test-scores (ideational apraxia) correlate 0.80 in a group of 50 patients with brain damage on the left side.

3 Anatomical Correlates of the Apraxias

The theoretical status of the Liepmann apraxias is intimately tied to the cerebral localization of the cerebral lesions producing them. In fact, Liepmann's most significant contribution may well have been his establishment of the motor apraxias as a consequence of left-hemisphere lesions and his demonstration of the importance of the corpus callosum for communicating motor programming functions from the left to the right hemisphere.

3.1 Liepmann's Studies

Particularly notable was Liepmann's (1900, 1906) success in predicting the lesions of his patient, the *Regierungsrat,* whose right-sided apraxia was attributed correctly to the pathological separation of the left "sensomotorium" (pre- and post-central gyri) from the rest of the brain through focal lesions in the left supramarginal gyrus, in the left frontal lobe, and in the corpus callosum.

Remarkable also was Liepmann's (1905b, 1908) analysis of apraxic symptoms in 89 right-handed patients at the Dalldorf Sylum in Berlin. In this relatively brief article, entitled *The left hemisphere and action,* Liepmann reported on the results of his neurological investigation of 42 left-sided and 41 right-sided hemiplegics, evaluating the behavior of the nonparetic limbs as a function of the hemisphere assumed to be affected by reason of the (contralateral) side of paresis. Liepmann was able to show that, of the 42 left hemiplegics (right-hemisphere pathology), disturbances in the performance by the right (ipsilateral) hand of any of the required acts were rare and major deficits nonexistent. Among the 41 right hemiplegics (left-hemisphere pathology), almost exactly 50% (20) of the patients presented substantial apraxia of the left (ipsilateral) hand. Although 14 of the 20 apraxic patients also had severe motor aphasia, their apraxia was as marked when required to imitate the examiner's motions as when the act was to be elicited verbally. That the apraxic patients *understood* what was to be done was indicated by comments (or gestures in the aphasics) from the patients themselves to the effect that they could imagine what they were supposed to do but couldn't do it.

Liepmann's article thus established apraxia as a disturbance distinct from aphasia, asymbolia, and agnosia, and as resulting virtually exclusively from left-hemisphere pathology. Liepmann was, in fact, able to show that the lesions were largely in the anterior parietal lobe, not only because of the associated right hemiparesis, but also due to the association with motor aphasia. That neither paresis nor motor aphasia could be considered a requisite concomitant of, and therefore immediately responsible for, the apraxia of these patients was shown by the fact that half of the right (and all the left) hemiplegics revealed little apraxia, and 6 of 20 apraxic patients were nonaphasic while 4 right-hemiplegic nonapraxic patients had severe motor aphasia.

3.2 Apraxia as a Disconnexion Syndrome: Liepmann's Cases

Taken together with the observation that motor apraxia occurs almost exclusively as a result of left-hemisphere lesions, Liepmann's analysis suggests that the programming of motor acts of *either* hand takes place in the left hemisphere. Presumably, the left-hemisphere lesions postulated by Liepmann also caused an apraxia of the *right* hand of his right-hemiplegics, but an analysis of possible right-sided apraxia was, of course, prevented by the paralysis of the right upper limb. This places the "center" for programming complex motor behavior of either hand in the left hemisphere, an analogy to the left-hemispheric dominance for the motor programming of speech to which Liepmann made frequent explicit reference ("One can call apraxia the aphasia of the extremities", Liepmann, 1905b, p. 2322).

In another paper ("On the function of the corpus callosum in action and relationships of aphasia and apraxia to intelligence"), Liepmann's (1907, 1908) analysis made explicit the disconnexionistic interpretation of left-sided apraxia. In this paper, he makes clear that apraxia of the left hand can be caused, with or without paralysis of the right hand, by interruption of the fibers of the corpus callosum leading from the motor command centers of the left hemisphere to the motor centers of the right (cf. Figure 1). If the corpus callosum alone is severed by lesion

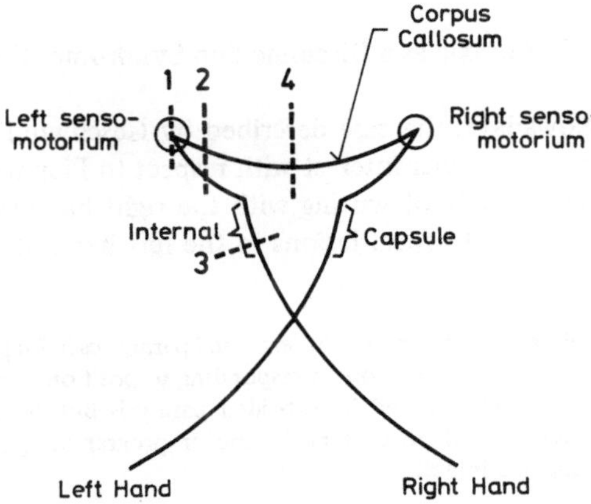

Figure 1. Anatomical relationships and assumed lesion sites related to the classical apraxias. *1* Cortical lesion; *2* Supracapsular lesion; *3* Capsular lesion; *4* Lesion of the corpus callosum. (After Liepmann, 1908, p. 59)

(Figure 1, No. 4), only the left hand will be apraxic, and neither paralytic. A lesion at No. 2, interrupting both interhemispheric fibers of the corpus callosum and the descending motor fibers to the right arm, will paralyze the right hand and leave the left hand apraxic, due to the disruption of fibers running to the motor control centers of the right hemisphere. Lesions of the internal capsule of the left hemisphere (No. 3) will paralyze the right hand without producing an apraxia of the left. Thus Liepmann makes quite explicit the significance of the interhemispheric fibers of the corpus callosum for the direction of behavior – an assumption that, as Geschwind has repeatedly pointed out (Geschwind, 1965, 1975; Geschwind & Kaplan, 1962), was lost to the English-speaking neurological community until revived by the investigations of Geschwind and Kaplan (1962) and Gazzaniga, Bogen, and Sperry (Bogen, 1969; Gazzaniga, 1964, 1966; Gazzaniga, Bogen, & Sperry, 1967; Sperry, 1961, 1966; Sperry & Gazzaniga, 1966; Sperry, Gazzaniga, & Bogen, 1969).

The validity of Liepmann's contention that the so-called sympathetic motor apraxias of the left hand, as well as aphasia, are the results of lesions of either the left hemisphere or the corpus callosum (or both) has been confirmed in a number of investigations. However, evidence will be presented in the next section that may limit somewhat the generality of Liepmann's conclusions. Three cases were described by Liepmann (1907, 1908) himself supporting his contention that the motor behavior of both upper limbs is under the control of the left hemisphere and that the action of the left limbs is programmed indirectly via the corpus callosum. One of these, the famous Ochs, (Liepmann & Maas, 1907) presented a severe left-sided apraxia and right-sided paralysis, as was observed in many other cases (Liepmann, 1905b). On postmortem examination of the brain of the patient, however, it turned out that, although sympathetic apraxia of the left hand was due to a lesion of the corpus callosum that spared the internal capsule, the right-sided paralysis was due to a *sub*-capsular lesion of the pyramidal tract at the level of the pons running to the right arm, which could have had no influence on the praxic behavior of the left hand. Liepmann also discussed two other papers describing cases similar to his own.[2]

3.3 Apraxia as a Disconnexion Syndrome: Recent Evidence

Nowadays, the case described by Geschwind and Kaplan (1962; Geschwind, 1965) is of particular interest with respect to Liepmann's theoretical position. This patient was capable of writing with the right hand (hence was neither hemiplegic nor agraphic due to focal lesions in the left hemisphere), but not with the left. The patient

2 Presumably some of the *left*-sided paralytics in Liepmann's study (1905b) had *right* hemisphere lesions at a position corresponding to position 2 and others corresponding to position 3, both of which resulted in left-sided paralysis but *neither* in *right*-sided apraxia because both the left-hemisphere centers for motor programming and their connections to the right hand remained intact.

carried out verbal commands with the right hand flawlessly. "When asked to perform actions with his left hand," however, "he made many errors and even the correct responses were performed with a considerable delay. Thus, when asked to draw a square, he drew a circle. When asked to draw a triangle (which he had done earlier with his right hand without hesitation), he paused, asked if the examiner meant the figure with three corners, and finally drew the figure. When asked to point to the examiner with his left index finger, he pointed to his own eye with his left index and middle fingers . . . When asked to show how he would brush his teeth with his left hand, he on one occasion made the motions of lathering his face and on another went through the motion of combing his hair" (Geschwind & Kaplan, 1962; 1974, p. 31). In addition, this case revealed many other symptoms of hemisphere disconnexion, such as inability to write with the left hand, to name objects, numbers, and letters held in the left hand, and to identify objects placed in his left hand by selecting corresponding ones, drawing pictures of them, or writing their names with the right hand.

Interestingly, this patient did *not* reveal a classic motor (ideomotor) apraxia of the left hand. In the words of the investigators, "When it was made clear to him that he was to show how the object placed in his hand was to be used, the subject (with eyes closed) proceeded with little hesitation to manipulate the item correctly, usually giving simultaneously an incorrect verbal account. Thus, given a hammer, he made hammering movements correctly while saying 'I would use this to comb my hair with it.' Given a key, he went through the motions of inserting it into a lock and turning it but said that he was 'erasing a blackboard with a chalk eraser.' Holding a pair of scissors correctly, he made cutting movements in the air but said that 'I'd use that to light a cigarette with.' When his eyes were open he correctly identified the objects in his left hand" (Geschwind & Kaplan, 1962; 1974, pp. 29–30).

Geschwind and Kaplan concluded that their patient did not show any signs of the classic parietal-lobe syndrome of either hemisphere, but rather acted as if his two hemispheres were functioning nearly autonomously. As long as the behavior eliciting stimulus and response elicitation were confined to a single hemisphere, the patient performed correctly. But as soon as the stimulus was directed to one hemisphere and a response required from the hand directed by the other hemisphere, performance was generally incorrect. These findings provided strong support for the assumption that this patient's principal lesion lay in the corpus callosum. At postmortem examination, following the first publication (Geschwind & Kaplan, 1962) of their findings, Geschwind (1974, p. 40) was able to verify the general predictions of the behavioral analysis.

The possible control of right hemisphere motor processes from the left hemisphere has also been taken up by Gazzaniga et al. (1967) in their study of dyspraxia following section of the corpus callosum and anterior commissure in nine epileptic patients. Earlier studies (Gazzaniga, Bogen, & Sperry, 1962, 1963, 1965; Gazzaniga & Sperry, 1967), which were concerned with visual, tactile, and language functions in the same patients, showed that the separated hemispheres functioned quite independently with respect to most mental or gnostic activities. Speech, writing, and calculation, plus visual gnosis for the right visual half-field and stereognosis of the right hand, were found to be lateralized in the dominant left hemisphere, while stereognosis of the left hand and visual gnosis of the left visual half-field were found to be restricted to the right (minor) hemisphere.

A special attribute of this investigation, in consonance with many others by the same research group, was the use of tachistoscopic methods of presentation of visual stimuli as elicitors of motor responses. Thus "instructions" could be presented separately to the left and right visual areas of the brain by projecting briefly flashed verbal and nonverbal visual stimuli in the right and left visual half-fields, respectively.

When the effects of visual input to the right visual half-field and tactile stimulation to the right hand on the responses of the right hand were analyzed, no obvious motor impairment was observed in the performance of the right hand in tasks such as pointing to a spot of light, copying simple geometric shapes, blind manipulation of objects, and – of particular relevance to classic notions of the motor apraxias – the mimicking of hand, thumb, and finger postures in response to sketches presented to the right visual half-field.

Similarly all praxic functions of the *left* hand were accomplished successfully when sensory information was projected to the right hemisphere. "Fine individuated finger movements and good differential mimicking of hand, thumb, and finger postures are readily demonstrated under these conditions . . . Objects like cigarettes, keys, glasses, eating utensils, etc., . . . are manipulated correctly and with highly refined delicate movements of fingers and thumb" (p. 609). These findings are particularly notable because they provide evidence that minor hemisphere motor control of the left hand is at least as sophisticated as the major hemisphere motor control of the right hand, and may even be somewhat superior in controlling praxic tasks of a constructional nature.

When hemisphere control of the *ipsilateral* hand was studied, left-hemisphere control of the left hand appeared to be markedly superior to right-hemisphere control of the right hand. Neither, however, approached the sophistication of contralateral control of either hand. In particular, the control of fine movements of the ipsilateral fingers involved in writing and positioning the hand and fingers was lacking, although tasks could be performed well so long as gross arm and trunk motions could be utilized to accomplish them. The fact that some ipsilateral control was demonstrated, however, does complicate the interpretation, for example, of the effect of left-hemisphere lesions, as analyzed by Liepmann, on apraxia of the left hand.

Performance was poorest under those conditions in which the right hand was to be controlled by the right hemisphere. Sometimes simple geometrical figures projected to the left visual half-field or placed in the left hand could be drawn by the right hand, but gross errors were more common. The authors suggest that the ipsilateral control of the right hand by the right hemisphere may also have been influenced by the *contralateral* control by the left hemisphere. For example, a triangle flashed to the right hemisphere could be drawn by the right hand so long as a competing figure was not flashed to the left hemisphere. If another figure was introduced in the right visual half-field, the right hand drew *it* rather than the one flashed to the left visual half-field.

3.4 Apraxia and Left-Hemisphere Function

The evidence from the investigations of Geschwind & Kaplan (1962), Gazzaniga et al. (1967), and Sperry et al. (1969) of dyspraxia in split-brain patients is clear: the right-hemisphere motor regions are fully capable of programming and initiating sophisticated left-handed motor behavior even though the right hemisphere is fully severed from neural contact with the left (dominant) motor centers. This fact would suggest that Liepmann's concept of left-hemisphere dominance, not only for language, but also for nonverbal motor programming, and not only for the right hand but also for the left, cannot be considered tenable. The issue, however, is not settled. In fact, there is overwhelming neurological evidence that *right*-hemisphere lesions do *not* (or extremely rarely) produce apraxia of the nonparetic left hand, let alone the right hand. Although Foix (1916) claimed that right parietal lesions could cause left-sided apraxia, Geschwind (1965) shows that Foix was probably in error, particularly with respect to Liepmann's cases. Furthermore, Hécaen and Gimeno Alava (1960) have analyzed 16 cases of whom three were their own, in which patients with apraxia of the left side had been reported to show clinical evidence of right hemispheric lesions. Of these, according to Geschwind (1965), 11 almost certainly also had lesions of the left hemisphere. Of the remaining five, two were known to be left-handed and only one of the others was shown at post-mortem examination to have had no involvement of the left hemisphere, suggesting that left-sided motor apraxia due to right-sided lesions is extremely rare.

Investigations of the effect of right- (and left-) hemisphere lesions on the apraxia of the *ipsilateral* hand are overwhelmingly consistent with Liepmann's results. For example, Ajuriaguerra, Hécaen, and Angelergues (1960) found not a single case of apraxia in their 151 cases with right-sided post-Rolandic lesions, although 48 cases of apraxia (23%) were found among 206 cases with left-sided lesions. In addition, 10 cases of apraxia were observed among 55 cases with bilateral lesions. DeRenzi, Pieczuro, and Vignolo (1968) studied ideomotor and ideational apraxia in 45 right- and 160 left-hemisphere patients using Morlaas' (1928) definition of ideational apraxia in terms of the demonstration of actual use of objects. They found that none of the right-hemisphere patients scored below the cut-off score of a control group on the ideomotor test, while 45 left-hemisphere patients, of whom all but four were also aphasic, did so. Almost identical results were obtained on the object-use test of ideational apraxia.

In another thorough study, DeRenzi et al. (1980) investigated imitation and demonstration-of-use apraxia in 80 right-hemisphere and 100 left-hemisphere patients, all of whom were right-handed. In this investigation the limb ipsilateral to the lesion was used in all tests. Consonant with the investigation of Kimura and Archibald (1974), half of the imitation tests were "symbolic", and therefore familiar, gestures, the other half "nonsymbolic", and hence unfamiliar gestures. Apraxia was found in 50% of the left-hemisphere patients, all but two of whom were also aphasic. In contrast with previous investigations, 20% (16) of the right-hemisphere patients were also found to be apraxic, of whom, however, only four revealed motor behavior that deviated markedly from that of the controls, and none produced errors approaching the most severely disturbed left-hemisphere aphasics. There is the possibility, however, that the tasks in this experiment were somewhat more difficult than is usually the case[3], although a separate analysis of symbolic and nonsymbolic gestures ruled out the possibility that nonfamiliar tests were particularly difficult for the right-hemisphere cases (Kimura & Archibald, 1974). Kimura's recent comparison of right-hemisphere patients with elderly healthy controls provided still more confirmation of the lack of disturbance, in this case in both simple manual postures and multiple hand movements, both of which were essentially meaningless or nonsymbolic, in right-hemisphere patients. DeRenzi et al. (1980) consider their results to be supportive of Liepmann's claim of the priority of the left hemisphere for the programming of motor behavior in view of the fact that "failure to initiate gestures is much more frequent and severe in the left limb following left-sided damage than in the right limb following right-sided damage" (p. 9).

In her investigation of left-hemisphere control of both oral and brachial (manual) movements, Kimura (1982) not only replicated previous findings that simple and complex apraxias are found almost only in left-hemisphere patients, but also studied the effect of anterior and posterior lesions *within* the left hemisphere on both manual postures, which did not involve sequences of actions, and on multiple hand movements, in which patients had to imitate sequences of hand positions. Both tasks were those used by Kimura and Archibald (1974). Both anterior (frontal) and posterior (parietal) patients were impaired in the tasks involving manual sequences, whereas only the parietal patients revealed significant deficits on the execution of hand postures. Again, manual apraxias are almost always associated with aphasia, and when not, are usually found in patients who have had a transient aphasia that has cleared. Conversely, aphasia without apraxia is found most often in patients with temporal-lobe lesions, rarely after anterior lesions, and almost never in patients with parietal lesions. From these and other findings related to the motor control of speech, Kimura concludes that the left parietal lobe plays a significant role in programming both oral and manual movements which may then be enacted via the left frontal region.

3 It may be recalled that Liepmann (1905b) himself found motor apraxia in almost exactly 50% of his right-sided paretics. However, little if any apraxia was observed in the left-sided paretics.

4 Apraxia as a Disturbance of Learned Behavior

In the clinical examination of both ideomotor and ideational apraxia, it is customary to require the patient to carry out various tasks (many of which have been described in previous sections of the present paper), which may be considered to belong to the patient's behavioral repertoire and whose accomplishment is usually the result of many years of practice. As a result, many investigators, beginning with Liepmann (1900) and including contemporary authors (e.g., Ajuriaguerra & Tissot, 1969; Geschwind, 1975) have restricted the definition of apraxia to disturbances of learned behavior. However, evidence has also been accumulating that apraxics are as affected in performing unfamiliar tasks as they are in carrying out complex movements that may be considered elements of their highly practiced behavioral repertoire (cf. Kerschensteiner et al., 1975).

With a mind to testing the involvement of prior experience in dyspraxia, Kimura and Archibald (1974; see also Kimura, 1982) compared performance by left- and right-hemisphere patients of familiar tasks involving relatively complex sequences of movements, with the performance of unfamiliar tasks that also included motor sequences, and tasks that involved *positioning* the hand or fingers but did not require the integration of successive motions. Their findings indicated that the left-hemisphere patients were significantly disturbed on *both* familiar and unfamiliar tasks that involved motion *sequences* but were not disturbed on tasks involving mere positioning of the hand or fingers. Kimura and Archibald conclude that motor deficits are associated with left-hemisphere lesions, in agreement with most authors, but that such deficits are not restricted to learned, purposeful, voluntary, or symbolic tasks, but rather are best characterized by disturbances of behavior sequences of whatever form.

In Kimura's most recent study (1982), a possible clue to the role of the effects of left-hemisphere brain lesions on the production of highly practiced motor sequences can be gleaned from her analysis of the effects of anterior and posterior lesions on oral motor behavior. On the basis of her data and many prior investigations, it can be assumed that Broca aphasics (with anterior lesions) have substantial difficulties in *initiating* both speech and other oral movements. Although both parietal and frontal patients are capable of *re*producing (shadowing) overlearned, multisyllabic sequences, neither group is capable of reproducing sequences of arbitrary oral movements not involving speech. For *single* oral tasks, whether or not they involve speech, only the frontal patients are disturbed. In manual functions, parietal patients are disturbed in performing both single hand postures and multiple manual tasks, whereas frontal patients are disturbed only in multiple manual tasks. These results, plus a very high correlation between errors of multiple (non-speech) oral and multiple manual tasks in left parietal patients, suggest not only an intimate association between motor processes in bucco-facial and manual functions, but also that the parietal lobe plays a critical role in the integration of unfamiliar, unpracticed motor sequences of whatever kind. The fact that frontal patients, although greatly disturbed in the production of spontaneous speech, are capable of shadowing highly familiar short phrases and that parietal patients can produce fluent, if defective, speech, suggests that the frontal speech area is critical for both the initiation and production of highly overlearned speech articulations, whereas the parietal lobe plays a critical role in the planning or programming of less probable, and hence more informative, word sequences, which are then transmitted to the frontal lobe for production.

In accordance with Kimura's ideas, it may be argued that the planning and "articulation" of manual behavior follow a similar principle: the pre-central motor areas are involved in the initiation and performance of highly practiced manual behavior, while the parietal lobe, consistent with the definition of ideational apraxia, is involved in the planning of manual motor sequences adapted to the (arbitrary) task at hand.

5 Conclusions

5.1 The Paradox of Disease of the Left Hemisphere

When the results of the studies on split-brain patients (Gazzaniga et al., 1967; Sperry et al., 1969), in whom we may assume essential integrity of both hemispheres, are compared with the vast majority of studies of patients with unilateral brain-damage, whose lesions may or may not have affected callosal fibers, a paradox may be observed that does not appear to have a simple explanation. Unilateral lesions of the left hemisphere produce motor apraxias in 20–50% of the affected patients, whereas unilateral lesions of the right hemisphere, even though they may be quite extensive, rarely result in apraxia of the right hand. Furthermore, section of all or critical portions of the corpus callosum (e.g., Liepmann & Maas, 1907; Geschwind & Kaplan, 1962) can result in a marked left-sided apraxia, even when the right hemisphere is essentially intact. On the other hand, section of the corpus callosum results in no apraxia, even in the left hand, so long as the response-eliciting stimulus is available to the hemisphere contralateral to the acting limb. How is one to resolve this paradox?

Several possibilities have been considered:

1. The motor apraxias are a language-dependent disorder (Ettlinger, 1969). Many aspects of the research to date would tend to support this contention. As Ettlinger shows, motor apraxia was found in split-brain patients (Gazzaniga et al., 1967) only when they were required to make responses with the *left* hand to *oral* commands. Even though the right hemisphere was able to "read" the names of familiar objects, verbal commands flashed to the right hemisphere were ineffective in eliciting responses, such as "squeeze", "point", or "knock" in *either* hand. We may assume, then, that language abilities of the right hemisphere play a negligible role in apraxia and certainly none for the *right* hand. Since *non*-verbal instructions to the right hemisphere appear to be fully adequate for generating accurate and

efficient behavior of the left hand under circumstances where the participation of the left hemisphere in guiding the left hand can be ruled out, one can only conclude that, except for verbal initiation of behavior, the left hemisphere need not be involved in the motor programming of the left hand.

This hypothesis is given further support by the overwhelming evidence from the fact that, in most studies, the majority of left-hemisphere apraxics are also aphasic. This observation was made by Liepmann, who found that 14 of his 20 apraxic right-sided hemiplegics were also aphasic, and has been repeated over and over again until the present time (DeRenzi et al., 1980; Kimura, 1982).

2. Apraxia is due to a disruption of the motor programming centers, which lie in the left hemisphere for both speech and non-speech behavior. This hypothesis, which is essentially that of Kimura (1982; Kimura & Archibald, 1974), is supported by the predominance of left-hemisphere apraxia and by the fact that most apraxics are also motor aphasics. Therefore, the involvement of aphasia in apraxia is best understood as the consequence of the fact that motor control of the limbs and neural control of speech production both represent aspects of the control of integrated muscular sequences, the programming of which may be attributable to similar and related neural mechanisms.

Neither of these hypothesis would appear, however, to be tenable in view of the paradoxical nature of the split-brain data. If the left hand of split-brain patients is fully capable of performing a wide variety of motor tasks without the participation of the (disconnected) left hemisphere and under circumstances in which the left hemisphere cannot "know" what instructions the right hemisphere has received, then why should a *lesion* of the left hemisphere disrupt the control by the right hemisphere of the motor behavior of the left hand, even though, as in the case of (nonverbal) imitation tasks and demonstration-of-use tasks, the (intact) right hemisphere is fully informed about the task to be performed by the left hand?

5.2 Priority of Action Control as a Function of the Left Hemisphere

An explanation of apraxia of the left hand in patients with damage to the left side of the brain which appears to be consistent with almost all the data may lie after all in a modified version of the concept of ideational apraxia as originally conceived by Liepmann. The concept should include the possibility that the principal function of the left hemisphere is the exercise of *priority of action control* in those situations in which it is called upon to do so.

Various pieces of evidence support this possibility:

1. One of the peculiarities of the behavior of Liepmann's *Regierungsrat,* who had a left-hemisphere lesion *without* right-sided paralysis, was that, when asked to perform a task, he was at first not capable of responding because he was right-handed and his right hand was apraxic. When told "Do it with the left hand," he was generally capable of carrying out the required act without further instruction. This suggests that the right hemisphere was fully capable of *directing* the appropriate

action of the left hand, but was not capable of *initiating* it – a function assigned by long habit to the left hemisphere in this right-handed patient.

2. In virtually every description I have been able to find of the behavior of the left hand of patients with damage to the left hemisphere, irrespective of the type of instruction and the modality by which it is issued, the errors that occur are usually errors of *commission* rather than errors of *omission.* Rather than doing what it is told to do, the left hand does *something else.* The "something else" is generally something no less meaningful or functional (*zweckmaessig:* in Liepmann's sense) than what it is supposed to do, but is simply different (not purposeful, or *zweckgemaess*). This suggests that the *right* hemisphere is quite capable of executing integrated behavior of the left hand, but receives distorted instructions from the (damaged) left hemisphere as to what kind of behavior it should direct the left hand to carry out.

3. When the left hemisphere is disconnected from the right, as in the split-brain patients of Gazzaniga et al. (1967), the right hemisphere is just as capable of directing the actions of the left hand as the left hemisphere is capable of directing the actions of the right (cf. Geschwind & Kaplan, 1962). In this case, however, the right hemisphere *cannot* be influenced by the priority of action of the left hemisphere since it is detached from it with the result that the right hemisphere proceeds with its own business without hindrance. But when the *right* hand receives ipsilateral instructions from the *right* hemisphere, as it can under very restricted circumstances, it is capable of carrying out at least rudimentary acts *so long as the left hemisphere does not receive conflicting instructions.* When split-brain patients are told to respond with the right hand to instructions projected to the right hemisphere (left visual half-field), but a conflicting stimulus is projected to the *left* hemisphere (right visual half-field), the right hand cannot ignore the left hemisphere instructions and does its bidding. This is of course because of the more powerful crossed corticospinal connections of the left hemisphere with the right hand. Unfortunately, Gazzaniga et al. do not appear to have carried out the contralateral control experiment, which would seem to be a crucial experiment for this explanation. Assuming that the left hemisphere is dominant in these right-handed split-brain patients in the sense of retaining priority of action, then ipsilateral instructions by the left hemisphere to the left hand should be less affected by interference from (irrelevant) instructions to the right hemisphere than was found to be the case for irrelevant instructions from the left hemisphere to the right hand, which was supposed to be under ipsilateral control by the right hemisphere.

4. There have been a few reports of *right*-hemisphere apraxia in left-handed patients, which are also consistent with the priority explanation. The most readily accessible is that by Heilman, Coyle, Gonyea, and Geschwind (1973) of a 62-year-old, left-handed male who suffered a left-sided hemiparesis 6 days prior to being seen by the authors. His speech comprehension and production were excellent, suggesting left-hemispheric dominance for speech. However, *spontaneous* writing with his (practiced) right hand was extremely poor and he even had difficulty writing his own name. In addition, he had a severe apraxia of the right arm, as tested by

pantomiming tasks to verbal instructions, demonstration of the use of real objects, and imitating the gestures of the examiner. He was able to perform none of these tasks adequately. The authors interpret their findings as showing that this patient was left-hemisphere dominant for oral speech and auditory comprehension but right-hemisphere dominant for nonverbal motor behavior, *including writing,* even though he had been taught to write with his right hand as a child. As a result, the patient displayed both agraphia and a severe motor apraxia of the right hand, deriving from a large lesion of the right hemisphere, which was dominant for motor behavior but subordinate for oral language, which in turn was undisturbed by the lesion.

5. A fact frequently referred to but little understood is that patients who reveal the severest motor apraxias during a neurological examination, in which the patient is required to perform a wide variety of acts in an essentially irrelevant context, are often if not usually capable of performing the same acts flawlessly in circumstances in which they are normally called for. The patient who is incapable of crossing himself on command or in imitation of the examiner crossing *him*self may do so spontaneously and perfunctorily when entering church. Under these circumstances, however, it must be assumed that the initiation of the act of crossing oneself derives from entirely different (subcortical?) neural sources which are under greater sensory or emotional control. To put the matter somewhat differently, the act of crossing oneself in church assumes a reflex-like quality that obviates the necessity of being "initiated" by cognitive commands.

6. Geschwind (1975) recalls Liepmann's suggestion that the hemisphere controlling the preferred hand is a storehouse of learning derived from the acquisition of motor skills. As a consequence the right hemisphere is either completely dependent upon the left for carrying out learned movements (which is certainly true for language, perhaps less so for motor behavior), or its own motor learning must be supplemented by the superior experience of the left. This would explain why right hemiplegics with damage to the left side of the brain have so much more trouble adapting to the use of their left arm and hand than do patients with just a broken preferred arm, who learn very quickly to use their nondominant limbs relatively well.

7. Finally, a large amount of evidence, including that of DeRenzi et al. (1980), suggests that left-hemisphere lesions result not only in erroneous behavior but also in marked hesitancy in carrying out the instructions of the examiner. Right-hemisphere lesions, on the contrary, are often associated with a remarkable spontaneity of behavior of the right hand, bordering on impulsivity. This would support the notion of the left hemisphere as initiator of action. According to this interpretation, the right hemisphere may be considered primarily as a reactor, whose function is that of the reception and analysis of higher-order sensory stimulation from the sense orders and the implementation of motor instructions from the left hemisphere.

The assumed control of the actions of the left hand by the left hemisphere differs, of course, from the control of motor speech by the left hemisphere in view of the fact that the actual motor commands to the left hand most likely arise in the right hemisphere, whereas motor commands in speech certainly derive largely, if not entirely, from the left hemisphere. Therefore, control by the left hemisphere of actions of the left hand must be considered to be a two-stage process, in which the left hemisphere both programs and initiates the implementation of the motor plan, which is then communicated to the right hemisphere for execution. If either the planning or the initiation function of the left hemisphere is disrupted through neural disease, both hands will either not react at all, or perform incorrectly. That the dominance of the left hemisphere of the control of the left hand is largely a matter of learning and practice is evidenced by the effect of handedness: left-hemisphere lesions in left-handed patients are far less likely to affect the behavior of the left hand than in right-handed patients.

Nevertheless, it must still be asked why the right hemisphere does not take over control of the actions of the left hand when the behavior of the right hand has been disrupted or paralyzed by left-hemisphere lesions. There appear to be two plausible interpretations, neither of which has received much attention, but either or both of which may be involved. As Paillard (1960, 1982) has argued, largely on the basis of neuroanatomical and neurophysiological evidence, both the parietal and the frontal cortex receive extensive afferents from the basal ganglia and limbic structures, and the basal ganglia in particular may well be involved in the generation and/or execution of motor plans. As a result of the overlearning and automatization of behavior of the right hand (in right-handed persons), the functional integration of the cortex and subcortical structures may be better established in the left than in the right hemisphere, and such integration may be critical for the timing, muscle selection and force regulation necessary for the performance of motor acts (Paillard, 1982). Analogously, the corticocortical connections involved in the interpretation of verbal instructions or the visual impressions of the motions of the examiner to be shadowed in a typical test of apraxia are very likely better developed in the left hemisphere of right-handed patients than they are in the right hemisphere, with the result that sensory information received simultaneously by both hemispheres receives priority in the left hemisphere even of left-hemisphere patients, and results in defective motor instructions to the right hemisphere. The strongest evidence for this assumption is still Liepmann's *Regierungsrat,* whose right-sided apraxia, due to left-hemisphere lesions, was alleviated by partial callosal lesions that may have released the obligation of the right hemisphere to the left. When instructed verbally to perform motor acts with the left hand, it may be assumed that the right hemisphere was able to take over the task of programming and initiating the motor commands to the left hand. Similarly, in patients with complete transsection of the corpus callosum (described by Geschwind and Kaplan, and by Gazzaniga, Bogen, and Sperry), both of whose hemispheres may be assumed to have received verbal or visual instructions when presented orally or via the central retina, the left hemisphere is dominant in planning and executing the motor programs of action which are then carried out by the right hand, thereby avoiding a conflict between "correct" behavior of the right hand and "erroneous" behavior of the left.

The actual nature of left-hemisphere priority of action control, which may lie in principle either in the selection, initiation, or execution of left-handed action, or in some combination of these, is difficult to specify. Since the right hemisphere of right-handed split-brain patients appears to be able to select, initiate, and execute actions of the left hand in the absence of left-hemisphere control, the latter would appear to be optional with respect to all motor functions of the left hand. A possible solution to the question of left-hemisphere control of the left hand may lie in the effect of restricted lesions of patients with a partially sectioned corpus callosum. Kleist (1912) suggests that the fibers of the corpus callosum may be divided into three sections: those that connect cortical fields of the frontal cortex, those that connect the sensorimotor fields of the central gyri, and those that connect the cortex of the supramarginal gyrus, the three divisions representing roughly equal portions of the corpus callosum. Kleist cites Liepmann as suggesting that it is the fibers of the central, sensorimotor third of the corpus callosum that may be held responsible for transmitting the motor programs of the left hemisphere to the right for control of the left hand. Kleist sees partial support for Liepmann's view in the fact that the case of Liepmann and Maas (1907), who presented a left-sided apraxia due to a focal lesion in the corpus callosum, was shown at autopsy to have retained the major portion of the splenium. Conversely, Hartmann's (1907) second case of left-sided sympathetic apraxia due to callosal lesions, in whom the *anterior* portion of the corpus callosum proved to be intact, would appear to exonerate fibers connecting the frontal cortex of the two hemispheres from responsibility for transmittal of motor instructions from the left to the right hemisphere. Taken together, the two cases of Liepmann and Maas and of Hartmann provide at least indirect evidence that it may be the central section of the corpus callosum, containing fibers connecting the central, sensorimotor gyri, which transmit motor programs from the left to the right hemisphere.

Kleist supports this argument with evidence of sympathetic apraxia in the *left* hand of Liepmann's *Regierungsrat.* Although far less severe than the apraxia of the right hand, it was nevertheless characterized by deficits of the simple motor or innervatory form, which was assumed to have been transmitted by the central motor area, and which may be considered to be roughly of the form of executive commands. Kleist suggests finally that the fine structure of behavior lost in motor or innervatory apraxia is that form of behavior which is most dependent upon the effects of prolonged motor learning (in the right hand of right-handed persons). Assuming that the highly overlearned motor sequences of single motor acts are involved in the performance of all motor behavior, however complex, the role of the left hemisphere in executing such behavior of the right hand, and sympathetically in the left, may be critical when the (inexperienced) left hand must take over the role of the right in patients with lesions of the left hemisphere.

That left-sided sympathetic apraxia due to lesions of the corpus callosum can also take other forms is suggested by Liepmann's own observations as well as by those of more recent cases of complete transsection of the corpus callosum as previously mentioned, in whom sympathetic apraxia of the left hand was complete.

Clarification of the role of fibers in restricted segments of the corpus callosum in transmitting motor commands from the left to the right hemisphere must await additional investigations of sympathetic apraxia in patients with partial lesions of the corpus callosum.

It is clear from the preceding discussion that a detailed analysis of individual cases of apraxia with localizable focal lesions would be helpful in clarifying both the nature of apraxia and the motor mechanisms of manual action. There is reason to expect that modern neurological methods of identifying the position of brain pathology will facilitate the elucidation of many interesting problems of higher-order motor control and their relationship to cognitive processes in general.

Acknowledgments. The help of Wolfgang Prinz and an anonymous reviewer in clarifying several important points of this paper is gratefully acknowledged.

References

Ajuriaguerra, J. de, Hécaen, H., & Angelergues, R. Les apraxies: Variétés cliniques et latéralisation lésionelle. *Revue neurologique,* 1960, *102,* 566–594.

Ajuriaguerra, J. de, & Tissot, R. The apraxias. In P. Vincken & G. Bruyn (Eds.), *Handbook of clinical neurology* (Vol. 4). Amsterdam: North-Holland, 1969.

Bogen, J.E. The other side of the brain. I: Dysgraphia and dyscopia following cerebral commissurotomy. *Bulletin of the Los Angeles Neurological Societies,* 1969, *34,* 73–105.

DeRenzi, E., Motti, F., & Nichelli, P. Imitating gestures: A quantitative approach to ideomotor apraxia. *Archives of Neurology,* 1980, *37,* 6–10.

DeRenzi, E., Pieczuro, A., & Vignolo, L.A. Ideational apraxia: A quantitative study. *Neuropsychologia,* 1968, *6,* 41–52.

Eggert, G.H. *Wernicke's works on aphasia: A sourcebook and review.* The Hague: Mouton, 1977.

Ettlinger, G. Apraxia considered as a disorder of movements that are language dependent: Evidence from cases of brain bi-section. *Cortex,* 1969, *5,* 285–289.

Finkelnburg, F.C. Niederrheinische Gesellschaft in Bonn. Medizinische Section. *Berliner klinische Wochenschrift,* 1870, *7,* 449–450, 460–461.

Foix, C. Contribution à l'étude de l'apraxie idéomotrice. *Revue neurologique,* 1916, *1,* 285–298.

Freud, S. *On aphasia: A critical study* (E. Stengel, trans.). New York: International Universities Press, 1953. (Originally published, 1891)

Gazzaniga, M.S. Cerebral mechanisms involved in ipsilateral eye-hand use in split-brain monkeys. *Experimental Neurology,* 1964, *10,* 148–155.

Gazzaniga, M.S. Visuomotor integration in split-brain monkeys with other cerebral lesions. *Experimental Neurology,* 1966, *16,* 289–298.

Gazzaniga, M.S., Bogen, J.E., & Sperry, R.W. Some functional effects of sectioning the cerebral commissures in man. *Proceedings of the National Academy of Sciences,* 1962, *48,* 1765–1769.

Gazzaniga, M.S., Bogen, J.E., & Sperry, R.W. Laterality effects in somesthesis following cerebral commissurotomy in man. *Neuropsychologia,* 1963, *1,* 200–215.

Gazzaniga, M.S., Bogen, J.E., & Sperry, R.W. Observations on visual perception after disconnexion of the cerebral hemispheres in man. *Brain,* 1965, *88,* 221–236.

Gazzaniga, M.S., Bogen, J.E., & Sperry, R.W. Dyspraxia following division of the cerebral commissures. *Archives of Neurology,* 1967, *16,* 606–612.

Gazzaniga, M.S., & Sperry, R.W. Language after section of the cerebral commissures. *Brain,* 1967, *90,* 131–148.

Geschwind, N. Disconnexion syndromes in animals and man. *Brain,* 1965, *88,* 237–294, 585–644. Reprinted in: N. Geschwind, *Selected papers on language and the brain.* Dordrecht, Holland: D. Reidel, 1974.

Geschwind, N. The apraxias: Neural mechanisms of disorders of learned movements. *American Scientist,* 1975, *63,* 188–195.

Geschwind, N., & Kaplan, E. A human cerebral deconnection syndrome. *Neurology,* 1962, *12,* 675–685. Reprinted in N. Geschwind, *Selected papers on language and the brain.* Dordrecht, Holland: D. Reidel, 1974.

Hartmann, F Beiträge zur Apraxielehre. *Monatsschrift für Psychologie und Neurologie,* 1907, *21,* 97–118, 248–270.

Hécaen, H. Suggestions for a typology of the apraxias. In M.S. Simmel (Ed.), *The reach of the mind.* New York: Springer, 1968.

Hécaen, H., & Gimeno Alava, A. L'apraxie idéo-motrice unilatérale gauche. *Revue neurologique,* 1960, *102,* 648–653.

Heilman, K.M. Ideational apraxia – a redefinition. *Brain,* 1973, *96,* 861–864.

Heilman, K.M., Coyle, H.M., Gonyea, E.F., & Geschwind, N. Apraxia and agraphia in a left hander. *Brain,* 1973, *96,* 21–28.

Kerschensteiner, M., Poeck, K., & Lehmkuhl, G. Die Apraxien. *Aktuelle Neurologie,* 1975, *2,* 171–178.

Kimura, D. Acquisition of a motor skill after left-hemisphere damage. *Brain,* 1977, *100,* 527–542.

Kimura, D. Left-hemisphere control of oral and brachial movements and their relation to communication. In D.E. Broadbent & L. Weiskrantz (Eds.), The neuropsychology of cognitive function. *Philosophical Transactions of the Royal Society of London,* 1982, *B 298,* 135–149.

Kimura, D., & Archibald, Y. Motor functions of the left hemisphere. *Brain,* 1974, *97,* 337–350.

Kleist, K. Der Gang und der gegenwärtige Stand der Apraxieforschung. *Ergebnisse der Neurologie und Psychiatrie,* 1912, *1,* 343–452.

Lehmkuhl, G., & Poeck, K. A disturbance in the conceptual organization of actions in patients with ideational apraxia. *Cortex,* 1981, *17,* 153–158.

Liepmann, H. Das Krankheitsbild der Apraxie ("motorischen Asymbolie") auf Grund eines Falles von einseitiger Apraxie. *Monatsschrift für Psychiatrie und Neurologie,* 1900, *8,* 15–44, 102–132, 182–197. (Reprinted by Karger, Berlin, 1900)

Liepmann, H. *Über Störungen des Handelns bei Gehirnkranken.* Berlin: S. Karger, 1905. (a)

Liepmann, H. Die linke Hemisphäre und das Handeln. *Münchener medizinische Wochenschrift,* 1905, *52,* 2322–2326, 2375–2378. (Reprinted in Liepmann, 1908) (b)

Liepmann, H. Der weitere Krankheitsverlauf bei dem einseitig Apraktischen und der Gehirnbefund auf Grund von Serienschnitten. *Monatsschrift für Psychiatrie und Neurologie,* 1906, *19,* 217–243. (Reprinted by Karger, Berlin, 1906)

Liepmann, H. Über die Funktion des Balkens beim Handeln und die Beziehungen von Aphasie und Apraxie zur Intelligenz. *Medizinische Klinik,* 1907, Nos. 25 & 26. (Reprinted in Liepmann, 1908)

Liepmann, H. *Drei Aufsätze aus dem Apraxiegebiet.* Berlin: Karger, 1908.

Liepmann, H., & Maas, O. Fall von linksseitiger Agraphie und Apraxie bei rechtsseitiger Lähmung. *Journal für Psychologie und Neurologie,* 1907, *10,* 214–227.

Meynert, Th. *Klinische Vorlesungen über Psychiatrie auf wissenschaftlichen Grundlagen für Studierende und Aerzte, Juristen und Psychologen.* Vienna: Braumüller, 1890.

Morlaas, J. *Contribution à l'étude de l'apraxie.* Paris: Amédée Legrand, 1928.

Paillard, J. The patterning of skilled movement. In J. Field, H.W. Magoun, & V.E. Hall (Eds.), *Handbook of physiology. Section 1: Neurophysiology,* (Vol. 3, pp. 1679–1708). Bethesda, MD: American Physiological Society, 1960.

Paillard, J. Apraxia and the neurophysiology of motor control. In D.E. Broadbent and L. Weiskrantz (Eds.), The neuropsychology of cognitive function. *Philosophical Transactions of the Royal Society of London,* 1982, *B 298,* 111–134.

Pick, A. *Studien über motorische Apraxie und ihr nahestehende Erscheinungen; ihre Bedeutung in der Symptomatologie psychopathischer Symptomenkomplexe.* Leipzig and Vienna: Franz Deuticke, 1905.

Poeck, K. Neuropsychologische Symptome ohne eigenständige Bedeutung. *Aktuelle Neurologie,* 1975, *2,* 199–208.

Poeck, K., & Lehmkuhl, G. Das Syndrom der ideatorischen Apraxie und seine Lokalisation. *Nervenarzt,* 1980, *51,* 217–225.

Sperry, R.W. Cerebral organization and behavior. *Science,* 1961, *133,* 1749–1757.

Sperry, R.W. Brain bisection and mechanisms of consciousness. In J.C. Eccles (Ed.), *Brain and conscious experience.* New York: Springer, 1966.

Sperry, R.W., & Gazzaniga, M.S. Language following surgical disconnection of the hemispheres. In C.H. Millikan & F.L. Darley (Eds.), *Brain mechanisms underlying speech and language.* New York: Grune and Stratton, 1966.

Sperry, R.W., Gazzaniga, M.S., & Bogen, J.E. Interhemispheric relationships: The neocortical commissures; syndromes of hemispheric disconnection. In P.J. Vinken & G.W. Bruyn (Eds.), *Handbook of clinical neurology,* (Vol. 4). Amsterdam: North-Holland, 1969.

Warrington, E.K. Constructional apraxia. In P.J. Vinken & G.W. Bruyn (Eds.), *Handbook of clinical neurology,* (Vol. 4, Chap. 4, pp. 67–83). Amsterdam: North-Holland, 1969.

Wernicke, C. *Der aphasische Symptomencomplex: Eine psychologische Studie auf anatomischer Basis.* Breslau: Cohn & Weigert, 1874.

Wilson, S.A.K. A contribution to the study of apraxia with a review of the literature. *Brain,* 1908, *31,* 164–216.

Summary. The history and current status of the classic motor apraxias as originally described by Hugo Liepmann are discussed in relation to the diseased hemisphere, the locus of lesion, the effect of learning, and hemisphere control. The emphasis of Liepmann and Geschwind on the importance of disconnexion syndromes for motor apraxia is discussed in relation to the question of left-hemisphere control of motor action in right-handed persons. The paradox of adequate right-hemisphere control of the motor behavior of the left hand in patients whose hemispheres have been disconnected by transection of the corpus callosum suggests that the dominance of the right hemisphere by the left in the control of motor behavior in intact right-handed people derives from a priority of motor control by the left hemisphere. This motor control is in turn largely attributable to the involvement of the left hemisphere in the control of highly overlearned motor behavior of right-handed persons.

4 A Motor-Program Editor

DAVID A. ROSENBAUM and EDWARD SALTZMAN

Contents

1 Introduction

Before the execution of a sequence of voluntary movements, a memory representation with instructions for the entire sequence is thought to be established in the central nervous system. There are three main sources of support for the existence of such a memory representation, or *motor program,* as it has been called. First, animals whose sensory feedback systems are impaired can often perform movements skillfully (see Evarts, Bizzi, Burke, Delong, & Thach, 1971; Glencross, 1977; Keele, 1968; Summers, 1981). Second, anticipatory errors in movement production (e.g., Spoonerisms in speech) indicate that information about segments of a sequence may often be available in memory well before the time of their execution (see Fromkin, 1973, 1980; Garrett, 1975, 1982; Norman, 1981). Third, the time to initiate a sequence of movements can depend on characteristics of the sequence, such as its timing (Klapp, 1978; Rosenbaum & Patashnik, 1980a,b) or length (e.g., Henry & Rogers, 1960; Sternberg, Monsell, Knoll, & Wright, 1978).

Despite the evidence for motor programs, little is known about how they are created, structured, and implemented. In this paper, we attempt to shed light on these issues by proposing a new model for data reported elsewhere (Rosenbaum, Saltzman, & Kingman, 1984). Those data came from choice reaction-time (RT) experiments in which human subjects were required to choose as quickly as possible between two sequences of finger-tapping responses such as those used in piano

Cognition and Motor Processes
Ed. by W. Prinz and A.F. Sanders

playing or typing. To account for the data we reported, we proposed a Hierarchical Decisions model, which is reviewed below. A difficulty with this model, which was not apparent when we proposed it, is that it fails to account for some recent results obtained in choice RT studies similar to ours. In the present paper, we outline an alternative model that can accommodate all the findings alluded to above as well as results from serial choice RT studies. The model is presented in greater detail and tested further in a forthcoming paper (Rosenbaum, Inhoff, & Gordon, Note 1).

2 The Hierarchical Decisions Model

2.1 Background Data

The model proposed by Rosenbaum et al. (in press) was designed to account for data from two experiments. The first had three choice conditions. In one, if one choice signal appeared on a screen the subject was required to press a button with the left index finger, then to press a button with the left ring finger, and then to press a button with the left middle finger. (We denote these three responses *i, r, m,* respectively.) In the same condition of the experiment, if the other possible choice signal appeared the required response sequence was right index, followed by right ring, followed by right middle (which we denote *I, R, M,* respectively). Both sequences consisted of $n = 3$ responses. In another condition of this experiment, the two possible sequences consisted of $n = 2$ responses: one of the sequences was *ir;* the other was *IR*. Finally, in another condition, the two possible sequences each consisted of $n = 1$ response: *i* versus *I*. For a given subject, a given choice signal was always mapped to the same hand throughout the experiment, and all subjects performed in the $n = 1$, $n = 2$, and $n = 3$ conditions. Subjects were always instructed to minimize the time from the appearance of the choice signal until completion of the required sequence.

The results of greatest interest concern response latencies. The time for the first response, T_1, increased linearly with n. The time between the first response and the second, T_2, was much shorter than T_1, but was longer for sequences of length $n = 3$ than for sequences of length $n = 2$, the magnitude of the difference being 33 msec, which was not statistically different from the slope of the best-fitting linear function relating T_1 to n (37msec). The time between the second response and the third, T_3, was approximately equal to T_2 in 3-response sequences.

In the second experiment reported by Rosenbaum et al., the main question was how the time to produce a sequence of length n depended on the length n' of the other possible sequence. In the first experiment n and n' were always equal, so it was impossible to assess the contributions of n and n' separately. In the second experiment the conditions studied were $n = 1, n' = 1$ (herewith 1-1), $n = 1, n' = 3$ (herewith 1-3), and $n = 3, n' = 3$ (herewith 3-3). As in the first experiment, choices were always made between hands, and the 1-response sequences and 3-response sequences were the same as those used in Experiment 1 (i.e., the 1-response sequences were *i* and *I*, and the 3-response sequences were *irm* and *IRM*). In all other respects, the second experiment was conducted in the same way as the first, although different subjects participated.

The main findings from Experiment 2 also concern response latencies. For required sequences of length $n = 1$, T_1 was longer in condition 1-3 than in condition 1-1. T_1 was also longer for required sequences of length $n = 3$ than for required sequences of length $n = 1$, but was only slightly shorter in condition 3-1 than in condition 3-3. T_2 and T_3 were not reliably different from one another, nor were they reliably affected by the value of n'.

2.2 The Model

The Hierarchical Decisions model, which we proposed to account for the above results, is depicted in Figure 1. Panel A shows the structures assumed to be used when subjects choose between sequences *irm* and *IRM*. It is assumed that each of the sequences has associated with it a higher-order, or *control*, element. Choosing between the two sequences is assumed to entail a choice between these two control elements, followed by a choice of the first response within the sequence that is initially chosen, followed by execution of that response. Production of the second response is assumed to entail a choice of the second response within the sequence that is initially chosen, followed by execution of that response. Production of the third response is assumed to entail a choice of the third response within the initially chosen sequence, followed by execution of that response. The timing results of the first experiment can be accounted for by assuming that the time to make a choice at any given choice point depends only on the number of alternatives at that point.[1]

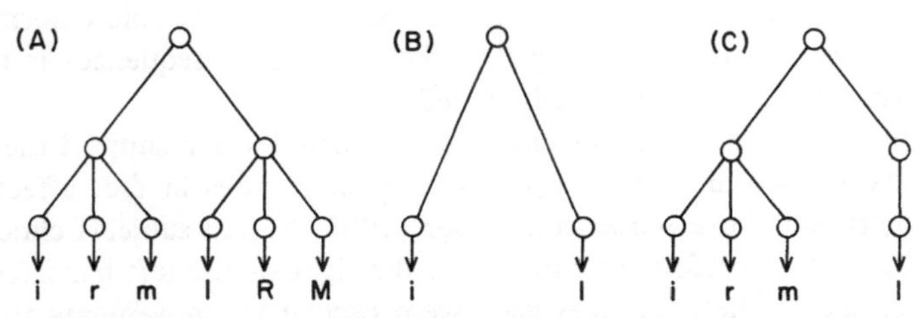

Figure 1. Alternative choice procedures assumed by the Hierarchical Decisions model for different choice situations. Panel A: two sequences of length $n = 3$; Panel B: two sequences of length $n = 1$; Panel C: one sequence of length $n = 3$ and one sequence of length $n = 1$. (Reprinted with permission from Rosenbaum et al., in press)

1 Fundamentally the same model, with the same assumptions about timing spelled out in more detail, has been offered by Rosenbaum, Kenny, and Derr (1983) to account for timing effects in the rapid production of memorized finger patterns initiated without reaction-time pressure.

Panel B shows the structures assumed to be used when subjects choose between responses *i* and *I*. Here, an initial choice between the two hands is presumed to be unnecessary because of an assumption about economy of choices, namely, that choices are always made within the lowest level possible. When subjects choose between *i* and *I*, they can choose between elements at the "finger" level and can avoid making prior choices at the "hand" level.[2]

Panel C illustrates the structures assumed to be used in choosing between sequences *irm* and *I* (or equivalently *i* and *IRM*). Here another assumption of the Hierarchical Decisions model is illustrated. The assumption is that decisions are always made *within* the same functional level, never *between* functional levels. If this assumption were not made, it would be possible to choose between a *hand* (for the 3-response sequence) and a *finger* (for a 1-response sequence). Instead, by virtue of the assumption, if the left hand is one of the elements entering into a choice, the alternative element must be the right *hand,* not a *finger* (e.g., the index finger) of the right hand. On the basis of this assumption, we can account for the fact that more time was needed to initiate a 1-response sequence when it was paired with a 3-response sequence than when it was paired with another 1-response sequence. Likewise, the assumption allows us to account for the fact that approximately the same amount of time was needed to initiate a 3-response sequence regardless of whether it was paired with another 3-response sequence or a 1-response sequence.

2.3 A Difficulty with the Model

Although the Hierarchical Decisions model can account for the results of the experiments reviewed above, it makes an incorrect prediction which was not apparent at the time that the paper by Rosenbaum et al. was completed. The prediction is that, all other things being equal, the similarity of two response sequences should have no effect on the time to choose between them. The Hierarchical Decisions model makes this prediction because it assumes that the sole determinant of the time to choose between two (equally likely) response sequences is the way in which the responses are hierarchically coded.

Some data have recently been reported which support the idea that the similarity of two alternative response sequences does in fact affect the time to choose between those sequences. Heuer (1982a,b) had subjects choose between response sequences performed with either the right or the left hand (as was the case in our studies). The sequences used were oscillatory movements that required rapid, repeating displacements of one hand in the vertical or horizontal direction. Heuer found that when subjects chose between vertical movement sequences or between horizontal movement sequences, their choice RTs were faster than when they chose

2 This assumes that there really are functionally differentiated *finger* and *hand* levels. Evidence for such an assumption is found in a study by Miller (1982).

between vertical and horizontal movement sequences. Hence, subjects in Heuer's experiments could initiate one of two possible movement sequences more quickly if the movement sequences were similar than if they were dissimilar.

In our laboratory we have also recently obtained evidence for a response similarity effect in choice RTs (Rosenbaum & Saltzman, Note 2; see also Rosenbaum et al., Note 1). We found that subjects were faster to initiate a sequence such as *iim* when the other possible sequence was *IIM* than when the other possible sequence was *IMM.* Similarly, the sequence *imm* took less time to initiate when the other possible sequence was *IMM* than when the other possible sequence was *IIM.* In general, RTs were shorter for similar than for dissimilar sequences.

Klapp (Note 3) has also collected data consistent with the hypothesis that response similarity influences choice RT. He found that when subjects chose between a short-duration manual response (a Morse code "dit") and a short-duration vocal response (saying "dip"), the time for the manual response was shorter than when the alternative vocal response was a long-duration "daap." Using other pairs of short and long manual and vocal responses, Klapp consistently obtained shorter latencies for manual responses that shared the same duration classification as the alternative, vocal response. This finding, like the findings of Heuer and Rosenbaum et al. mentioned above, support the hypothesis that choice RTs are inversely related to the similarity of the responses between which a choice must be made. However, this hypothesis does not follow in any obvious way from the Hierarchical Decisions model; therefore a more comprehensive model is needed.

3 The Motor-Program Editor Model

We would now like to propose an alternative to the Hierarchical Decisions model. The new model is meant to account for the similarity effects just summarized, as well as the data that the Hierarchical Decisions Model could explain. The main ideas in the new model are these:

1. In a choice RT task, the subject prepares for the reaction signal in part by preparing a serially ordered set of motor subprograms. Each subprogram has only the features shared by the two possible response alternatives at the corresponding serial position. Thus, if a choice must be made between sequences *irm* and *IRM,* the initially prepared motor program consist of three subprograms. The first designates an index finger as well as other motor features that are certain for the first serial position. The second designates a ring finger as well as other motor features that are certain for the second serial position. The third designates a middle finger as well as other motor features that are certain for the third serial position. What is missing from each subprogram in this choice context is the "hand feature" (i.e., left or right) of the corresponding to-be-required response (as well as other features that may depend on hand identity). In general, the missing features from any initially prepared subprogram are those features that distinguish the alternative possible responses in the corresponding serial position.

2. Once the choice reaction signal is identified, those motor features that could not be specified in advance are now specified.

3. The time required to specify all the features needed to define a forthcoming response sequence is assumed to depend on the number of feature specifications that must be completed.

3.1 Accounting for Previous Results

Consider how the above assumptions can be used to account for response similarity effects in choice RT studies. The greater the similarity of two responses, the larger is the number of features shared by their corresponding subprograms. Consequently, within a single subprogram that must later be edited to allow for production of one of two possible responses, the number of features that can be specified in advance increases with the similarity of the two responses; in other words, the number of features remaining to be specified after the reaction signal, and so the time needed for editing, decreases with the similarity of the responses. The Editor model says, therefore, that choice RTs are shorter for similar response alternatives than for dissimilar response alternatives because less editing is needed when the response alternative are similar.

How can the Editor model account for the results reported by Rosenbaum et al.? Consider the first experiment, where it was found that choice RTs increased linearly with the number of responses in the required sequence. This finding can be explained by saying that as the number of required responses increases, more subprograms require feature specifications following identification of the choice signal. In particular, specification of "hand" (left or right) has to occur three times when subjects choose between sequences *irm* and *IRM*, twice when subjects choose between *ir* and *IR*, and once when subjects choose between responses *i* and *I*.

Now consider how the Editor model can explain the results of the second experiment. Recall that in that experiment subjects made choices between 1- and 3-response sequences. Suppose, as a corollary of the first assumption of the Editor model, that whenever subjects choose between two sequences of unequal length they set up a motor program consisting of the larger number of subprograms that might be necessary. Thus, if subjects had to choose between a sequence of length $n = 3$ and a sequence of length $n = 1$, they set up a motor program consisting of three subprograms. If the 3-response sequence is called for, the hand identity of the three required subprograms is supplied to the initially readied program after the choice signal is identified[3]. Note that this is also what was assumed to occur in producing a three response sequence when the other possible response sequence also had three responses, which accounts for the fact that T_1 was found to be approx-

3 This statement assumes that the subprograms in serial positions 2 and 3 were incompletely specified at the time the choice signal was presented.

imately equal for 3-response sequences paired with 3-response sequences and for 3-response sequences paired with 1-response sequences.

Now consider what happens if instead of the 3-response sequence the 1-response sequence is required. Two editing operations are then necessary. One is specifying the hand of the first subprogram; the other is cancelling the second and third subprograms. If cancellation is achieved by simply truncating the entire program – an assumption supported in some recent experiments (Rosenbaum, Inhoff, & Gordon, Note 1) – the number of editing operations, and hence the time required for these editing operations, would be less than when a 3-response sequence was called for, as we found.

Finally, consider the finding that the choice RT to produce a single response when it was paired with another single response was shorter than the choice RT to produce a single response when it was paired with a 3-response sequence. According to the Editor model, all that is required in choosing between two single responses such as *i* and *I* is to specify the hand of the initially prepared subprogram. However, in choosing between a single response and a 3-response sequence both hand specification and subprogram cancellation are required.[4]

3.2 Implications for Sequential Effects

The Editor model has important implications for the analysis of sequential effects in choice RT experiments. For the purpose of describing these implications, we first pose a question that has not been widely discussed before, to the best of our knowledge: What happens to motor programs after they have been used?

A priori there are two fates that a just-used motor program could experience. One is that it could be lost from memory as soon as it has been implemented. The other is that it could be retained in memory. As one of us has argued elsewhere (Rosenbaum, 1980), it seems preferable to preserve just-used motor programs, for if the features of a movement sequence that has just been performed are similar to the features of a movement sequence that is forthcoming, it might be possible to make use of the motor program for the earlier sequence in setting up the motor program for the sequence to follow. This could be accomplished by *editing* the just-used program, e.g., by changing those features distinguishing the previous movement sequence

4 T_j data, for $j > 1$, can be explained with the Editor model by assuming that features for subprograms in serial positions 2 and 3 are not always specified during the T_1 interval, but instead remain to be specified after the first response has been produced. If this assumption is accepted, one must also assume that the mean value of T_1 represents a mixture of times to specify 2, 3, . . . or n features, where serial position 1 must have all of its features specified in order to be performed, but features for later responses can be specified later if there is pressure to minimize T_1. If an attempt is always made to minimize T_1, then as n increases T_1 will tend to increase but so will the probability that features for a given serial position need to be specified after the first response has been performed. (This predicts that T_1 should be inversely related to later response times if the subject is allowed to vary T_1 freely.)

from the movement sequence to follow. Alternatively, if the new motor program were set up "from scratch," which is what would happen if the previous motor program disappeared from memory, much more programming would be needed. (This idea can be appreciated easily by anyone who has become adept at using a computerized text editing system. With such a system, revising a body of text can be achieved by simply making necessary changes. By contrast, revising a manuscript in a conventional fashion –that is, with a system with no memory – often requires reduplication of the entire manuscript or major parts of it.)

In view of the potential benefit of editing previously used motor programs to allow for the production of forthcoming responses, the question that arises is whether motor programs do in fact persist in memory after they have been used. One way of addressing this question is to ask whether there are response-related sequential effects in serial RT tasks or, for that matter, in other tasks requiring successive movements. If, in an RT task, the time to perform one motor response depends on characteristics of its predecessor, it can be argued that information pertaining to the predecessor is present in memory when the second response is prepared. As it turns out, the evidence for such sequential effects is profuse (see Kirby, 1980; Kornblum, 1973).

Do the data on sequential effects support the hypothesis that programming of a forthcoming movement sequence is achieved by editing the program for the last response? For the sake of discussion, let us assume that such editing consists of changing those features that distinguish the last response from the one to come. We have already suggested that it takes a constant amount of time on average to supply a feature to an incomplete, prereadied motor program (in a choice RT task). Suppose that *changing* a feature also takes a constant amount of time. The consequence of this assumption is that on average as more features need to be changed (i.e., the more dissimilar two successive responses are) the longer the feature-changing process should take.

Data from analyses of sequential effects in choice RT experiments support the above prediction. A principal source of support is the "response-repetition" effect, where the time to perform a given response is much shorter if that same response was performed in the preceding trial than if some other response was performed in the preceding trial (Bertelson, 1963, 1965). The Editor model predicts this result, of course, because if two responses are the same in successive trials, no feature changes are needed to prepare the second response (assuming that the motor program for the first response is not altered in anticipation of a possibly different response in the next trial).

There is also support for the more general proposal that the RT for a given response increases with the number of features distinguishing the motor program for that response from the motor program for the preceding response. One source of support for this prediction is the fact that response repetitions tend to be quicker than any other kind of response transition in serial choice RT tasks. This outcome is what one would expect given that the fewest number of features need to be changed in response repetitions.

A related piece of evidence is that transitions between fingers of different hands take longer than transitions between fingers of the same hand in serial choice RT tasks (Kornblum, 1973; Rabbitt, 1968). Moreover, as Rabbitt and Vyas (1970) discovered, transition times between different hands are somewhat longer if different fingers are used (e.g., *I* followed by *m*) than if the same fingers are used (e.g., *I* followed by *i*). These effects are the sort one would expect if the number of motor features distinguishing successive responses affects the time to make the transition between the responses.

More evidence for this kind of effect comes from the observation (Rabbitt, 1966) that in serial choice RT tasks involving the two hands and the two feet, transition times are faster *within* limbs on different sides of the body (e.g., left hand followed by right hand) than *between* limbs on different sides of the body (e.g., left foot followed by right hand). Presumably, more features need to be changed in using the right hand after the left foot than in using the right hand after the left hand.[5]

The suggestion that emerges from the above discussion is that motor programming of successive movement sequences may be achieved by continual re-editing of a single motor program. An appealing property of this hypothesis is that the mechanisms used for controlling response transitions are seen as fundamentally the same as the mechanisms used for choosing between alternative possible responses.

Acknowledgment. Preparation of this chapter was supported in part by a grant from the Mellon Foundation and grant BNS-8120104 from the National Science Foundation.

References Notes

1. Rosenbaum, D.A., Inhoff, A.W., & Gordon, A. *Choosing between movement sequences: A hierarchical editor model.* Manuscript submitted for publication, 1983.
2. Rosenbaum, D.A., & Saltzman, E. *Levels of choice in the production of movement sequences.* Paper presented at the Twenty-third Annual Meeting of the Psychonomic Society, Minneapolis, MN, November 1982.
3. Klapp, S.T. Personal communication, December 1982.

5 One result that appears to contradict the prediction of the Editor model is that transition times are slowest for ipsilateral, between-limb changes (e.g., right foot followed by right hand). However, Blyth (1963, 1964) found that substitution errors involving ipsilateral limbs are by far the most common in four-limb choice tasks. It is possible, therefore, that subjects exercise special caution before producing a response that uses a different limb but the same side of the body as the preceding response. Blyth's finding, therefore, is not inconsistent with the Editor model.

References

Bertelson, P. S-R relationships and reaction-times to new versus repeated signals in a serial task. *Journal of Experimental Psychology,* 1963, *65,* 478–484.

Bertelson, P. Serial choice reaction-time as a function of response versus signal-and-response repetition. *Nature,* 1965, *205,* 217–218.

Blyth, K.W. Ipsilateral confusion in 2-choice and 4-choice responses with the hands and feet. *Nature,* 1963, *199,* 1312.

Blyth, K.W. Errors in a further four-choice reaction task with the hands and feet. *Nature,* 1964, *201,* 641–642.

Evarts, E.V., Bizzi, E., Burke, R.E., DeLong, M., & Thach, W.T., Jr. Central control of movement. *Neurosciences Research Program Bulletin,* 1971, *9,* 1–169.

Fromkin, V.A. (Ed.) *Speech errors as linguistic evidence.* The Hague: Mouton, 1973.

Fromkin, V.A. (Ed.) *Errors in linguistic performance: Slips of the tongue, ear, pen, and hand.* New York: Academic Press, 1980.

Garrett, M.F. The analysis of sentence production. In G.H. Bower (Ed.), *Psychology of learning and motivation,* (Vol. 6). New York: Academic Press, 1975.

Garrett, M.F. Production of speech: Observations from normal and pathological language use. In A. Ellis (Ed.), *Normality and pathology in cognitive functions.* London: Academic Press, 1982.

Glencross, D.J. Control of skilled movements. *Psychological Bulletin,* 1977, *84,* 14–29.

Henry, F.M., & Rogers, D. Increased response latency for complicated movements and a "memory drum" theory of neuromotor reaction. *Research Quarterly,* 1960, *31,* 448–458.

Heuer, H. Binary choice reaction time as a criterion of motor equivalence. *Acta Psychologica,* 1982, *50,* 35–47. (a)

Heuer, H. Binary choice reaction time as a criterion of motor equivalence: Further evidence. *Acta Psychologica,* 1982, *50,* 49–60. (b)

Keele, S.W. Movement control in skilled motor performance. *Psychological Bulletin,* 1968, *70,* 387–403.

Kirby, N. Sequential effects in choice reaction time. In A.T. Welford (Ed.), *Reaction times.* London: Academic Press, 1980.

Klapp, S.T. Reaction time analysis of programmed control. In R. Hutton (Ed.), *Exercise and sports sciences reviews,* Vol. 5. Santa Barbara, CA: Journal Publishing Affiliates, 1977.

Kornblum, S. Sequential effects in choice reaction time: A tutorial review. In S. Kornblum (Ed.), *Attention and performance,* (Vol. 4). New York: Academic Press, 1973.

Miller, J. Discrete versus continuous stage models of human information processing: In search of partial output. *Journal of Experimental Psychology: Human Perception and Performance,* 1982, *8,* 273–296.

Norman, D.A. Categorization of action slips. *Psychological Review,* 1981, *88,* 1–15.

Rabbitt, P.M.A. Times for transitions between hand and foot responses in a self-paced task. *Quarterly Journal of Experimental Psychology,* 1966, *18,* 334–339.

Rabbitt, P.M.A. Repetition effects and signal classification strategies in serial choice-response tasks. *Quarterly Journal of Experimental Psychology,* 1968, *20,* 232–240.

Rabbitt, P.M.A., & Vyas, S.M. An elementary taxonomy for some errors in laboratory choice RT tasks. *Acta Psychologica,* 1970, *33,* 56–76.

Rosenbaum, D.A. Human movement initiation: Specification of arm, direction, and extent. *Journal of Experimental Psychology: General,* 1980, *109,* 444–474.

Rosenbaum, D.A., Kenny, S., & Derr, M.A. Hierarchical control of rapid movement sequences. *Journal of Experimental Psychology: Human Perception and Performance,* 1983, *9,* 86–102.

Rosenbaum, D.A., & Patashnik, O. A mental clock setting process revealed by reaction times. In G.E. Stelmach & J. Requin (Eds.), *Tutorials in motor behavior.* Amsterdam: North-Holland, 1980. (a)

Rosenbaum, D.A., & Patashnik, O. Time to time in the human motor system. In R.S. Nickerson (Ed.), *Attention and performance VIII.* Hillsdale, NJ: Erlbaum, 1980. (b)

Rosenbaum, D.A., Saltzman, E., & Kingman, A. Choosing between movement sequences. In S. Kornblum & J. Requin (Eds.), *Preparatory states and processes.* Hillsdale, NJ: Erlbaum, 1984.

Sternberg, S., Monsell, S., Knoll, R.L., & Wright, C.E. The latency and duration of rapid movement sequences: Comparisons of speech and typewriting. In G.E. Stelmach (Ed.), *Information processing in motor control and learning.* New York: Academic Press, 1978.

Summers, J.J. Motor programs. In D.H. Holding (Ed.), *Human skills.* Chichester, Wiley, 1981.

Summary. We suggest that motor programs for forthcoming movement sequences are prepared by making changes in motor programs for preceding movement sequences. Specifically, the model says that programming a new movement sequence is accomplished by taking the program for the preceding movement sequence and changing those features that distinguish the old sequence from the new. The model also says that in choice reaction-time (RT) tasks, subjects prepare for response alternatives by setting up a motor program with features shared by the alternatives; after the choice signal is presented, features needed to define the required response are added to the initially readied program. This model is shown to account for data from serial and discrete choice RT tasks.

5 Eye Movement Control During Reading: The Effect of Word Units

GEORGE W. MCCONKIE and DAVID ZOLA

Contents

1 Eye Movements in Reading

In recent years psychologists have shown a renewed interest in eye movement in reading (see reviews by Levy-Schoen & O'Regan, 1979; McConkie, 1983; Rayner, 1978a). This work has been motivated by more than a simple curiosity about the nature of eye movement control. Rather, eye movement data are regarded as having the potential for testing theories about the ongoing perceptual and language processing taking place during reading. As people read, a great deal of variability is exhibited in how far they move their eyes and in how long their eyes remain centered on different locations in the text. There is general faith in, and some evidence for, the notion that this variability reflects differences in the nature of the perceptual and cognitive processes occurring at different locations in the text. It is assumed that if we could discover the ways in which mental processes influence eye movement behavior, then we would be able to use eye movement records to infer the nature of the processing occurring at different places in the text. In effect, the eye movement pattern would then become a language by which the brain communicates some of its activities to the psychologist. The hope that this can be achieved is a strong motivator for research on eye movement control in reading (Just & Carpenter, 1980; McConkie, Hogaboam, Wolverton, Zola, & Lucas, 1979).

Cognition and Motor Processes
Ed. by W. Prinz and A.F. Sanders

During reading the eyes execute a rapid series of saccadic movements averaging about six to ten letter positions in length. They occur at the rate of three or four per second, with each saccade taking the eyes to a different location and providing the reader with a clear perception of a new region of text. How the mind decides where to send the eyes on each saccade has been a matter of speculation among psychologists for decades [i.e., Dodge's (1907) argument for the involvement of peripheral vision]. For some time it was believed that learning to establish a regular rhythm of saccadic movement was an oculomotor skill that contributed to skilled reading. However, attempts to improve reading through oculomotor training proved fruitless.

Hochberg (1970) gave strong credence to the distinction between foveal and peripheral vision in his formulation of a dual eye guidance system. He postulated a peripheral search guidance mechanism that communicates information to the oculomotor system about where the eyes must be moved for clearest visibility of detail, and a cognitive search guidance mechanism that affords hypotheses about where to look in order to gain further needed information for reading. Recent research has provided clear evidence that readers use some peripheral information in determining where the next fixation will be located (McConkie & Rayner, 1975; O'Regan, 1980; Rayner, 1978b; Zola, 1981). O'Regan (1981; see also Rayner, 1979), in attempting to account for where the eyes are sent during reading, stated the "Convenient Viewing Position Hypothesis," suggesting that the eyes tend to go to centers of words, and, if that fails, corrective action is sometimes required, taking the eyes to a more optimal position. Rayner and McConkie (1976) described a range of alternative ways in which the guidance of eye movements might occur during reading and argued for a moment-to-moment control in response to ongoing mental processes taking place. Shebilske (1975) opted for a more delayed form of control, one reflecting the amount of buffered information available from prior fixations. Finally, Levy-Schoen (1981) has suggested that eye guidance in reading is based on a learned oculomotor routine which moves the eyes in a basic left-to-right pattern along one line of text and on to the next. However, this routine can be influenced and even overridden by momentary mental events occurring during reading. Such modulation would lead to the variability seen in eye movement records.

Thus, at present there is controversy both about the nature of the information used in deciding where to send the eyes next (i.e., visual information from fovea or periphery, central information from a basic oculomotor pattern, information concerning the contents of a buffer, and hypothesis about upcoming text and/or information from other ongoing processes involved in the perception and comprehension of the text) and how soon the information can be brought to bear on influencing where the eyes will be sent.

To gain further insight into the nature of eye movement control in reading, we have collected eye movement records from a number of college students as they read a short passage about the early history of Alaska. For the present paper, forward saccades from this data set were analyzed to yield descriptive information about the influence of three variables on the likelihood of any given letter in the text being the recipient of the next fixation:

1. How far the letter was from the prior fixation;
2. The length of the word the letter is in;
3. The letter's serial position in the word.

2 A Set of Eye Movement Data

As subjects have come to our laboratory to participate in other studies, we have typically had them read a 417-word passage taken from a high-school level encyclopedia. Its readability is estimated for 16-year-old students. Thus, it was relatively easy reading for the college students who participated in our research. However, they were told that they would be given questions after reading the passage, implying that they should read carefully.

The text was displayed on a cathode-ray tube (CRT) one line at a time in normal upper and lower case type. The subject was able to call for each successive line by pressing a button that changed the text within a few msec. The CRT was about 68 cm from the subjects' eyes, a distance at which 4 letter positions occupied one degree of visual angle. Maximum line length was 73 letter positions. As subjects read, their eye position was monitored every millisecond using an SRI Dual-Purkinje Eyetracker.

The analyses to be described were based on data from 51 subjects providing a total of nearly 20 000 saccades. From these data we selected each forward saccade that was preceded by a forward saccade, where these movements did not represent a rereading of the line or part of the line of text, and where neither saccade nor the fixation between them was contaminated by eyetracking failures (i.e., blinks, or loss of track). This procedure resulted in a reduced data set of approximately 9200 forward saccades.

2.1 Relationships to Prior Eye Movement Events

The degree to which individual saccades are independently controlled was explored by correlating the length of each saccade in the data set with the length of the preceding forward saccade. The correlation of $r = 0.21$ proved to be due almost entirely to individual differences in average saccade length. The average correlation within subjects was $r = 0.05$. In addition, the correlation between the length of a saccade and the duration of the prior fixation was $r = -0.0001$. These data are in agreement with prior reports (Andriessen & deVoogd, 1973; Rayner & McConkie, 1976) and are suggestive of independent control of individual eye movements. Such data give encouragement to the notion that each saccade reflects stimulus or processing characteristics present at the time immediately preceding that saccade, rather than more general influences existing over longer periods of time (for a differing opinion, see Shebilske, this volume).

2.2 Distribution of Saccade Length

Data concerning saccadic eye movements can be conceptualized in either of two ways: either as the likelihood of making saccades of different lengths, or as the likelihood of the eyes going to different locations in the text. In this paper, we have adopted the second perspective, and explore the effects of the three variables listed above on the locus of the next fixation. Our strategy has been to examine the effect of each of these variables while controlling for the others and to allow for some indication of their possible interaction.

A frequency distribution of saccade lengths in the selected data set is shown in Figure 1. The mean saccade length is 7.20 character positions, and the standard deviation of the distribution is 2.90. The distribution also has a median value of 6.87. This distribution gives a general indication of the likelihood of fixating a letter next which lies different distances from the present fixation location. The interpretation of this distribution, however, depends upon the view one takes of the nature of eye movement control. For example, it might be taken as indicating the result of oculomotor learning: the average distance that readers have learned to cast their eyes and the normal variability induced by various cognitive factors (Levy-Schoen, 1981). Or it might be taken to indicate the range of distances to words that are anticipated in reading and must be fixated next in order for visual confirmation to occur (Hochberg, 1970). Or it might be taken as indicating the range of distances

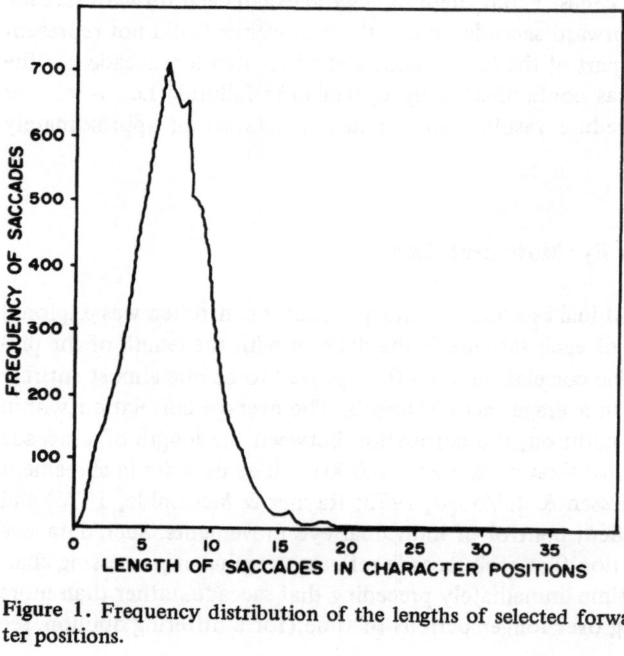

Figure 1. Frequency distribution of the lengths of selected forward saccades, given in character positions.

at which perception or identification fail, and thus, added visual clarity is required for reading to continue (McConkie, 1979; O'Regan, 1979b). Thus, the proper interpretation of this distribution is an issue that has not yet been resolved.

3 The Importance of Words

3.1 Word-Unit Influences

The most perceptually obvious structure in the stimulus array of a page of text is its arrangement in lines and the subdividing of lines into words. A very important question asks whether visual characteristics of a word influence the likelihood of the next fixation being attracted to letter positions in the word. There is some evidence that this information is used in determining future fixation location: there are fewer fixations in large blank regions (Abrams & Zuber, 1972–73) and spaces between sentences (Rayner, 1975a), and more fixations on the centers of words than on their beginnings and ends (O'Regan, 1981; Rayner, 1979; Zola, 1981).

To assess more accurately the degree to which the eyes tend to be attracted to certain letter positions in a word, we partitioned our data according to the location of the fixation with respect to different letter positions in words of different length. For instance, all fixations were found which were located three or four letter positions to the left of the first letter of a 5-letter word. Then the proportion of times that the next fixation fell on that letter was calculated. A similar proportion was obtained for each of the other letter positions of a 5-letter word; in each case, this statistic represented the likelihood that the letter would be fixated immediately following a fixation lying three or four character positions to the left of it. The results are graphically presented in Figure 2. This figure also shows similar proportions when the prior fixation was five or six letter positions to the left of each letter position. Thus, these curves indicate the degree to which letter position in a word influences the likelihood that a given letter will be fixated next when distance and word length are held constant. Figure 3 presents similar data for 3-, 5- and 7-letter words when the prior fixation was five or six character positions before the letter. Figure 4 presents the same data when the previous fixation was nine or ten character positions in front of the current fixation.

The curves show a strong influence of letter position in a word. If a position is within 8 to 10 letter positions of the present fixation location, it is most likely to be fixated next if it is slightly left of the center of a 5- or 7-letter word. Letter positions further away than this are benefited more by being closer to the beginning of the word. When 3-letter words are involved, however, the likelihood of a letter being fixated is greater if it lies immediately to the right of the word with a general favoring of end letters over beginning letters, even if the letter lies as much as 9 to 10 positions to the right of the prior fixation location.

While some letter positions are clearly preferred, it is equally important to notice that there are still many fixations at other positions, including the space

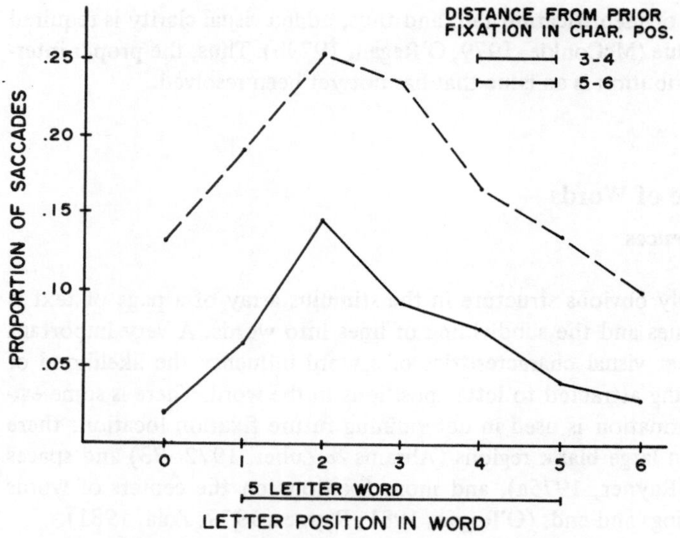

Figure 2. Likelihood of fixating different letter positions of a 5-letter word when the prior fixation was located 3 or 4 and 5 or 6 letter positions to the left.

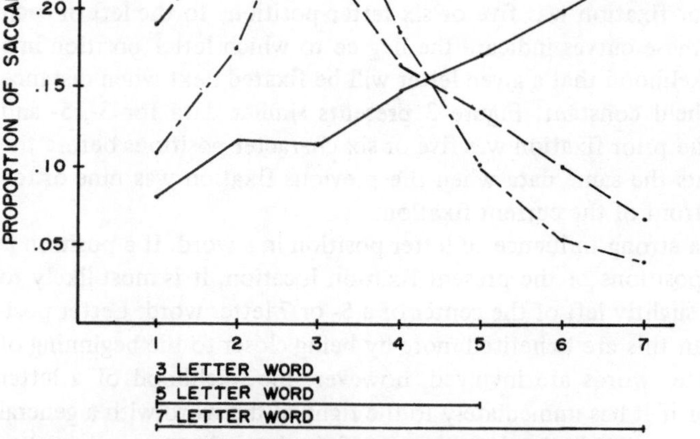

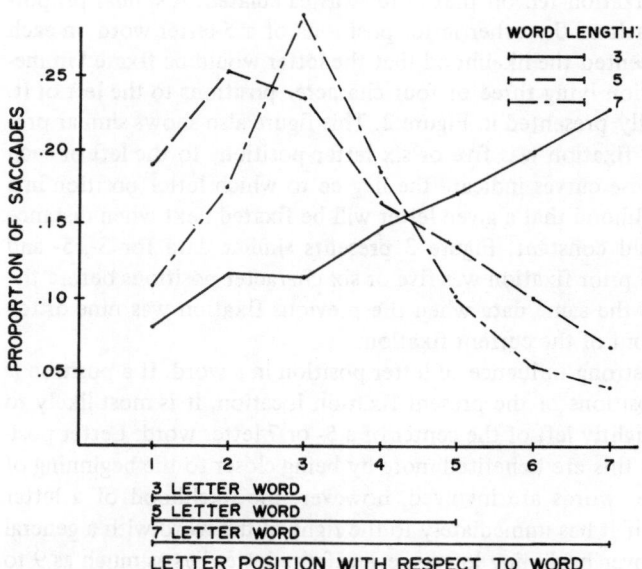

Figure 3. Likelihood of fixating different letter positions of 3-, 5- and 7-letter words when the prior fixation was located 5 or 6 letter positions to the left.

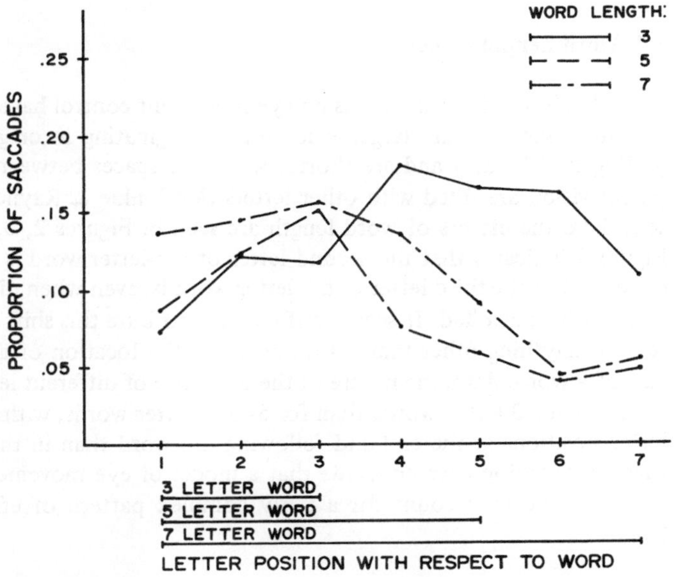

Figure 4. Likelihood of fixating different letter positions of 3-, 5- and 7-letter words when the prior fixation was located 9 or 10 letter positions to the left.

before or after a word. These observations raise two questions. First, why are certain letter positions preferred over others? Second, why aren't more fixations drawn to those locations?

In response to the first question, certain positions could be preferred because of an eye movement control algorithm that seeks these locations (i.e., go slightly left of center in the next word), or because some other determiner of fixation location correlates with word position (e.g., relative perceptibility of letters of larger sub-word units).

How one regards the second question concerning the spread of fixation locations depends on the answer given to the first. If the eye movement control system seeks to center the eyes at certain word locations, the existence of fixations at other locations must indicate either that there is error in the control or that the region sought is sufficiently large that a relatively broad region represents a hit, or both (Rayner, 1979). On the other hand, if the real basis for eye movement guidance simply correlates with word position, then the fact that this correlation is not perfect is the basis for a spread of fixation locations. This latter possibility then serves to motivate further research aimed at seeking a more fundamental basis for eye movement control. At present, we can only conclude that a model of eye movement guidance must predict a greater likelihood of fixating certain locations in words than others, and that the pattern varies with distance of those locations from the present fixation location.

3.2 Word Length Effects

Several effects of word lengths on eye movement control have previously been documented: saccades are larger when either originating in or going to longer words (O'Regan, 1979a,b) and are shorter when the spaces between words lying in peripheral vision are filled with other letters (McConkie & Rayner, 1975). In the present data, the effects of word length are seen in Figures 2, 3, and 4. For example, Figure 3 indicates that the second letter of a 5-letter word is most likely to attract the eyes, but the third letter of a 7-letter word is, even when distance from the prior fixation is controlled. It seems difficult to attribute this shift of where the eyes are sent to anything other than an influence of the location of the beginning and the end of a word. Also, the nature of the influence of different letter positions is quite different for 3-letter words than for 5- or 7-letter words, with greater attraction for letter positions at the end and following the word than in the middle of it. From these observations, we conclude that a model of eye movement control in reading must be able to account for a fairly complex pattern of effects related to word length.

3.3 Word Identification Effects

In addition to such stimulus configuration factors as word length and letter position, previous research has also demonstrated that factors related to the identifiability of words influence where the eyes go. Erroneous letters in words to the right of the fixation location can shorten saccades (McConkie & Rayner, 1975; O'Regan, 1980; Rayner, 1975b; Zola, 1981), though this occurs in a relatively narrow region (McConkie & Underwood, in preparation). Also, there is a tendency to fixate the word "the" (thought to be more perceptible due to its high frequency in the language) less than other 3-letter words (O'Regan, 1979a,b).

In the present data there are patterns which seem most easily explained by assuming that whether a word is previously identified or not influences the likelihood of letters in the word receiving the next fixation. For example, Figure 5 presents the frequency distribution of making saccades of different lengths from fixations located one letter position in front of a word. When fixating immediately prior to words of length 5 and 7, the distribution of saccade lengths appears to be bimodal. The dip between the modes comes at about the region between the words in these two cases. The results suggest that at times the word immediately to the right of the fixation was identified, in which case the eyes were sent to the next word beyond it. At other times, the word to the right was not identified and was then the locus of the next fixation. Interestingly, this bimodality disappears when the words lie just three letter positions to the right of the fixation (see Figure 6). The data suggest that most of the time the word to the right was not identified on that fixation, thus requiring it to be fixated next.

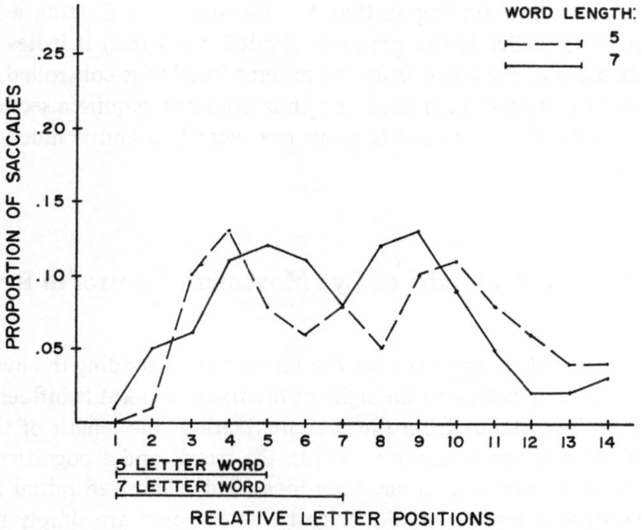

Figure 5. Frequency distribution of lengths of saccades following fixations one letter position prior to 5- and 7-letter words respectively.

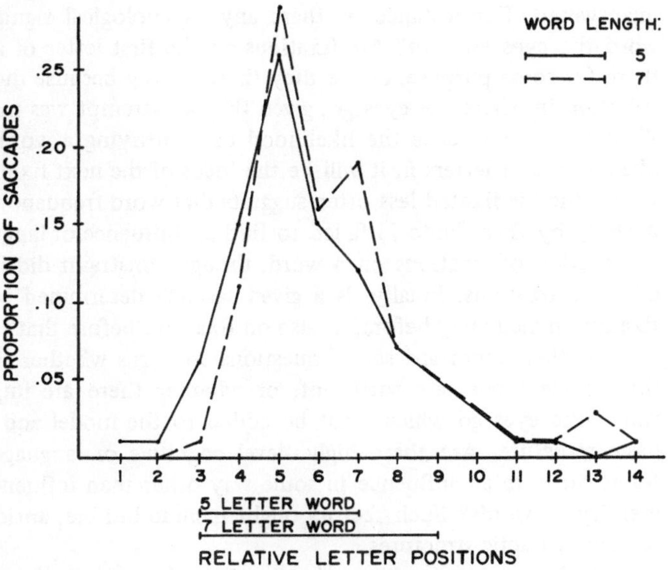

Figure 6. Frequency distribution of lengths of saccades following fixations three letter positions prior to 5- and 7-letter words respectively.

Another finding is that the likelihood of fixating a letter position is much lower if it lies in the presently fixated word than if it lies in the next word when distance of the letter from the present fixation is controlled. Apparently the fixated word is usually identified and thus does not require a second fixation; whereas the next word to the right is often not identified and is much more likely to require a fixation.

4 Toward a Model of Eye Movement Control in Reading

At present, it appears that the likelihood of sending the eyes next to some particular letter position to the right of the fixation point is influenced by stimulus factors (i.e., its distance from the present fixation, the length of the word it is in, and the letter position it occupies within the word) and a cognitive factor (i.e., whether or not the word it is in has been identified). Thus, an initial model of eye movement control in reading would suggest that the eyes are simply sent to the next unidentified word while reading carefully. Furthermore, where the eyes are sent is strongly influenced by location preferences that are a complex function of word length and distance.

This simple model appears to be capable of accounting for most present observations about forward saccades made during reading. Still, it leaves many questions unanswered. For instance, is there any psychological significance to where in a word the eyes are sent? Are fixations on the first letter of a 7-letter word placed there for some purpose, or are they there simply because there is some chance distribution in where the eyes go, given that an attempt was made to fixate a word? What factors influence the likelihood of identifying a word, thus influencing the likelihood that letters in it will be the locus of the next fixation? The fact that the word "the" is fixated less often suggests that word frequency might have an effect. A study by Zola (Note 1) failed to find an influence of language constraint on the distribution of fixations on a word, though constraint did influence the duration of those fixations. Finally, is a given saccade determined by information on the fixation immediately before, or also on fixations before that?

Another important set of questions concerns whether the factors included in this simple model are sufficient, or whether there are important influences on where the eyes go which must be added to the model and which will change its basic structure. Are there higher-level cognitive or language factors that will be found to have an influence in some way other than influencing the likelihood of identifying words? Such factors might include buffers, anticipations of upcoming text, or syntactic structures.

Finally, nothing has been said here about the control of regressive saccades, of for forward saccades during rereading of the text, or of the factors determining how long the eyes will remain in a location before moving on. Even less is known about these aspects of eye movement control in reading. They all require much more investigation.

Our message here is quite simple. Eye movements in reading reflect the moment-by-moment brain state changes induced by an interaction of the stimulus pattern and the task of comprehending. The underpinnings for a model of control of forward movements in reading involve influences due to word identification, word length and letter position the word, and distance from the current location. Our current hypothesis suggests that the reader may simply send his eyes to the next unidentified word with positioning in that word based upon its length and its distance from the point of fixation. It is from this perspective that we will continue our efforts to understand the eye guidance system in reading.

Reference Note

1. Zola, D. *The effect of redundancy on the perception of words in reading.* (Tech. Rep. No. 216) Urbana, IL: University of Illinois, Center for the Study of Reading, 1981.

References

Abrams, S.G., & Zuber, B.L. Some temporal characteristics of information processing during reading. *Reading Research Quarterly,* 1972–73, *8,* 42–51.

Andriessen, J.J., & deVoogd, A.H. Analysis of eye movement patterns in silent reading. *IPO Annual Progress Report,* 1973, *8,* 30–35.

Dodge, R. An experimental study of visual fixation. *Psychological Review,* 1907, *8,* 1–95.

Hochberg, J. Components of literacy: Speculations and exploratory research. In H. Levin & J.P. Williams (Eds.), *Basic studies on reading.* New York: Basic Books, 1970.

Just, M.A., & Carpenter, P.A. A theory of reading: From eye fixations to comprehension. *Psychological Review,* 1980, *87,* 329–354.

Levy-Schoen, A. Flexible and/or rigid control of oculomotor scanning behavior. In D.F. Fisher, R.A. Monty, & J.W. Senders (Eds.), *Eye movements: Cognition and visual perception.* Hillsdale, NJ: Erlbaum, 1981.

Levy-Schoen, A., & O'Regan, K. The control of eye movements in reading. In P.A. Kolers, M.E. Wrolstad, & H. Bouma (Eds.), *Processing of visible language.* New York: Plenum Press, 1979.

McConkie, G.W. On the role and control of eye movements in reading. In P.A. Kolers, M.E. Wrolstad, & H. Bouma (Eds.), *Processing of visible language.* New York: Plenum Press, 1979.

McConkie, G.W. Eye movements and perception during reading. In K. Rayner (Ed.), *Eye movements in reading: Perceptual and language processes.* New York: Academic Press, 1983.

McConkie, G.W., Hogaboam, T.W., Wolverton, G.S., Zola, D., & Lucas, P.A. Toward the use of eye movements in the study of language processing. *Discourse Processes,* 1979, *2,* 157–177.

McConkie, G.W., & Rayner, K. The span of the effective stimulus during a fixation in reading. *Perception and Psychophysics,* 1975, *17,* 578–586.

O'Regan, K. Saccade size control in reading: Evidence for the linguistic control hypothesis. *Perception and Psychophysics,* 1979, *25,* 501–509. (a)

O'Regan, K. Moment to moment control of eye saccades as a function of textual parameters in reading. In P.A. Kolers, M.E. Wrolstad, & H. Bouma (Eds.), *Processing of visible language.* New York: Plenum Press, 1979. (b)

O'Regan, K. The control of saccade size and fixation duration in reading: The limits of linguistic control. *Perception and Psychophysics,* 1980, *28,* 112–117.

O'Regan, K. The "convenient viewing position" hypothesis. In D.F. Fisher, R.A. Monty, & J.W. Senders (Eds.), *Eye movements: Cognition and visual perception.* Hillsdale, NJ: Erlbaum, 1981.

Rayner, K. Parafoveal identification during a fixation in reading. *Acta Psychologica,* 1975, *39,* 271–282. (a)

Rayner, K. The perceptual span and peripheral cues in reading. *Cognitive Psychology,* 1975, *7,* 65–81. (b)

Rayner, K. Eye movements in reading and information processing. *Psychological Bulletin,* 1978, *85,* 616–660. (a)

Rayner, K. Foveal and parafoveal cues in reading. In J. Requin (Ed.), *Attention & performance VII.* New York: Erlbaum, 1978. (b)

Rayner, K. Eye guidance in reading: Fixation locations within words. *Perception,* 1979, *8,* 21–30.

Rayner, K., & McConkie, G.W. What guides a reader's eye movements? *Vision Research,* 1976, *16,* 829–837.

Shebilske, W. Reading eye movements from an information-processing point of view. In D. Massaro (Ed.), *Understanding language.* New York: Academic Press, 1975.

Summary. This chapter deals with the control of forward saccadic eye movements in reading. Currently there is some controversy both about the nature of the information used in deciding where to send the eyes next and how soon the information can be brought to bear on influencing where the eyes will be sent. Analyses of a set of eye movement data that deals with the interplay between eye guidance and word pattern information are described. The conclusion is that the likelihood of forward saccades taking the eyes to a particular letter position is a function not only of the distance of that position from the prior fixation, but also of the word length and the letter position in the word which that position occupies. An hypothesis is advanced which suggests that, in reading, the eyes are simply sent to the next unidentified word with location preferences in the word being a complex function of length and distance.

II Motor Contributions to Perception and Cognition

6 Motor Theories of Cognitive Structure: A Historical Review

ECKART SCHEERER

The (motor) theory is so simple and so easy to present that every one is glad to believe it. The only question that any one cares to raise is how much of it will the known facts permit one to accept.
Pillsbury (1911, p. 84)

Contents

1 Introduction

In this review an attempt is made to sketch the history of motor theories in the fields of perception and cognition. The term "motor theory" has never been well defined; in the absence of an accepted definition, it will be used to denote a theory which maintains that movement or its sensory consequences are a constituent element of a mental process, such that the process would not exist or would have different qualities, had the motor component been absent. The review will be restricted to visual perception and cognition, and emphasis will be placed on eye movements. A complete inventory of motor theories will not be given; rather, I intend to describe three lines of thought to which most motor theories can be related.

2 The Classic Tradition: Movement as a Basis of Space Perception

In the early 18th century, perception was considered in a way that has dominated much of perceptual theorizing up to the 20th century. The perspective is tied to the

Cognition and Motor Processes
Ed. by W. Prinz and A.F. Sanders

distinction between sensation and perception (cf. Neumann, 1972). Sensation is taken as an affection or modification of the mind, while perception refers to the knowledge of objects external to the mind. The question then arises: how does the mind come to know external objects on the basis of sensations? The first motor theories of perception were formulated in this context. A case in point is Berkeley's "New Theory of Vision" (1709), in which the perception of depth was explained by convergence movements, and two-dimensional perception by tactile impressions that, in turn, are dependent on hand movements when exploring objects.

The ultimate motive behind the sensation/perception dichotomy was the Cartesian dualism between mind and matter. Mind was seen as an entity to which spatial attributes cannot be applied, while matter was defined in terms of extension. How, then, can the mind come into contact with the spatially defined world of matter?

This metaphysical problem forms the base of the first systematic motor theory of space perception, which was formulated by Lotze (1846, 1852). While for Berkeley, tactile impressions possessed an intrinsically spatial quality that could be transferred to vision, Condillac (1754) denied that even tactile impressions had intrinsically spatial attributes. Lotze follows him in that he formulated a theory of secondary "spatialization" for touch as well as for vision. However, only the latter will be considered here, because Lotze's theory of *tactile* "local signs" was far less influential than his theory of *visual* local signs. According to Lotze, the spatial ordering of our visual experience arises from a combination of two sources of information. On the one hand, we have a purely qualitative system of visual sensations; on the other hand, we have a system of eye movements and the muscular feelings accompanying them, which vary on an intensive dimension. Neither of these systems, in itself, has a spatial quality. The two systems are associatively connected because whenever a point outside the fovea is stimulated, the stimulus evokes a color sensation and simultaneously, on a purely reflex basis, a tendency to move the eye such that the stimulus is brought into foveal vision. The combination of color sensation and muscular feeling is then taken by the mind as a system of signs for spatial relations in the outer world.

Following a long tradition of psychological historiography, Lotze's theory is usually discussed in terms of the empirism-nativism debate, with Lotze taking the empiricist stand and Stumpf and Hering, for instance, the nativist position. I doubt, however, that this sorting of the theoretical positions is entirely correct. As a matter of fact, Lotze appealed to a natural spatial tendency of the mind to explain the fact that the combination of color sensations and eye movements is interpreted in a spatial fashion (Lotze, 1852, pp. 334ff.). Thus, space "as such" is not the product of experience. The real issue with which Lotze was concerned is brought out by the following quotation:

> The intuition of the real location of external objects can only be obtained by means of a recreation, rather than a mere apprehension, of spatiality. Everywhere extension is replaced by intensity, and on the basis of the latter the mind has to reconstruct a new spatial world (Lotze, 1852, p. 328).

Thus, it might be more fruitful to situate Lotze's theory with reference to the controversy between constructivist and direct-realist theories of perception, which of late has again attracted the attention of many psychologists (cf. Ullman, 1980). Lotze's theory clearly belongs to the constructivist camp, as is true for all eye-movement theories of perception.

Lotze's theory was not entirely original: it represented a crystallization of theoretical trends that had gained considerable popularity at that time. Nevertheless, compared with his immediate predecessors and contemporaries, Lotze made two assumptions that are specific to him and, at the same time, set the framework in which most motor theorists in the 19th century were destined to move.

The first assumption concerns his insistence on the *reflex nature* of the eye movements which supply local signs. In the 18th century, the motor basis of perception was generally sought in the practical activity of everyday life. At the beginning of the 19th century, we notice a certain tendency to rely on eye (and head) movements, rather than on tactile exploration, to account for the genesis of two-dimensional visual perception (e.g., Gruithuisen, 1810, pp. 49ff.); but the eye movements that were invoked were those occurring voluntarily, in free viewing situations. It was Lotze who replaced the freely moving eye by a reflex apparatus that responds to bright stimuli in the periphery of the visual field. His conception determined the main stream of motor theorizing, although some less well-known psychologists continued the older tradition for a while (e.g., George, 1854; Cornelius, 1861).

Lotze's move to a reflex conception had two motives (apart from the fact that he was able to draw upon increasing solid knowledge about the structure of the nervous system, knowledge that had made possible the formulation of the reflex-arc model). Firstly, Lotze thought that the association between color sensations and eye movements was established in the nervous system, rather than in the mind, and he was convinced that all bodily processes could be reduced to mechanical laws. Secondly, owing to their deterministic nature, the reflex movements could supply a stable system of local signs.

The second peculiarity of Lotze's theory is that it replaced actual eye movements by mere *tendencies to eye movement*. If the basic premisses of the theory are accepted, such a move is absolutely necessary to account for the fact that different locations in the visual field can be perceived simultaneously. Herbart (1825) had assumed that only one point in the visual field could be perceived in any moment, but Volkmann's (1846) pioneer tachistoscopic investigations showed that this assumption was false. Consequently, Lotze complained about the exaggerated role some of his contemporaries assigned to eye movements; instead, once the system of local signs has been acquired, he thought both that it is available simultaneously throughout the visual field, and that eye movements do not contribute to the perception of visual size, form, and movement.

Lotze's theory contains most of the core assumptions made by motor theorists in the second half of the 19th century. The best known proponents of this type of theorizing are Helmholtz and Wundt. What follows is a brief outline of how they tried to solve some of the problems inherent in Lotze's theory.

A standard objection against Lotze was that his theory rested on purely metaphysical grounds. Although later motor theorists were firmly convinced that they had liberated the theory from its metaphysical underpinnings, one may doubt whether or not they had, in fact, succeeded. Along with most psychologists of their time, both Wundt and Helmholtz believed that sensation had only two attributes – quality and intensity – and that the spatial properties of experience must be supplied by an additional process. As late as 1898, Wundt stated this principle as follows: "Whenever an idea of extension occurs, sensation and movement must have cooperated" (Wundt, 1898/1921, p. 389).

Traces of the Cartesian mind-body dualism are clearly visible in this statement. From a psychological standpoint, it faced the difficulty that the sensory and motor components of spatial perception could not be demonstrated by means of introspective analysis. This problem haunted Wundt when he began his work on sensory perception around 1860; he solved it by formulating the "principle of creative synthesis". Applied to the problem of space perception, the principle states that, owing to an active mental process, spatial attributes have an emergent quality that is caused by, but cannot be deduced from, the combination of intensity and quality. Thus, local signs are *complex*. They are also *genetic*, in that they do not result from separate associations between single sensations and movements but, rather, from a slowly evolving adaptive process which supplies the basic dimensions of spatial experience, not the precise space values of single points in the visual field. Both modifications of the local-sign doctrine were directed against what Wundt considered the "naive" associationism of the British empiricists, which had been taken over by Helmholtz.

Another difficulty of 19th century motor theories concerns basic uncertainty about the nature of the motor signal involved in perceptual processes. The sensualist bias which prevailed at the time led most investigators, especially those in the British tradition, to accept the "muscle sense" as the source of movement-produced signals. However, general reliance on eye-movement tendencies made it difficult to maintain that relevant sensory information resulted from actual muscle contractions. Also, evidence was accumulating about the different perceptual effects of active and passive eye movements.

Some historians of science believe that Helmholtz developed his theory of innervation feelings because at the time there was no information about the presence of sense organs in the muscles. However, it must be understood in the above-mentioned context, and at any rate, the idea that movements are judged by the central motor impulse or by an effort of the will was not absolutely new at the time. The real innovation introduced by Helmholtz becomes apparent in the following quotation:

> The direction of seen objects relative to our body is *judged* by means of feelings of innervation of the eye muscles, but it is constantly *controlled* by the effect produced by the innervations, that is, by the shift of the (retinal) images (Helmholtz, 1867, p. 801).

Thus, Helmholtz postulated that the central motor discharge is compared with the movement of the retinal image, rather than with the contractions of the eye muscles, and, as a result, his theory qualifies as an early version of the "reafference principle" (von Holst & Mittelstaedt, 1950).

However, the "reafference" part of Helmholtz's theory escaped the attention of his contemporaries. As a result, his theory was discussed with reference to the presence or absence of proprioceptive information about the position of the limbs – a question which is quite irrelevant to it. This emphasis led to the downfall of the theory, because evidence was accumulating that appreciation of the position of the limbs is impossible in the absence of proprioceptive information (cf. the discussion in Fröbes, 1923, pp. 158ff.). By the end of the century, the notion of innervation feelings had generally fallen into disrepute, except with Wundt. He developed a modified version of the theory, in which innervation sensations were taken as a corollary discharge (*Miterregung*) spilling over from motor centers to neighboring sensory centers (Wundt, 1908–1911, Vol 2, pp. 37ff.).

The motor theory of local signs is not tied necessarily to Helmholtz's efferent theory, but it had the same fate: at the end of the century, nobody believed in it except Wundt. The following arguments – most of which go back to Stumpf (1873) – were usually adduced against it: it is difficult to understand why a spatial (rather than any other) attribute should emerge from a combination of quality and intensity; introspection knows nothing of sensations arising from eye movements; the accuracy of eye movements is much lower than the accuracy of spatial discrimination; and finally, the discrimination of simultaneously visible colors depends logically on their separate localization, rather than vice versa (see Fröbes, 1923, p. 281, for these arguments).

The downfall of the motor theory of local signs had the result that the dogma of the non-spatiality of sensation was no longer accepted universally. Even former students of Wundt (e.g., Külpe, 1893, pp. 30ff.) were willing to count extension or localization among the basic attributes of sensation. However, this did not necessarily lead to the *total* demise of motor theories in vision. A case in point is Ebbinghaus's position. He thought that the basic dimensions of visual space are inborn, and, thus, that there is no necessity for the acquisition of local signs on a motor basis. On the other hand, the segmentation of the visual field into identifiable objects is based on eye movements, a point which applies especially to the appreciation of the length and direction of lines (determined by the effort needed to move the eye from one endpoint of a line to the other). Most optical illusions depend on this, and Ebbinghaus was in a position to quote actual eye-movement data from Judd's (1905) pioneer investigations of the Müller-Lyer illusion (Ebbinghaus-Dürr, 1913, p. 88). However, at least in Germany, he and Wundt were the last psychologists, for a long time, to give serious consideration to a motor explanation of optical illusions, not mention visual perception in general. The profound change of the theoretical "climate" is evident in the fact that the very editor of the second (and posthumous) volume of Ebbinghaus's handbook passed the judgment that, among all theories of optical illusions, the eye-movement theory is the only one that certainly is untenable (Ebbinghaus-Dürr, 1913, pp. 117ff.)!

Universal rejection of motor theories resulted not so much from an accumulation of negative evidence, but rather from a general "paradigm shift": motor processes had been invoked wherever a straightforward explanation along the lines of sensualist elementarism had failed; once the psychology of sensory elements had been relinquished, they were simply no longer needed.

3 Motor Theories of the Higher Mental Processes

While the end of the 19th century was characterized by general scepticism toward motor approaches to perception, we find during the same period an increasing tendency to apply motor explanations to higher mental processes. We also find, for the first time, a theorist who endeavors to explain conscious awareness in terms of motor processes.

Space limitations prevent more than a passing reference to the "partial motor theories" of the time: Ribot's (1889) motor theory of attention, the James-Lange theory of emotion (James, 1890), and the theory of thinking developed, by Titchener (1909), in response to the challenge of sensualism put forward by the Würzburg school. All these theories share two properties. First, they focused on the *adaptive* function of motor processes, whereas the Lotze tradition was concerned with their *representative* function. Second, they depended on the principle of *peripheralism* – a result of the replacement of the psychological definition of sensation by a physiological definition (cf. Neumann, 1972).

A much more inclusive motor theory was formulated, at the end of the last century, by Hugo Münsterberg. Like many former students of Wundt, Münsterberg had an ambivalent attitude to the central core of Wundt's system, i.e., to the *apperception doctrine.* Wundt taught that apperceptive connections were not subject to the principle of psychophysical parallelism. In addition, he considered apperception, an internal volitional activity, to be the prototype of mental life in general. He thus violated two basic principles accepted universally by the psychology of his time: physiological reductionism and sensualism. On the other hand, the apperception doctrine stressed the active character of mental life, an aspect which rendered it attractive to many German psychologists – including Münsterberg – who shared Wundt's background in German idealism, in particular the action-based philosophy of Fichte. Thus, in Münsterberg's own words, his action theory was destined to "inherit the strict psychophysical conceptions" of the association doctrine and, at the same time to follow the example of the apperception doctrine in "taking account of the active side of mental life, of attentional and inhibitory phenomena" (Münsterberg, 1900, p. 527).

Münsterberg subscribed initially to a peripheralist version of the motor theory which had been proposed by Ribot (e.g., Münsterberg, 1889); in its final form, however, his theory assumed a *centralist* flavor. For a sensation to gain access to consciousness, it must be accompanied by a central motor discharge. In addition to the traditional attributes of quality and intensity, all sensations vary in the attributes of vividness and evaluation. Neural excitation varies in two dimensions only: the location of the neural pathway and the strength of the excitation. How can *four* dimensions of sensation be represented by *two* dimensions of excitation? Intensity and quality are determined by stimulus parameters, thus, it is natural to assume that they are represented by the parameters of afferent excitation. The only way to represent vividness and evaluation, then, is in the efferent pathway. Whether or not a sensation is consciously experienced depends on its vividness and, therefore, on the strength of the efferent impulse elicited by it.

Another notable feature of the action theory was its strong reliance on *inhibitory processes,* which, in turn, was justified by its motor orientation. Münsterberg was probably the first psychologist to state a principle that is maintained by many motor theorists: "Only such ideas are incompatible to us which are connected with antagonistic motor impulses" (Münsterberg, 1891, p. 105). That is, opposition or incompatibility are not a matter of sensations per se, but accrue to them only through the movements or actions that are evoked by them. For Münsterberg, the prototype case of incompatibility in the motor domain was the antagonism between flexor and extensor muscles. To give a physiological basis to his action theory, he appealed to the antagonistic arrangement of subcortical motor centers which, he considered, were the only locus of inhibition in the central nervous system. The latter point poses a certain difficulty for the theory, because vividness was assumed to be a conscious attribute, and the cortex was taken as the only locus of conscious processes. Accordingly, Münsterberg was led to postulate the existence of subcortical-cortical feedback loops, i.e., he assumed that the activity of cortical areas was regulated by impulses originating from subcortical centers. The idea has a distinctively modern flavor, but despite Mach's early speculations about lateral inhibition it rested on the faulty assumption that there are no inhibitory interactions between afferent neural impulses. Neither did it take account of the complexity of central connective relations and cortical inhibition.

Münsterberg's theory is perhaps the closest approximation to behaviorism that was possible within a psychology that still defined itself as the science of mental life. Although Münsterberg used the concept of action to designate his theory, it is, in fact, a theory of movements rather than of purposeful acts. The reason is that Münsterberg adhered to a strict epistemological division between scientific psychology and the humanities. All aspects of human experience that relate to active striving and to the realization of values were to be dealt with by the humanities; scientific psychology was restricted to the study of conscious contents, as registered by a passively onlooking ego (*vorfindendes Ich*).

Münsterberg had developed his theory while he was in Germany, but, after his move to the United States, it provided new impetus to the wave of motor theorizing in American psychology. Indeed, in 1910, Walter Pillsbury felt obliged to raise a dissenting voice and to urge caution against assimilating all of consciousness to movement (Pillsbury, 1911).

Pillsbury analyzed several pitfalls of the motor theories then in vogue. Against the *peripheralist* version (which then was coupled with a Titchener-type sensualist bias) he pointed out that an attempt to reduce all qualities of sensation to movements is unlikely to succeed, because the number of qualitatively different sensations is much higher than that of movements. But even if it succeeded, it would not provide a fundamental breakthrough in psychology, because movements are available to consciousness via kinesthetic sensations, and these are still sensations, after all. The centralist version, represented by Münsterberg, was regarded by Pillsbury as a theoretical manoeuver designed to render the motor theory untestable. Another domain of motor theories were *dynamic effects,* such as selective attention and perceptual organization. Against an explanation of selective effects in terms of motor preparation, Pillsbury objected that it fails to explain why the "motor traces" are in a state of preparation; if this task is taken seriously, sensory excitations and previous habits must be invoked, and an explanation in purely motor terms becomes impossible. The only form of motor theory with which Pillsbury felt genuine sympathy is one which asserts that the grouping of qualities into objects and events is made by actions. But, this theory cannot explain why a movement should assume cognitive value outside the context in which it has originally occurred. Why should eye movements that occur when I think of a triangle have greater cognitive value than an image of the triangle? A similar problem poses itself when *recognition memory* is explained by the evocation of movements previously made to the stimulus: the movements represent a recognition problem in themselves – they are difficult to remember and to discriminate, the more so because many movements go entirely unnoticed.

Against all attempts to explain consciousness by movements alone, Pillsbury insisted that

> all action is sensory-motor. . . Consciousness is made up of sensation understood with reference to movement, and of movement in the light of and under the control of sensation (Pillsbury, 1911, p. 99).

At several points, Pillsbury's address reads like a prophecy. He thought that motor theories are popular because they are easy to understand; he praised them for having brought psychology into contact with practical life. On the negative side, he noted that "to say that all functions are to be explained by movements would be meaningless if function meant nothing more than movement" (Pillsbury, 1911, p. 95). The latter, of course, is exactly what Watson did, and the attractiveness of the behaviorist program had exactly the reasons Pillsbury gave for the popularity of motor theories.

There are good reasons, then, to assume that early behaviorism was simply motor theorizing driven to its extreme, i.e., the logical consequence of a theoretical trend that had been developing during the 20 years preceding the formulation of the behaviorist program. More specifically, Watson took over the peripheralist version of motor theorizing. Among the objections he raised against the image concept is that it was defined as a "centrally excited sensation" and, thus, came into conflict with the unbounded determinism on which Watson wanted to base psychology as a "branch of natural science" (Watson, 1914, pp. 16ff.). However, when Watson and his fellow behaviorists carried through their program, we notice a certain paradox: theirs was a "partial motor theory" in the sense outlined above. They developed a motor theory for the thinking process only, while perception, the traditional domain of motor theorizing, was not subjected to an analysis in motor terms. This remained true even in the later developments of behaviorism. In Hull's theory, for example, the only mechanism related to perception is afferent neural interaction; efferent mechanisms are not mentioned in this context.

It is a truism of psychological historiography that, during the reign of behaviorism, the *image* had no place in psychology. But this received opinion seems to overlook the fact that, at least until the 1930s, there existed a line of thought which upheld the central place of the image in psychology. One exponent of this line of thought was Margaret Flow Washburn. She developed a "motor theory of consciousness" and was particularly concerned about the application of the motor theory to what she called the "complexer mental processes" (Washburn, 1916).

Washburn's theory had both a physiological and a psychological part (Abel, 1927). Her physiological theorizing was, in many respect, similar to Münsterberg's theory. However, she was concerned not only with central motor impulses but also with their effects on the muscles, effects which become available to consciousness in the form of kinesthetic sensations: clearly, a concession to the peripheralism of Titchener, her teacher. Another deviation from Münsterberg was that Washburn made access to consciousness dependent on the ratio of excitation to inhibition in the motor discharge. Münsterberg had assumed that vividness was dependent on the strength of motor excitation only, a position which seemed counterintuitive in view of the fact that the most readily evoked movements are automatic, i.e., not accompanied by consciousness. According to Washburn (1916, p. 25) the difficulty can be removed by assuming that access to consciousness is an inverted-U function of the excitation/inhibition ratio. A medium value of the ratio produces what Washburn called "tentative movements", i.e., "actual slight contractions of the muscles which the larger movements would involve" (Washburn, 1916, p. 26). When tentative movements occur slowly and with delays we are in an attentive state. Attention, then, is a motor phenomenon; because it is the indispensible precondition for the formation of associations, association also depends on motor processes.

The excitation/inhibition ratio determines our awareness not only of peripherally excited sensations but also of "centrally excited sensations", i.e., images. In an image, the sensory effects of past stimulations are revived. The image is formed by a process akin to Pavlovian conditioning: when a motor center is excited, all sensory centers that have recently (or frequently) discharged into the motor center are excited by the lowering of synaptic resistances (Washburn, 1914). The properties of images are determined mainly by the delay between the motor impulse and the actual occurrence of the movement: the longer the delay, the more extensive the spread of excitation from motor to sensory centers and, consequently, the more detailed and inclusive the image. Because images result from association and association depends on attention, i.e., a motor process, images always contain a kinesthetic component, which is fused into a modality-specific component. Images, then, are the "conscious accompaniments" of tentative movements, but not all tentative movements are accompanied by images; imageless processes occur when the movements are automatized or when the associative process is obstructed by incompatible movements.

At this point, we have already reached the psychological part of Washburn's theory, the part concerned with the structure and function of movement systems. Washburn's concept of tentative movements was, of course, the theoretical heir of the movement tendencies that all motor theories require. But we notice a significant change in emphasis. The earlier theorists had been concerned with single movements and movement tendencies derived from them, and they had interpreted movement tendencies in a purely retrospective fashion: they were supposed to derive from overt movements, but they were not related to future movements or actions. Washburn's insistence on the occurrence of tentative movements in delayed action shows that she was much more interested in their prospective function: "only in delay can an action-problem be formulated, can the realization and anticipation of something to be done materialize" (Abel, 1927, p. 96). Once the term "image" has been dropped, Washburn's descriptions of tentative movement and its role in the formation of movement systems sound very similar to the constructs that were introduced by later behaviorists, such as "fractional antedating goal response" or "vicarious trial and error".

Unlike the behaviorists of her time, Washburn was prepared to meet the Gestalt psychologists on their own field, the study of perception. She accepted the Gestalt laws but felt that they were not adequately explained by Köhler and his associates. For Washburn, the organizing and unifying factor in perception is given by the potential actions that one can perform on an object. They are the essence of the "thing-character" by which perceptual configurations are constituted (Washburn, 1926).

There are, however, other aspects of the "thing-character" which escaped Washburn's attention. One of them is the "objectivity" of things, in the sense that the knowledge of things presupposes the establishment of a dualism between the subject as the knower and the object as the known. This aspect of the object category is the focus of interest in the next line of motor theorizing we shall review.

4 The Genetic Approach to Motor Theorizing

At the risk of oversimplification, the motor theories reviewed so far may be characterized as follows: in the first type, which originated with Lotze, motor processes were seen as instruments of *representation,* while in the second type, they were seen as instruments of *adjustment.* The most general feature of the theories to be reviewed next is the elimination of the opposition between adjustment and representation. If we employ "cognition" as a generic term for veridical representation, then we can say that *cognition is the product of previous adjustment,* and, at the same time, it is *a precondition for future adjustments.* The only way to prove this general thesis is by way of genetic analysis, a task which was not acknowledged by the other types of motor theories, despite lip-service paid to the genetic principle.

The evolutionary-genetic approach to psychology was initiated by Herbert Spencer. True to the tradition of British empiricism, he assigned the impressions of the "muscle sense" a central role in his psychology. In his theory of visual space perception, he belongs to the tradition which, in Europe, was represented by Lotze; that is, he deduced the spatial attributes of perception from a combination of non-spatial sensations and eye movements (Spencer, 1855/1886, pp. 180ff.). However, contrary to the Lotze tradition, Spencer extended the motor approach to the perception of "primary" (i.e., non-spatial) attributes of objects in the external world. The experience of resistance is a primary attribute of objects, and, while it is usually signaled by a combination of pressure and muscle sensations, from a genetic standpoint, priority belongs to the muscle sensations (Spencer, 1855/1886, pp. 239ff.).

Muscle sensations and the experience of resistance also play an important role in the differentiation of subject and object. According to Spencer, a *preliminary* subject/object differentiation is supplied by the fact that "before reasoning begins . . . . consciousness divides into the vivid and the faint aggregates" which roughly correspond to perception ("vivid aggregate") and memory or imagination ("faint aggregate"). *Complete* subject/object differentiation requires the intervention of an active exploration of our own body, an exploration which reveals a constant connection between the inner experience of force and modifications in that part of the "vivid" aggregate called "my own body", and the identity of these modifications with those modifications in the rest of the "vivid" aggregate which are not provoked by our own activity. Thus, the separate and continuing existence of the "vivid" aggregate (representing the outer world) and the "faint" aggregate (representing our inner experience) was conceived via a transformation of the *feeling* of force, originating from the experience with our own body, into the *idea* of force, originating from experiences with external objects (Spencer, 1855/1886, pp. 508ff.).

Spencer's ideas were expanded into a systematic motor theory by the Russian physiologist Sechenov, who himself acknowledged his indeptedness to Spencer (Sechenov, 1878/1956, pp. 275ff.). Outside his own country, Sechenov is known almost exclusively for his discovery of central reflex inhibition and as a precursor of the later "reflexological" movement initiated by Bekhterev. But when Sechenov insisted that the reflex was the prototype of all mental activity, he did not mean to reduce all mental life to reflexes. Rather, he was motivated by his dissatisfaction with what he regarded as the principal defect of psychology in his time, viz., that it treated consciousness as a self-encapsulated entity which was not under the causal influence of the environment or of the brain. Against this defect, he set the principle that every mental act originates and ends in the outside world and that, in addition, it has a central phase, which Sechenov identified with consciousness. As far as the methodology of psychology is concerned, Sechenov strongly urged a genetic approach, which he practiced in his main work on psychology, the treatise on the "Elements of Thought" (Sechenov, 1878/1956).

Sechenov was a motor theorist by virtue of his principle that the muscles are not only organs of physical work but also instruments for the cognition of external reality. At the level of sensory cognition, movements have a threefold influence: they *improve* the conditions of perception; they *divide* a continuous sensation into a chain of separate perceptual acts; and they provide a *link* between different perceptual acts.

To understand this statement, we must acknowledge that Sechenov, like William James several years after him, pictured the world of the newborn child as a chaos of sensations. In this chaos we have, however, a stream of stronger sensations corresponding to the more powerful impressions. The first motor reactions of the body intensify the already stronger sensations by bringing the sense organs into a suitable position; this serves to segregate more clearly the stream of sensations into a strong and a weak part[1]. At the same time, movements transform the continuous stream of sensations into a variable chain; the best example is in vision, where each new fixation brings a new group of sensations into sight. Finally, a connection between the successive links in the chain is made by the muscular sense. Again, vision affords the best example: movements of the head and of the eyes signal the spatial relation between the impressions received in each fixation.

Thus, movements perform simultaneously both an analytic and a synthetic function, and it is these functions that transform the initial chaos of sensations into perception. Obviously, Sechenov, here, paid tribute to the tradition according to which movements transform sensation into perception. But notice an important shift of emphasis. "Classic" theorizing about perception, in the sense outlined above, had been of the "enrichment" type: a constant core of sensations is supplemented by associations, unconscious inferences, etc., and thus is transformed into perception. In opposition to this general trend, Sechenov's views on perception belong to the class of differentiation theories.

A second difference is that Sechenov invoked the muscle sense not only for the area of perception but also for the field of abstract thinking. In this connection, he relied on certain precedents established by Spencer, but he gave them an original twist by appealing to the role of locomotor movements: because the walking movements are repeatable, periodical, and equidistant, they establish the number system and the basic units of measurement in time and space. The proposal sounds curiously naive, but it should be noted that appeals to rhythm as the basic organizing princi-

1 Sechenov seems to have taken over the idea of separating experience into a "strong" and a "weak" part from Spencer, but he uses it in a different sense. Spencer's distinction was the heir to the traditional dichotomy between the outer senses (= vivid aggregate) and the inner sense (= faint aggregate). For Sechenov, both "strong" and "weak" sensations are elicited by the outer world, and the criterion for distinguishing them was, apart from stimulus strength, their biological relevance.

ple of mental activity were common at the time; it should also be remembered that Piaget (1947/1966) considers rhythm to be the biological prototype of reversibility, the defining characteristic of operational thinking.

Finally, as a physiologist Sechenov also dissented from the accepted opinions of his time. His constant reference to the "muscle sense" was not so much a matter of sensualist dogma, but a consequence of his insight into the necessity of reafferentiation for the regulation of movements. He was, perhaps, the first to understand that certain movement disorders, such as ataxia, are the result of a loss of reafferent stimulation – an insight which half a century later (although without being credited to Sechenov) helped to develop the new discipline of cybernetics (Rosenblueth, Wiener, & Bigelow, 1943). In brief, although Sechenov talked about the reflex *arc*, in reality he accepted the principle of the reflex *circle*. In this respect, he anticipated the basic discovery of the next motor theorist I shall review.

In the movement known as *American functionalism*, the principle that the basic unity of all activity must have a circular structure is usually credited to Dewey (1896), but its implications were explored most energetically by James Mark Baldwin. There is no evidence that Baldwin was influenced by Sechenov when he formulated his concept of the circular reaction (Baldwin, 1895/1898); rather, they both derived their basic orientation, independently from each other, from Spencer. In fact, according to Baldwin's principle of the circular reaction, the function of movements is almost opposite to that which Sechenov had assigned to it: for Baldwin, movement tended to repeat or at least to maintain the current state of stimulation, while for Sechenov it produced new stimuli. Nevertheless, Baldwin was able to reach the same general conclusion as Sechenov: movement is the foremost instrument of cognition of the outer world, a conclusion that speaks to the vast flexibility of motor theorizing.

One problem with Baldwin's conception is that maintenance of a stimulus situation is not necessarily the same as forming an image of the situation. In fact, in their most primitive form, circular reactions cannot have a cognitive function: the organism would be oriented exclusively toward the result produced by its own movements, thus rendering it incapable of evaluating the precision with which his movements reproduce a given stimulus situation. According to Baldwin, the cognitive function of movements arises only as the result of a sequence of developmental steps, in which the initial orientation towards results gives way first to an orientation towards the movement itself, and finally to the comparison between the movement and the result produced by it. As an outcome of the comparison process, the perceived model and the muscular sensations are fused into a new response which can be executed in the absence of a visible model.

The real importance of this developmental sequence is that it results in the acquisition of the inner-outer dualism which is basic to cognition. Baldwin's argument runs as follows. When circular responses have reached the stage at which a movement and its result can be compared with each other, their repetition assumes the character of a "try-it-again" activity: the movement is repeated until it has produced the desired result. The subjective counterpart of this "persistent imitation" is a feeling of effort. This is registered along with the actual movements that give rise to it, and, thus, the child comes for the first time to compare a subjective state with an event that is registered via the sense organs. As a result, he emerges from an initial "projective" stage, in which inside and outside are not yet separated, into a "subjective" stage, in which he applies the inner-outer dualism to himself, but not yet to other persons. However, the child observes the similarity between his own movements and that of people he is trying to imitate. Consequently, he achieves the conclusion that, in other people too, the movement which he observes is an outside manifestation of an inner effort: this marks the attainment of "ejective" stage of personality development[2].

Baldwin developed this reasoning to substantiate his general thesis that the self is a social product. In the more modest context of the present review, its importance lies elsewhere. It signals the emergence of a new type of motor theory that, in comparison with earlier motor theories, has two properties. First, movements subserve the transition from perception to mental images, rather than from sensation to perception. This meant a break with the earlier tradition, according to which percepts and images had been subsumed under one catergory ("ideas", *Vorstellungen*), and images were considered to be "faint copies" of previous percepts. According to the new approach, images are different from percepts because they can be distinguished from the objects to which they refer; for percepts, such as distinction is impossible. Second, the movements involved in the formation of mental images are thought to be imitative, and the image is considered to be an "internal imitation" of the object. This means, on the one hand, that the relationship between image and object can be described in terms of similarity and, on the other hand, that the process of image formation is embedded in a social context[3].

Although very influential at the time, Baldwin's ideas rapidly fell into oblivion. They have been re-discovered only in the last few years as an anticipation of the Piagetian program of a genetic epistemology. In fact, it is obvious that Piaget's views on the role of imitation in the acquisition of symbolic activities are very similar to what Baldwin had to say about the topic.

According to Piaget (1945/1962), the development of mental images results from the child's ability for deferred imitation, an ability that is acquired at the end

---

2 Baldwin's use of the term "projective" differs from the one to which we are accustomed, which is influenced by psychoanalysis and ultimately derives from the problem of the "spatialization" of originally non-spatial experience. A "projection" in the modern sense becomes possible only in the Baldwinian "ejective" stage, because it presupposes the distinction between the self and other people. The term "ejective", which according to Romanes (1884, p. 16) was coined by Clifford, was quite commonly used at the end of the 19th century.

3 Note that the imitative nature of the image, in the context of Baldwin's theory, does not necessarily imply that it has a social genesis. In line with an old tradition (cf. Scheerer & Schönpflug, 1984), but contrary to modern usage, Baldwin applies the term "imitation" not only to social interactions, but in the broadest imaginable sense, in fact, virtually synonymous with any kind of interaction with the environment (Baldwin, 1895/1898, p. 224). The most important form of nonsocial imitation is self-imitation, i.e., the repetition of circular responses.

of the sensorimotor stage of intelligence. Among intelligent adaptations in general, imitation is a special case in that it is characterized by a preponderance of accommodation, while in symbolic play we find a preponderance of assimilation. The very fact that accommodation and assimilation are temporarily separated from each other allows the differentiation between the sign and the significate. The mental image – the internal imitation of an object or action – acquires the function of a signifier, a function that is different from the object or action it signifies.

Is it correct to pick out Piaget's views on image formation to qualify him as a motor theorist, given that he derives *all* intelligence from sensorimotor activity and, thus, could be considered a motor theorist in a much broader sense? Beyond the level of sensorimotor intelligence, the mental image is the only part of the cognitive system that cannot be detached *entirely* from movements. Neither perception nor the mental operations have this quality. It is true that perception is, in a large measure, determined by centrations and couplings, which are realized by fixations and eye movement. But they operate on the primary perceptual field, and Piaget maintained that the latter is entirely independent from any motor process, once it has been detached from sensorimotor intelligence (Piaget & Inhelder, 1966/1979, p. 30). On the other hand, mental operations are emancipated from sensorimotor intelligence to the extent that their inner structure is independent from the actual movements through which they are realized. However, Piaget stressed that mental images are motor, in addition to their quasi-sensory character, and he placed great emphasis on experimental evidence purporting to show that imagining an object is regularly accompanied by eye movements similar to the ones used in exploring an object (Piaget & Inhelder, 1966/1979, pp. 17ff.).

Piaget's motive for insisting on the (at least partial) motor nature of the mental image was his desire to prove the thesis that there is no direct developmental sequence leading from percepts to images; rather, images evolve from sensorimotor intelligence through imitation.

Images share some of the sensory qualities of percepts, but this does not prove that they are "faint copies" of previous percepts. Their structure is different from that of percepts. The field effects that are so typical for perception, for instance, play at most a very subordinate role in images. Even the similarities between the eye movements subserving visual exploration and those subserving the evocation of visual images do *not* speak for a perceptual origin of the image. Rather, they have a common origin in that each eye movement related to the contour of an object has a strong imitative component (Piaget & Inhelder, 1966/1979, p. 19). At a more general level, the difference between images and perceptions is that the former are *symbolic* while the latter are only *representative.* The most important aspect of this distinction is that images use *schemata,* that is, they involve the construction of condensed and simplified internal models, while perception uses *schemes* – plans which rely on the use of common elements occurring in analogous modes of behavior.

5 The Historical Roots of Some Current Motor Theories

By way of conclusion, I shall attempt to trace current variants of motor theorizing to their historical sources.

Lotze's motor theory applies to the problem of *exocentric spatial localization,* i.e., the spatial ordering of the visual field independent of the position of the observer. Today, a strict version of the theory (which maintains that the spatial values of individual points in the visual field are acquired on the basis of eye movements) no longer has any followers.

There is a weak version of Lotze's theory, however, which – in Wundt's apt phrase – may be termed "the influence of eye movements on surveying the visual field" (Wundt, 1908–11, Vol. 2, p. 566). Here, eye movements subserve the "measurement" of the length and direction of lines. Historically, the preferred testing ground for this type of theory has been the study of geometrical-optical illusions.

From the beginning, the motor theory of optical illusions has been faced with two difficulties. The first is the necessity to appeal to unobservable events called "eye-movement tendencies" (because optical illusions are present under conditions were eye movements are impossible). The second is the objection that the movements of the eye, when inspecting an illusion-inducing figure, are the effect rather than the cause of the illusion. It seems fair to state that these difficulties still persist in the recent attempts to revive the Wundt-Ebbinghaus motor theory of optical illusions (Festinger, White, & Allyn, 1968; Virsu, 1971). However, they have brought about the discovery (or re-discovery; see Judd, 1905) that "erroneous" eye movements (e.g., overshoots in the Müller-Lyer figure) are followed by corrective saccades and that the practice-induced decrement of the illusions is obtained only under conditions permitting free scanning of the figure. Thus, eye movements have an *error-correcting,* rather than an *error-inducing* function. Accordingly, in the most recent systematic survey of optical illusions, eye movements are treated as the foremost instrument by which an observer "somehow garners information about the nature and extent of the illusion" (Coren & Girgus, 1978, p. 184).

Compared with the classic version of the motor approach to perception, this is a rather modest statement. Nevertheless, a certain similarity to the Lotze type of theorizing may be discerned. It consists in the idea that eye movements serve to re-establish a veridical link between perception and the outer world, a link which is destroyed (Lotze) or at least severely distorted (Coren & Girgus) in the afferent and central "stages" of the perceptual process.

The real domain of current debate about the role of eye movements in perception, though, is not the study of visual illusions arising from stationary patterns, but rather is the "Helmholtz problem" – the stability of *egocentric localization* despite retinal image shifts produced by eye movements. Helmholtz's solution to the problem has been revived by von Holst and Mittelstaedt (1950) and by Sperry (1950). Unfortunately, von Holst and Mittelstaedt developed their "reafference principle" on the basis of the optokinetic following-response of insects, which they considered to be an "optomotor reflex". As a result, their theory is criticized even today because of its alleged foundation in reflex physiology (e.g., Turvey, 1979). In fact, however, it was oriented *against* "classic reflex theory" (von Holst & Mittelstaedt, 1950, p. 464).

Among psychologists, the reafference principle has been accepted and expanded by Held (1961) and by Festinger and Canon (1965). Held's amplification involved the idea that the correlation between efferent and reafferent signals is not fixed, but is acquired in early life and can be changed through an adaptive process similar to the one envisaged by Wundt. Festinger criticized von Holst for not having formulated a comprehensive efferent theory of perception (Festinger, Ono, Burnham, & Bamber, 1967, p. 7), an oversight Festinger proposed to repair by relying on the concept of "efferent readiness". While such amendments are not quite in von Holst's spirit, they are more congenial to the classic theories: Held's proposal fits nicely with the empiricism of early motor theorists, and Festinger's with their reliance on eye-movement tendencies.

The *efferent* (Festinger) or *sensorimotor* (Held) approach to perception enjoyed considerable popularity until about 1975, but of late it has come under increasing attack by the followers of Gibsonian "direct realism" (see the discussion about the paper by Gyr, Wiley, & Henry, 1979). It is not the purpose of the present review to participate in the current debate (see Shebilske, this volume, p. 99); instead, some comments will be made on the extent to which the claims made by both efferent theorists and direct realists can be related to the historical prototypes of motor theories.

In Held's theory, a crucial condition for the acquisition of motor-based spatial localization is the *active* nature of the movement, and a special emphasis is placed on locomotion. The appeal to active movement bears a superficial similarity to Helmholtz's concept of the "effort of the will". The similiarity is, in fact, superficial. Helmholtz was simply not concerned with the perceptual effects of moving about in one's environment. Also, in Held's "passive" locomotion conditions, eye movements cannot be prevented, and they depend on an "effort of the will". Thus, it is doubtful that Helmholtz would have predicted the differential effects of active vs. passive locomotion reported by Held (1965).

A second point relates to the distinction between *saccadic* and *smooth pursuit* eye movements. It is occasionally argued that "compensation" of retinal image shifts by an efference copy is restricted to voluntary eye movements (i.e., saccades), while such compensation does not occur in smooth pursuit movements. The reasoning behind this argument is that smooth pursuit movements have the function of compensating exafferent motion and, thus, it would be absurd to compensate them in perception, because such compensation would lead to perceived rest in the presence of actual motion. While this reasoning must be taken seriously, it cannot be bolstered by appealing to the authority of Helmholtz. Saccades were discovered more than 10 years after the first edition of Helmholtz's *Physiological Optics;* moreover, Helmholtz adduced cases, in favor of his theory, which must have involved smooth pursuit movements. Finally, Helmholtzians of the early 20th century denied a differential effect of voluntary vs. unvoluntary eye movements on the central compensation of retinal image shifts (see Bischof, 1966, pp. 382ff., for a statement of the problem and a brief historical review).

From the standpoint of direct realism (e.g., Turvey, 1979), efferent theories are typically criticized for assuming that (1) afferent stimulation lacks structure, (2) the relationship between afference and efference is arbitrary, and (3) afferent stimulation is ambiguous and thus needs to be disambiguated by efferent processes, while efference, itself, presumably is always non-ambiguous. When applied to the classic motor theories of perception, these criticisms are certainly justified. In fact, classic motor theories made even stronger assumptions in the wrong direction (from the standpoint of direct realism). For instance, afferent stimulation was not held to be ambiguous but to be totally *meaningless,* and both Lotze and Helmholtz maintained that local signs could be acquired even if there was a totally random connection between points on the receptor surface and in the visual centers. The latter point is especially important because it shows that their central concern was almost a cosmological one: to explain how order can arise from an initial state of chaos.

Such radical claims have been abandoned by contemporary efferent theorists, but traces of the way of thinking underlying them are still visible. The methodology of studies aimed at demonstrating "efferent factors in perception" is often characterized by the use of severely impoverished visual stimulation (cf. Haber, 1979). At a theoretical level, contemporary efferent theorists differ from their classic predecessors in that it is no longer assumed that visual stimulation lacks structure in *principle,* but rather that it *may* lack structure, and that in such a case efference will be needed to make up for the structural deficit in afferent information. The design of Held and Rekosh's (1963) well known study is a very direct expression of this theoretical conviction: evidence for exclusively efference-based adaptation to prismatic displacement is sought by employing a visual stimulus composed of randomly arranged dots.

In the case of egocentric localization, then, we face the same trend as in the case of exocentric localization – efference is relegated to the status of a secondary principle operating under conditions of dubious ecological validity. In view of this situation, one is tempted to ask which type of theory is the more "radical" version of motor theorizing – the variants of "efferent" or "sensorimotor" approaches, or the theory of direct realism. While the perceptual effects of movements are treated in the former as a kind of nuisance variable which must be cancelled to make possible perception of a stable world, the movement of the perceiver is a primary precondition in the latter for the production of higher-order invariants underlying both the perception of static surfaces and the perception of one's own movements (Gibson, 1979/1982, pp. 76ff.). Thus, although perception is exclusively based on afferent information, afferent information, in turn, depends on movement. In this sense, direct realism may be considered to be the more "radical" motor theory of perception.

While there is a clear line leading from Lotze and Helmholtz to contemporary motor theories of perception, the same cannot be said of the two other types of motor theorizing reviewed in this paper. Accordingly, only a brief enumeration will be given of certain contemporary notions that can be traced to them.

If we look at the tradition starting with Münsterberg and ending in behaviorism, we note that in the field of attention it still has its followers. Basically, the late-selection theories of attention, where the "bottleneck" is placed at the response selection stage, follow up the intellectual trend started by Ribot and Münsterberg. To the same tradition belongs the idea that incompatibility between overt responses or internal action programs is the main condition limiting dual-task performance (e.g., Neisser, 1976, pp. 99ff.).

If behaviorism is taken seriously, it denies that there are any internal representations, and in this case it is not relevant to our topic. But the various forms of *cognitive behaviorism,* which have been antedated by Washburn's motor theory of consciousness, still have an important lesson for the contemporary cognitive psychologist, namely the idea that "tentative movements", "movement tendencies", and the like must have a positive function (besides being mere symptoms of an internalization process which, for one reason or the other, has remained incomplete). Perhaps Washburn's answer, that their function is planning and anticipation, is too general, but it seems to point in the right direction.

Finally, any motor theory of *imagery* belongs to the tradition started by Spencer and ending with Piaget if it meets the criterion that eye movements are not an epiphenomenon of the image itself (in the sense that an already evoked image can be scanned or explored as if it were a real object or picture), but rather are a precondition for the evocation of an image. Besides Piaget's theory, there are several other theories meeting this criterion, e.g., Hebb's (1968) theory of imagery and Noton and Stark's (1970) "scan path" theory. However, Piaget's theory seems to be unique in that it denies any direct isomorphism and genetical link between the "figurative" qualities of perception and of imagination. The relationship between imagery and perception has been given considerable attention in the recent "imagery debate" (see Kosslyn, Pinker, Smith, & Shwartz, 1979), but none of the participants has come up with a viable solution. A final observation is that, after an initial period of enthusiasm in the early 1970s, eye movements play little or no role in contemporary discussions of imagery. Perhaps it would still be a fruitful task for imagery research to discover to what extent differences between perceptual and imaginal representations are determined by the properties of motor representations.

References

Abel, Th.M. Washburn's motor theory: A contribution to functional psychology. *American Journal of Psychology*, 1927, *39*, 91–105.

Baldwin, J.M. *Mental development in the child and in the race* (1895). Translation: *Die Entwicklung des Geistes beim Kinde und bei der Rasse.* Berlin: Reuther & Reichard, 1898.

Berkeley, G. *An essay towards a new theory of vision.* Dublin: Pepyat, 1709.

Bischof, N. Psychophysik der Raumwahrnehmung. In W. Metzger (Ed.), *Handbuch der Psychologie,* (Vol. 1, Part 1, pp. 307–408). Göttingen: Hogrefe, 1966.

Condillac, E.B., de *Traité des sensations.* Paris, London: de Buré l'aîné, 1754.

Coren, S., & Girgus, J.S. *Seeing is deceiving: The psychology of visual illusions.* Hillsdale, NJ: Erlbaum, 1978.

Cornelius, C.S. *Die Theorie des Sehens und räumlichen Vorstellens, vom physikalischen, physiologischen und psychologischen Standpunkte aus betrachtet.* Halle/Saale: Schmidt, 1861.

Dewey, J. The reflex arc concept in psychology. *Psychological Review,* 1896, *3,* 357–370.

Ebbinghaus, H. *Grundzüge der Psychologie,* (Vol. 2), E. Dürr (Ed.). Leipzig: Veith, 1913.

Festinger, L., & Canon, L.K. Information about spatial location based on knowledge about efference. *Psychological Review* 1965, *72,* 373–384.

Festinger, L., Ono, H., Burnham, C.A., & Bamber, D. Efference and the conscious experience of perception. *Journal of Experimental Psychology Monograph,* 1967, *74*(4).

Festinger, L., White, C.W., & Allyn, M.R. Eye movements and decrement in the Müller-Lyer illusion. *Perception & Psychophysics,* 1968, *3,* 676–682.

Fröbes, J. *Lehrbuch der experimentellen Psychologie*, (Vol. 2, 2nd ed.). Freiburg/Brsg.: Herder, 1923.

George, L. *Lehrbuch der Psychologie.* Berlin: Reimer, 1854.

Gibson, J.J. *The ecological approach to visual perception* (1979). Translation: *Wahrnehmung und Umwelt.* München, Wien, Baltimore: Urban & Schwarzenberg, 1982.

Gruithuisen, F.v.P. *Anthropologie oder von der Natur des menschlichen Lebens und Denkens; für angehende Philosophen und Ärzte.* München: Lentner, 1810.

Gyr, J., Wiley, R., & Henry, A. Motor-sensory feedback and geometry of visual space: An attempted replication. *Behavioral and Brain Sciences,* 1979, *2,* 59–94.

Haber, R.N. When is sensory-motor information necessary, when only useful, and when superfluous? *Behavioral and Brain Sciences,* 1979, *2,* 68–70.

Hebb, D.O. Concerning imagery. *Psychological Review,* 1968, *75,* 466–477.

Held, R. Exposure-history as a factor in maintaining stability of perception and coordination. *Journal of Nervous and Mental Disease,* 1961, *132,* 26–32.

Held, R. Plasticity in sensori-motor systems. *Scientific American,* 1965, *213*(5), 84–95.

Held, R., & Rekosh, J. Motor-sensory feedback and the geometry of visual space. *Science,* 1963, *141,* 722–723.

Helmholtz, H. *Handbuch der physiologischen Optik.* Leipzig: Voss, 1867.

Herbart, J.F. *Psychologie als Wissenschaft, neu gegründet auf Erfahrung, Metaphysik und Mathematik,* (Vol. 2). Königsberg: Unzer, 1825.

James, W. *The principles of psychology.* New York: Holt, 1890.

Judd, C.H. The Müller-Lyer illusion. *Psychological Review Monograph Supplement,* 1905, *29,* 55–82.

Kosslyn, S.M., Pinker, S., Smith, G.E., & Schwartz, S.P. On the demystification of visual imagery. *Behavioral and Brain Sciences,* 1979, *2,* 535–481.

Külpe, O. *Grundriß der Psychologie.* Leipzig: Engelmann, 1893.

Lotze, H. Seele und Seelenleben. In R.Wagner (Ed.), *Handwörterbuch der Physiologie,* (Vol. 3, pp. 142–263). Braunschweig: Vieweg, 1846.

Lotze, R.H. *Medicinische Psychologie oder Physiologie der Seele.* Leipzig: Weidmann, 1852.

Münsterberg, H. Schwankungen der Aufmerksamkeit. In H. Münsterberg (ed.), *Beiträge zur experimentellen Psychologie,* (Heft 2, pp. 69–124). Freiburg/Brsg.: Mohr, 1889.

Münsterberg, H. Über Aufgaben und Methoden der Psychologie. *Schriften der Gesellschaft für psychologische Forschung,* 1891, *1,* 93–272.

Münsterberg, H. *Grundzüge der Psychologie.* Leipzig: Barth, 1900.

Neisser, U. *Cognition and Reality.* San Francisco: Freeman, 1976.

Neumann, O. Empfindung (II). In J. Ritter (Ed.), *Historisches Wörterbuch der Philosophie,* (Vol. 2, pp. 464–474). Basel, Stuttgart: Schwabe, 1972.

Noton, D., & Stark, L. Scanpaths in saccadic eye movements when viewing and recognizing patterns. *Vision Research,* 1970, *11,* 929–942.

Piaget, J. *La formation du symbole chez l'enfant* (1945). Translation: *Play, dreams and imitation in childhood.* New York: Norton, 1962.

Piaget, J. *La psychologie de l'intelligence* (1947). Translation: *Psychologie der Intelligenz,* (2nd ed.). Zürich: Rascher, 1966.

Piaget, J., & Inhelder, B. *L'image mentale chez l'enfant* (1966). Translation: *Die Entwicklung des inneren Bildes beim Kind.* Frankfurt/M.: Suhrkamp, 1979.

Pillsbury, W.B. The place of movement in psychology. *Psychological Review,* 1911, *28,* 83–99.

Ribot, Th. *La psychologie de l'attention.* Paris: Alcan, 1889.

Romanes, G.J. *Mental evolution in animals.* New York: Appleton, 1884.

Rosenblueth, A., Wiener, M., & Bigelow, J. Behavior, purpose, and teleology. *Phisolophy of Science,* 1943, *10,* 18–24.

Scheerer, E. & Schönpflug, U. Nachahmung. In J. Ritter & K. Gründer (Eds.), *Historisches Wörterbuch der Philosophie,* (Vol. 6). Basel, Stuttgart: Schwabe, 1984.

Sechenov, I. *The elements of thought* (1878). Translation in I. Sechenov, *Selected physiological and psychological works,* (pp. 265–401). Moscow: Foreign Languages Publishing House, 1956.

Spencer, H. *Principles of psychology* (1855). Translation: *Die Principien der Psychologie,* (Vol. 2). Stuttgart: Schweizerbart, 1886.

Sperry, R.W. Neural basis of the spontaneous optokinetic response produced by inverted vision. *Journal of Comparative and Physiological Psychology,* 1950, *43,* 482–489.

Stumpf, C. *Über den psychologischen Ursprung der Raumvorstellung.* Leipzig: Hirzel, 1873.

Titchener, E.B. *Lectures on the elementary psychology of the thought processes.* New York, London: Macmillan, 1909.

Turvey, M.T. The thesis of efference-mediation of vision cannot be rationalized. *Behavioral and Brain Sciences,* 1979, *2,* 81–83.

Ullman, S. Against direct perception. *Behavioral and Brain Sciences,* 1980, *3,* 373–415.

Virsu, V. Tendencies to eye movement and misperception of curvature, direction and length. *Perception & Psychophysics,* 1971, *9,* 65–72.

Volkmann, W.A. Sehen. In R. Wagner (Ed.), *Handwörterbuch der Physiologie,* (Vol. 3, pp 264–351). Braunschweig: Vieweg, 1846.

von Holst, E., & Mittelstaedt, H. Das Reafferenzprinzip (Wechselwirkungen zwischen Zentralnervensystem und Peripherie). *Die Naturwissenschaften,* 1950, *37,* 464–476.

Washburn, F.M. The function of incipient motor processes. *Psychological Review,* 1914, *21,* 376–390.

Washburn, F.M. *Movement and mental imagery: Outlines of a motor theory of the higher mental processes.* Boston: Houghton Mifflin, 1916.

Washburn, F.M. Gestalt psychology and motor psychology. *American Journal of Psychology,* 1926, *37,* 516–520.

Watson, J.B. *Behavior: An introduction to comparative psychology.* New York: Holt, 1914.

Wundt, W. Zur Theorie der räumlichen Gesichtswahrnehmungen (1898). Repr. in: *Kleine Schriften,* (Vol. 3, pp. 285–422). Stuttgart: Kröner, 1921.

Wundt, W. *Grundzüge der physiologischen Psychologie,* (6th ed.). Leipzig: Engelmann, 1908–1911.

Summary. Current motor theories of cognitive structure can be traced back to three types of traditions. The first centers around the role of motor processes in the transformation from sensation, which is considered to have no intrinsic spatial attributes, to spatial perception. It originated in British Empiricism and reached its zenith in the "local sign" theories and in the theory of innervation sensations. At the end of the last century, this type of motor theorizing was discarded almost universally. Instead, motor explanations were applied to "higher mental processes" such as attention and thinking, and various "motor theories of consciousness" were developed, according to which motor excitations are an essential precondition of conscious awareness. While behaviorism was the heir of this intellectual tradition, it was pursued during the reign of behaviorism until about 1940. A third line of thought owes its existence to the application of evolution theory to psychology. It was focused on the role of movement in the differentiation between the self and the environment, with special emphasis on mental images considered as internal imitations of objects. Today, the first tradition is represented by various "efferent" or "sensorimotor" theories of perception, the second has certain followers in the field of attention, and the third forms part of the Piagetian approach to mental imagery.

7 Context Effects and Efferent Factors in Perception and Cognition

WAYNE L. SHEBILSKE

Contents

1 Introduction

Context influences the appearance of visual stimuli and the meaning of linguistic stimuli, i.e., appearance and meaning are influenced by circumstances, conditions, and objects that surround a stimulus. The moon illusion exemplifies a context effect: The moon appears much larger and closer when it is at the horizon than when it is at its zenith even though it projects the same retinal image in both positions. The illusion is a context effect since different contexts (terrain and horizon sky versus zenith sky) induce different perceptions.

Context effects are important for understanding efferent factors in perception. Efferent innervations are neural messages going from motor control centers to muscles. Their direction is opposite that of afferent innervations, which go from

Cognition and Motor Processes
Ed. by W. Prinz and A.F. Sanders

peripheral receptors to central structures. Since the late nineteenth century, scientists have debated the plausibility of an efferent theory of perception, which states that efferent and afferent innervations interact to determine perceptions. Proponents of efference theories include Helmholtz (1806/1963), Coren (1981), and Matin (1982); opponents include James (1890/1950), Turvey (1977, 1979), and Bower (1975). Proponents have provided ample evidence that efferent factors can influence perceptions when other information is controlled. People can, for instance, see the direction of lights flashed briefly in reduced conditions that require judgments to be based upon motor information. In these reduced-information conditions, the perceptual system either uses efferent information directly, or it uses afferent motor signals that are contingent upon efferent innervations (Matin, 1972, 1982; Shebilske, 1976, 1977). The fact that people *can* use motor information when visual stimulation is reduced does not mean, however, that they *do* use it in everyday conditions when rich visual stimulation is available. Thus, the importance of efferent factors in perception may depend upon context. This paper will review experiments that analyze efferent factors in various contexts ranging from pointing at targets against a dark background to hitting a baseball in a fully illuminated, structured environment.

Context effects are also important for understanding efferent factors in cognition. Researchers have established a correlation between eye movements and underlying cognitive processes during pattern recognition, visual search, reading, and other tasks (cf. Stark & Ellis, 1981). The correlation suggests that cognitive processes interact with efferent commands to eye muscles. Some researchers have attempted to analyze this interaction at a physiological level (e.g., Brown, 1976). But most researchers analyze efferent factors in cognition at a process level. This paper will concentrate on a subset of the process-level research. Specifically, it will review experiments that use contextual manipulations to shed light on the way in which cognitive processes influence the control of eye movements.

2 Efference and Visual Perception

Researchers have used at least four distinct paradigms to study efferent factors in visual perception. They are: (1) correlational studies that establish a correspondence between eye movements and perception by recording eye movements during visual illusions, (2) experiments that isolate motor information by measuring perception in a reduced visual context, (3) experiments that manipulate motor information by inducing erroneous eye movements, and (4) experiments that examine the interaction between motor and visual information by inducing erroneous eye movements in various visual contexts. The first three paradigms have established important facts about visual-motor coordination. They have not, however, provided a basis for refuting the hypothesis that motor information is irrelevant when full visual information is available. In fact, this very hypothesis was put forth by Turvey

(1979), Matin, Picoult, Stevens, Edwards, Young, and McArthur (1982), and Bridgeman (Note 1). The fourth paradigm has produced results that are contrary to this hypothesis. Experiments from the fourth paradigm suggest that motor information influences visual perception in fully illuminated, well structured environments. Further experiments, and perhaps new paradigms, will be required to determine how visual and motor information interact.

This section will review the four paradigms, and then it will outline two opposing process models that are viable explanations of what we currently know about efferent factors in visual perception.

2.1 Observing Correlations Between Erroneous Eye Movements and Geometric Illusions

Coren (1981) reviews the history of the idea that sensory and motor functions interact to form conscious percepts (see also Scheerer, this volume, pp. 77–98). Early speculation held that perception of shapes (e.g., triangles, squares, circles, alphabetical characters) is primarily determined by muscular information. Accordingly, we see a shape because we get muscular impressions when our eyes trace out the shape while viewing a stimulus. Empirical support for this speculation came when researchers established a correspondence between erroneous eye movements and geometric illusions. When viewing the Mueller-Lyer illusion, for instance, people make longer scan paths when they view the wings-out version than when they view a line the same length in the wings-in version. Correspondingly, people see a longer line in the wings-out version. This support was temporarily undermined by the observation that the Mueller-Lyer illusion and other illusions are seen in stimuli that are flashed too briefly for eye movements to occur. The eye movement explanation of illusions gained popularity again with the suggestion that perceptions are determined not only by overt eye movements, but also by the readiness to issue efferent commands for eye movements (e.g., Festinger, Burnham, Ono, & Bamber, 1967).

This explanation makes a cause-effect statement: efferent readiness causes geometric illusions. The explanation cannot in principle be tested, therefore, by the paradigm of observing correlations between erroneous eye movements and illusions because correlational observations cannot establish cause-effect relationships. The correlational paradigm will always leave open the question of whether eye movements cause geometric illusion or whether the illusions cause the erroneous eye movements.

The following sections move away from paradigms that are locked into correlational observations, and they move toward paradigms based on experiments. They also move from questions about geometric illusions to questions about perception of visual direction.

2.2 Isolating Motor Information in Reduced Visual Contexts

Efferent factors have been implicated in visual direction constancy (VDC), which is the invariance of apparent egocentric direction despite changes in oculocentric direction (Shebilske, 1977). Visual direction constancy has two components: (1) dynamic VDC, which refers to the apparent rest of objects during saccadic eye movements, and (2) static VDC, which refers to the apparent invariance of egocentric direction of objects before and after eye movements. The two components are independent in the sense that people in darkroom conditions see lights streak in the opposite direction of eye rotation during saccades, but they see the same egocentric direction of the lights before and after the saccades (Shebilske, 1977; Bridgeman, Lewis, Heit, & Nagle, 1979). Experiments on static VDC have isolated contributions of (1) retinal information, which is "information in the visual system resulting from stimulation of the retina by light" (Matin, 1982, p. 4), and (2) extraretinal eye position information (EEPI), which is "any information regarding the position of his eye in the orbit that an observer has available which does not derive from stimulation of the retina by light" (Matin, 1982, p. 4). The neutral term EEPI is useful because, although efferent factors are involved in EEPI, they may not be the only factors involved (cf. Shebilske, 1976, 1978).

Several procedures have been used to isolate retinal information and EEPI, but we will discuss only one example in which subjects judge the relative positions of briefly flashed lights in the dark. One light flashes, an eye movement occurs, and a second light flashes during or after the eye movement. Because of the dark conditions, the only available retinal information is the location of the retinal images of the flashes. Results show that the perception of egocentric direction is not completely determined by retinal location. For example, suppose the retinal location of the first flash was on the fovea, the eye moved 30° to the right, and a few seconds later the other light flashed 30° to the right. Both flashes would fall on the fovea, but the subject would not see both flashes in the same egocentric direction. Instead, the subject would see the second flash close to its actual position 30° to the right of the first. The eye movement would be taken into account presumably by processing EEPI. In fact, results from this paradigm suggest that static VDC is determined by an algorithm that algebraically combines retinal location and EEPI.

Figure 1 presents some heuristic assumptions that greatly simplify a discussion of how this algorithm might work. The assumptions make perceived visual location a simple linear combination of EEPI and retinal location. The figure can be used to illustrate static VDC by comparing the two drawings. For instance, perceived visual location is 0° for both A′ (in the right-hand drawing) and A (in the left-hand drawing) even though eye position and retinal location is different in the two drawings. The figure also illustrates how an algebraic combination of retinal location and EEPI can explain the results of the preceding hypothetical experiment in which a light flashed on the fovea before and after a 30° eye movement. These lights correspond to A and B′ in the figure. As illustrated, A is seen at 0° and B′ at 30° even though images of both strike the fovea.

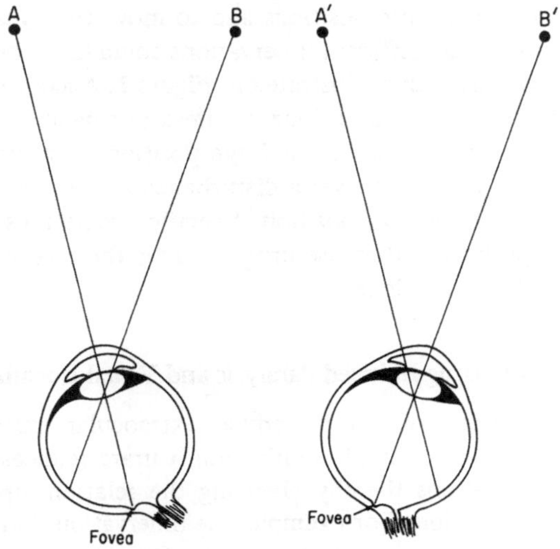

Figure 1. Simplifying assumptions of an algorithm for perceiving visual direction. The assumptions are: (a) Visual location of target [VL(T)]: A and A′ are defined as targets located at a spatial location of 0°, B and B′ as targets located 30° to the right of A and A′; (b) Retinal location (RL): the fovea is defined as a retinal location of zero. Deviations from the fovea are measured in degrees of visual angle. Target deviations to the right and left produce plus and minus changes in retinal location respectively; (c) Extraretinal eye position information (EEPI): the direction of gaze toward A and A′ is defined as zero. For other targets, EEPI is defined in terms of the visual angle between the targets and A or A′ (right = plus; left = minus). The algorithm is VL(T) = RL + EEPI. The algorithm specifies the locations of the targets in the figure as follows: VL(A) = 0° + 0° = 0°, VL(B) = 30° + 0° = 30°, VL(A′) = −30° + 30° = 0°, VL(B′) = 0° + 30° = 30°.

As mentioned earlier, the basic paradigms for isolating motor information during eye movements supports the existence of an algorithm like the one illustrated in Figure 1. These paradigms do not specify the source of EEPI, however. Experiments on the source of EEPI have used slightly different paradigms, which are discussed in the next section.

2.3 Manipulating Motor Information in Reduced Visual Contexts

Several review papers outline techniques used to manipulate motor information during eye movements in reduced visual contexts (e.g., Bridgeman, Note 1; Skavenski, 1976; Shebilske, 1977; Matin, 1982). Here we will review results from only two procedures: (1) inducing paralysis with drugs and (2) inducing minor motor anomalies (MMAs), which are errors in EEPI that can be induced by everyday conditions. Both procedures manipulate the relationship between efferent innervations and eye positions. Specific efferent innervation patterns are required to hold the eyes in specific positions and to move the eyes from one position to another (Collins, 1975). Efferent innervations could therefore be the source of EEPI in the static VDC algorithm illustrated in Figure 1. Accordingly, perceived visual location would remain veridical as long as there remained a one-to-one correspondence between efferent innervations and eye position. Conversely, perceived visual location would be in error whenever a disturbance altered the relationship between efferent innervations and eye position. Specific predictions along this line have been tested in experiments that use drugs to alter the relationship between efferent innervations and eye positions.

2.3.1 Drug-Induced Paralysis and Visual Localization

Injections of curare induce extraocular muscle paralysis, which increases with increased levels of curarization. Curare reduces muscular responsiveness to efferent innervations thereby changing the relationship between efferent innervations and eye position. For example, the innervation that held the eye at 40° before paralysis may only hold the eye at 30° after paralysis. The right-hand drawing of Figure 1 can be thought of as the condition existing before paralysis. Partial curarization would leave everything the same except that EEPI would no longer be veridical if it were based on efferent innervations. Specifically, efferent innervations would indicate an EEPI of 40°. As a result, the perceived visual location of B′ would be 40°. Leonard Matin and his co-workers have recently reported a series of experiments that support this prediction (Matin, 1982). In partially paralyzed observers, they found errors in the predicted direction on four separate measures of visual localization under darkroom conditions: (1) visually perceived median plane settings, (2) visually perceived eye-level horizontal settings, (3) auditory/visual matches of localization, and (4) pointing to a visual target. Thus, these results support the conclusion that efferent factors influence perception of visual direction. The same conclusion is supported in studies of MMAs.

2.3.2 Minor Motor Anomalies and Visual Location

Whereas Matin induced partial paralysis to alter the relationship between efferent innervations and eye position, other researchers have induced MMAs toward the same end. An MMA is erroneous EEPI that results in misjudgments of gaze direction and in illusions of visual direction. This MMA occurs after the eyes maintain an eccentric direction of gaze (Craske, Crawshaw, & Heron, 1975; Ebenholtz, 1976; Levy, 1973; Paap & Ebenholtz, 1976; Park, 1969) or after the head maintains a background tilt (Fogelgren & Shebilske, 1979; Shebilske & Fogelgren, 1977; Shebilske & Karmiohl, 1978). Many physiological factors could contribute to MMAs, but a major contributor is likely to be posttetanic potentiation, which is a temporary increase in muscle responsiveness after sustained stimulation. Posttetanic potentiation is consistent with many characteristics of MMAs including the direction of the effects (Shebilske, 1981). One can always predict the direction of MMA effects, therefore, by assuming that stimulated muscles have become hyperresponsive to efferent innervations. If the eye is held rightward, for instance, the muscles pulling the eye to the right become potentiated. This potentiation shifts the relationship between efferent patterns and eye positions such that any given efferent pattern holds the eye more *rightward* than it did before potentiation. In other words, according to the relationship that held before potentiation, the efferent pattern indicates that the eye is to the *left* of its actual position after potentiation. Thus, if efference influences EEPI as indicated in the curare studies, then in dark-room conditions subjects should see targets to the left of their actual position after maintaining a rightward eye position. This is in fact the obtained result, and in general subjects see visual direction illusions in the opposite direction as a potentiating stimulus (Table 1).

Another MMA is erroneous EEPI that results in misjudgments after convergence at less than 30 cm or beyond about 30 cm (Craske & Crawshaw, 1978; Ebenholtz & Wolfson, 1975; Ebenholtz, 1981). The direction of the effects are consistent with a potentiation explanation. The muscles that pull the eyes toward the nose became potentiated when observers fixate near objects. As a result, any given efferent pattern converges the eyes *more* than it did before potentiation. The efferent pattern therefore indicates that the eyes are *less* converged than they actually

Table 1. Changes in apparent visual direction or distance of a target after exposure to conditions that induce MMAs. Pretests are given before exposure; posttests are given after. Changes are posttest minus pretest shifts to the right (R), left (L), up (U), down (D), farther (F), or closer (C).

Inducing Condition	Change in Visual Direction or Distance
Eyes maintain an eccentric direction of gaze to the *right*	L
Eyes maintain an eccentric direction of gaze to the *left*	R
Head is tilted *back* and then returned to upright	D
Head is tilted forward and then returned to upright	U
Eyes maintain convergence on a target closer than about 30 cm	F
Eyes maintain convergence on a target farther than about 30 cm	C
Scanning targets from left to right as in reading English	L
Scanning targets from right to left as in reading Hebrew	R

are. Consequently, under darkroom conditions in which EEPI influences distance judgments, subjects see targets farther than their actual position. In general, subjects see distance illusions in the opposite direction as a potentiating stimulus (Table 1).

2.4 Manipulating Motor Information in Reduced and Structured Viewing Conditions

The results presented so far establish that efferent factors influence visual localization under darkroom conditions, but they leave open the possibility that efferent factors are irrelevant under more naturalistic conditions in which visual localization is judged in the context of a structured visual environment. Several theorists have, in fact, maintained that the visual system ignores efferent factors whenever visual information is available (e.g., Turvey, 1977, 1979; Bower, 1975). Evidence bearing on this issue comes from both the paralysis paradigm and the MMA paradigm.

2.4.1 Paralysis and Visual Localization in Reduced and Structured Contexts

Matin and co-workers conducted their curare experiments in both reduced and structured environments (Matin, 1982). They took four kinds of measures which they grouped into three categories as follows: (1) Visual localization (visually perceived median plane settings and visually perceived eye-level horizontal settings), (2) Sensory/motor localization (pointing to a visual target), (3) Intersensory localization (auditory/visual match of localization). They found that the illusions that they had observed under darkroom conditions did not occur in a structured context on their tasks of visual localization and sensory motor localization. Illusions were obtained in a structured environment, however, on their intersensory localization task.

Matin (1982) concluded that EEPI influences intersensory localization in a structured context, but it is not involved in visual localization when viewing is in a structured visual field.

The results and conclusions about context effects that are emerging from paralysis studies are at odds with those emerging from MMA studies. Let us turn to the conflicting results before discussing further conclusions.

2.4.2 MMAs and Visual Localization in Reduced and Structured Contexts

Shebilske (1977, 1981) performed MMA experiments in both reduced and structured environments. In one experiment, subjects pointed with their unseen hand at a visual target before and after MMAs were induced by having subjects hold fixation for 1 min on a target that was displaced 60° to the right or left of the median plane. Half the pointing tests were performed in darkroom conditions, the other half in a structured visual environment. In the structured context, subjects saw a different foreground and background at pretests and posttests. Testing in specific foreground-background contexts and order of testing in darkroom and structured conditions was counterbalanced across subjects. Head position was restrained during the induction period, but not during the pointing test.

Posttest minus pretest shifts in the direction of pointing indicated MMA effects. The shifts were in the opposite direction to the inducing stimulus, which agrees with the hypothesis that MMAs alter EEPI through the process of posttetanic potentiation (see Table 1). The shifts were 8.33° in darkroom conditions and 4.37° in structured conditions. The darkroom shift was significantly greater than the structured shift, and the latter was significantly greater than zero.

In another experiment, subjects pointed with their unseen hand at a visual target before and after MMAs were induced by having subjects hold fixation for 10 min on a target located 11 cm away in the median plane (Shebilske, Karmiohl, & Proffitt, 1983). Half the pointing tests were performed in darkroom conditions, the other half in structured environment. Again, visual context was changed between pretests and posttests, conditions were counterbalanced, and head position was restrained during induction but not during pointing tests.

Posttest minus pretest shifts in distance of pointing indicated MMA effects. Subjects pointed farther away on the posttests in agreement with the hypothesis that MMAs alter EEPI through the process of posttetanic potentiation (see Table 1). The shifts were a 6.34 cm overestimation of distance in reduced conditions, and a 2.31 cm overestimation of distance in structured conditions. (The actual average target distance was 33 cm with a range of ± 3 cm). The darkroom shift was significantly greater than the structured shift, and the latter was significantly greater than zero.

2.4.3 A Comparison of Paralysis and MMA Studies

Results from curare experiments and MMA experiments are in opposition. Both paradigms suggest that EEPI influences visual localization in the dark, but the two paradigms suggest different effects of a structured visual context. In the curare experiments a structured visual context eliminated EEPI effects on pointing; in the MMA experiments, a structured visual context reduced, but did not eliminate, EEPI effects on pointing. Reasons for the discrepancy will be difficult to determine until further investigations are made. Interestingly, however, a common pattern emerges when one looks at results from three laboratories (Bridgeman, Note 1; Matin, 1982; Shebilske, 1977, 1981).

Anomalous motor information disturbs performance in a structured environment on pointing tasks and other tasks that require subjects to indicate a target's egocentric direction without indicating the target's spatial location. The only exception came in a pointing task in which Matin and his co-workers reported that pointing measures may have been too imprecise to detect effects (Matin et al., 1982). This weak exception must be balanced against the fact that the same researchers observed significant effects of anomalous oculomotor information when subjects aligned a visual target with an auditory target in a structured environment. Like pointing, the auditory alignment task can be done by perceiving the egocentric direction of the two targets without necessarily knowing the spatial location of either one.

The other null results of anomalous motor information on perception in a structured environment came from tasks in which subjects had to indicate a target's spatial location (Bridgeman, Note 1; Matin et al., 1982). Shebilske and Karmiohl (Note 5) discuss how such tasks might lead to a cognitive suppression of the perceptual effects of anomalous motor information.

Curare experiments have confirmed the importance of efferent factors and have thereby contributed substantially to our understanding of perception. They could also have practical applications for patients who suffer from paralysis. But paralysis, whether induced by drugs or some other factor, is a pathology. In contrast, MMAs are natural events that occur in everyday situations, e.g., people maintain eccentric eye positions during conversations, they read and watch TV in recliner chairs, and they maintain fixations for extended periods on nearby objects during close handwork. The experiments reviewed so far include only MMAs induced under artificial laboratory conditions. Other experiments, however, have induced MMAs under natural conditions and the results are now reviewed before discussing the implications of studying naturally induced errors in EEPI.

2.4.4 Naturally Induced MMAs

The experiment reported earlier in this paper on MMAs in convergence provides a background. It was duplicated in all essential details except that a more natural task was used to induce MMAs (Shebilske et al., 1983). Instead of holding fixation on a target 11 cm away, subjects did three 5 min intervals of close handwork. They rested the hand that they used for the pointing task and used the other hand to thread a needle that was mounted in front of them. The task became an engaging game because subjects received 5 cents each time they threaded the needle, and they were encouraged to beat the current needle threading record. Subjects usually threaded the needle about 25 times in each 5 min period. The task had important characteristics of observers (1) choosing their own viewing distance, (2) moving their head freely, and (3) viewing in a well illuminated structured environment. These characteristics made the task similar to natural tasks such as draftsmen drawing detailed sketches, electrical engineers assembling circuit boards, and needleworkers sewing precise stitches.

Viewing distance was measured after 1.5, 3.0, and 4.5 min of each of the three handwork intervals. These nine measures taken on ten subjects averaged 21.3 cm. There was, however, between-subject variability in average viewing distance: the nearest was 14.7 cm, the farthest 33.8 cm. There was also within-subject variability: the average range across the nine within-subject measures was 7.6 cm; the smallest range was 2.6 cm (15.2–15.8 cm), and the largest 17.8 cm (12.7–30.5 cm).

Pointing tests in a reduced condition showed a 1.22 cm overestimation of distance on the posttest relative to the pretest. This shift was small in comparison with the 6.34 cm shift obtained in the experiment reported earlier, but the shift was statistically significant. Pointing tests in the same structured condition used in the experiment that was reported earlier showed no posttest minus pretest shifts. The handwork experiment suggests two conclusions: (1) A natural task can induce MMAs in convergence; (2) Some coping mechanism uses structured visual information to eliminate the distance illusion that would otherwise be associated with the amount of MMA induced by some natural tasks.

What is the nature of this coping mechanism? As mentioned earlier, some theorists hold that the perceptual system ignores efferent information whenever visual information is available. This process of ignoring might be thought of as a coping mechanism that guards against unreliable efferent information. Total dominance of visual information over motor information would explain the results of the handwork experiment, but the experiment reported earlier raises a question about this explanation. If subjects ignored efferent information in the handwork experiment, why did they process more distorted efferent information under identical test conditions in the experiment reviewed earlier? By raising this question, the earlier experiment argues against the idea that people cope with MMAs by ignoring efferent information. The earlier experiment suggests instead that some coping mechanism reduces the influence of unreliable efference by processing structured visual information in addition to efferent information. It remains to be seen whether or not this coping mechanism is always powerful enough to eliminate the distance illusions that would otherwise be induced by natural tasks. Future experiments may find natural tasks that produce more MMA distortions than the coping mechanism can completely eliminate.

Such experiments have already been performed for MMAs that distort perception of visual direction. In one experiment, University of Virginia baseball players were tilted back 45° while reading for 1 min before going to bat against a pitching machine (Shebilske, 1981). The prediction was that being tilted back would cause subjects to see the ball lower than it actually was during batting (see Table 1). This prediction was supported because batters swung lower after being tilted back, than they did after reading for 1 min in an upright position. A control experiment indicated that the manipulation had not altered the batters' ability to swing. The same manipulation did not cause subjects to swing lower at a remembered target position on a wooden stake, but it did cause them to swing lower at visual targets on the stake. Together the experiments suggest that batters swung lower at balls and targets because MMA distortions caused them to see things lower than they actually were.

The previously reviewed study on MMAs and visual direction suggested that a coping mechanism uses structured visual information to reduce the effect of unreliable efferent information. The baseball experiment gave this coping mechanism a chance to operate during a natural task in a fully structured environment. The mechanism did not completely eliminate the MMA distortion caused by the everyday task of reading in a reclined position. The batters did not notice any visual distortions, but the distortions were present nevertheless. The baseball experiment suggests, therefore, that, without realizing it, people may be continually forced to cope with visual distortions caused by unreliable efferent information.

2.5 Theoretical Implications of Context and Efferent Factors in Perception

The eye movement experiments reviewed here suggest that efferent factors influence perception even though efference is sometimes unreliable. In fact, the MMA experiments raise the possibility that the need to cope with unreliable efferent information has shaped both the phylogeny and ontogeny of the visual system. One evolutionary solution would have been to develop a system that ignores efference and uses afferent visual information in a structured environment. Gibson (1950, 1966, 1979) suggested this solution when he proposed that invariant structures of ambient light directly specify the perception of environmental events unequivocally in a structured environment (see Wolff, this volume, p. 121). However, MMA studies contradict Gibson's direct perception theory within the very conditions to which it is meant to apply, namely, an information rich, kinetic situation in which perceptions relative to body parts are possible (in addition to the MMA studies reviewed here, see Shebilske, Note 4, and Shebilske & Karmiohl, Note 5).

Shebilske and Karmiohl (Note 5) propose an alternative solution. They suggested that evolutionary pressures made sensorimotor errors negligible with respect to our survival by promoting a multilevel space perception system that processes both motor information and invariant light information. Accordingly, the system that takes into account motor information and the system that picks up light information from a structured environment are subroutines of higher-order processes that determine perception of visual direction. This solution represents a compromise similar to the one that took place in the development of the rod and cone receptor systems. The compromise is between evolutionary pressures during day vision when invariant light information is often available and evolutionary pressure during night vision when structured light information is severely reduced.

Shebilske and Karmiohl (Note 5) maintain that a multilevel model will result in a unified account of space perception just as multilevel models have provided a unified account of language processing (e.g., Forster, 1979; Schank & Abelson, 1977). An advantage of a multilevel space perception model is that investigators can import research techniques that have been used to articulate multilevel language models. The next section reviews some of those techniques.

3 Inferring Higher-Order Efferent Factors in Complex Cognitive Tasks

So far, we have discussed efferent factors within the saccadic motor control system itself. These factors are lower-order with respect to multilevel models of eye movement control (e.g., Fisher, 1979; Masse, Note 3, Rayner & McConkie, 1976; Shebilske & Reid, 1979). Multilevel models also postulate higher-order efferent factors, which are perceptual and linguistic processes that influence eye movements. The remainder of this paper will consider higher-order efferent factors: it will outline current research issues about them and explain how context effects relate to those issues.

Research discussed so far is concerned with how perceptions are altered by lower-order efferent factors. Noton and Stark (1971) raised questions about how perception, memory, and pattern recognition are altered by higher-order efferent factors, which were assumed to be generated by cognitive models present in the brain. The main thrust of research on higher-order factors, however, has approached the question from the other direction: it has concentrated on how higher-order efferent factors influence eye movements (Stark & Ellis, 1981). The current issues are well represented in research on reading eye movements.

3.1 A Multilevel Model of Reading Eye Movement Control

Figure 2 shows a multilevel model that provides a framework for discussing current research issues on higher-order efferent factors. The illustration brings to mind a linear model, but it is not intended as one in the sense that all levels of processing

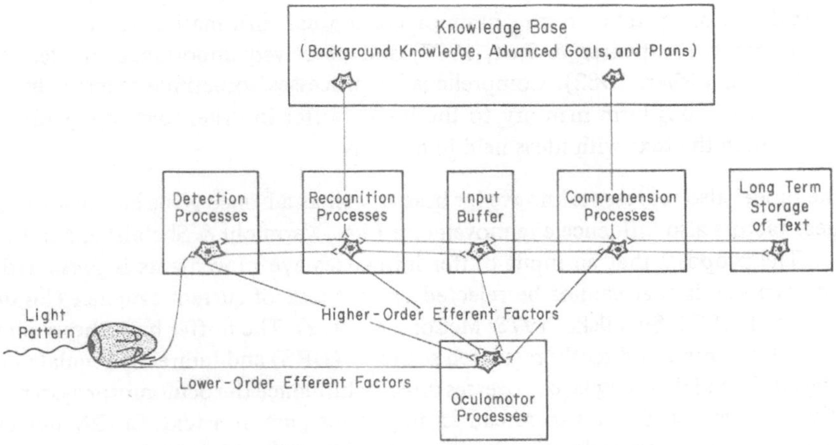

Figure 2. A multilevel model of reading eye movement control.

may operate in parallel and in noncanonical orders (Levelt, 1978; Newell, 1980; Winograd, 1972).

Overview of Model. Oculomotor processes control the direction and length of saccadic eye movements and the duration of fixations, but they do not completely determine eye movement patterns, which are also affected by higher-order efferent factors from four sources:

1. *Detection processes* discern visual features that the eyes extract from light patterns. Detection processes influence eye movements before visual features are analyzed for meaning. That is, detection processes direct the oculomotor system on the basis of visual characteristics per se. For instance, detection processes guide eyes according to spatial frequency during simple recognition tasks (Findlay, 1981), and they guide fixations away from blank spaces (Abrams & Zuber, 1972) and toward "convenient viewing positions", which could be determined by physical word length cues (O'Regan, 1981; Rayner, 1979; Rayner & McConkie, 1976).
2. *Recognition processes* map visual features onto linguistic units that correspond to letters, syllables, clauses, and sentences. Recognition processes direct the oculomotor system on the basis of linguistic information such as word frequency (Just & Carpenter, 1980; Kliegal, Olson, & Davidson, 1982; O'Regan, 1979).
3. An *input buffer* briefly stores the output of recognition processes making it available for further processing. The buffer directs the oculomotor system to slow down when the buffer is full and to speed up when it is almost empty.
4. *Comprehension processes* (a) link linguistic units with conceptual units, (b) map microstructural concepts, which correspond to words, clauses, and sentences, onto macrostructural concepts, which correspond roughly to summary statements that one would use in a topic outline, and (c) store a structured conceptual representation in long-term memory. Comprehension processes direct the oculomotor system on the basis of conceptual information such as thematic coherence (Carpenter & Just, 1977) and perceived importance of ideas (Shebilske & Fisher, 1982). Comprehension processes sometimes transfer information from long-term memory to the input buffer in order to compare ideas presented in the text with ideas held in memory.

The model also represents knowledge base as well as advance goals and plans because these factors also influence eye movements (e.g., Karmiohl & Shebilske, Note 2).

The proposal that an input buffer influences eye movements is presented as a null hypothesis that cannot be rejected on the basis of current evidence (Bouma & DeVoogd, 1974; Shebilske, 1975; McConkie, 1982). The buffer hypothesis allows a distinction between directly regulated scanning (DRS) and indirectly regulated scanning (IRS). In DRS, language processes directly influence the oculomotor system such as when a person decides to reread an important part of a text. In IRS, detection, recognition, and comprehension processes regulate the oculomotor system indirectly by determining the rate at which information is read into and out of the buffer.

By allowing for IRS, the buffer hypothesis has important theoretical implications because it stands at odds with the eye-mind assumption of Just and Carpenter (1980), which is the basis for an increasingly popular method of inferring processing time from "gaze durations." the eye-mind assumption holds that subjects look at a word while they process it so that processing time can be inferred from total gaze duration. In contrast, the buffer hypothesis implies that, during IRS, the comprehension processes operate on one word while the eyes look at another. Accordingly, gaze duration on a word does not reflect processing time of that word. As long as the buffer hypothesis remains viable, therefore, one must be cautious when interpreting models with "gaze duration" data (e.g., Just & Carpenter, 1980; Kliegal et al., 1982).

More generally, the multilevel model has implications for any attempt to establish a moment-to-moment link between eye movements and language processes. Those pursuing a moment-to-moment account of reading eye movements use local measures of eye movements, which Levy-Schoen and O'Regan (1979) defined as follows:

> local measures are made over a short time span and (are) taken in the vicinity of particular words of text . . . we expect them predominantly to reflect underlying processes that function over a short time span (p. 22).

According to the present model, those taking a moment-to-moment point of view must find ways to isolate lower-order efferent factors and to isolate the four sources of higher-order efferent factors. McConkie and Zola (this volume, p. 63) discuss an approach to this problem.

The model also has implications for those who take a more global point of view. They usually employ global measures, which Levy-Schoen and O'Regan (1979) defined as follows:

> Global measures . . . are essentially combinations of . . . local measures taken over a long time span . . . (they) would be expected to reflect higher level cognitive processes, relying upon information gathered over a long period of time (p. 22).

Those who take global measures are usually interested in studying processes that operate over longer time periods. But, according to the present model, global measures reflect combined influences of processes operating over a short time span (detection, recognition, and buffer processes) as well as processing operating over longer periods (comprehension processes). Those who take global measures to study comprehension processes (e.g., Mandel, 1979; Carpenter & Just, 1977; Shebilske & Fisher, 1982) must therefore isolate comprehension effects from effects of the other efferent factors.

3.2 Role of Context Effects in Analyzing Higher-Order Efferent Factors

Contextual manipulations can be used to isolate effects of comprehension processes on eye movements. Shebilske and Fisher (1983) outlined an example. They used eye movement records from good readers to measure reading rate in words per minute for behaviorally defined idea units, which usually consisted of one to three sentences (cf. Shebilske & Rotondo, 1981). They then computed correlations of reading rate and perceived importance of units, which had been judged by other subjects on a 10 point scale. This correlation was significant, suggesting that when good readers encounter important units they allocate additional time and resources to the comprehension processes that integrate units into a conceptual representation. This comprehension hypothesis implies a cause-effect statement: comprehension processes cause readers to modulate reading rate according to perceived importance. In general, cause-effect statements cannot be established by correlations, as mentioned earlier. In this case, the ambiguity of the correlation between reading rate and importance can be spelled out with respect to the multilevel model.

The comprehension hypothesis is only one of several possible explanations for a correlation between reading rate and importance of ideas. An alternative is that recognition processes account for the correlation. To illustrate this recognition hypothesis, suppose that authors used more low frequency technical terms and more complex syntax to express important ideas. These linguistic changes would slow down recognition processes, which would then slow down eye movements either directly or indirectly by reducing the rate of input into the buffer.

In the study described by Shebilske and Fisher (1983), the comprehension and recognition alternatives were tested. One group of students read ideas in their original context where reading rate had been modulated according to perceived importance. Another group read a subset of the same ideas in a context that was designed to equate the relative importance of ideas. In the original context, subjects read the important ideas significantly more slowly than unimportant ones. This difference in reading rate disappeared in the control context, however, suggesting that the words and sentences were equally easy to recognize.

The multilevel model suggests a generalization of this test. Whenever comprehension processes cause modulations in reading rate over some global unit, those modulations should depend upon how the global units are related to one another. Contextual manipulations that alter the relationship between units should therefore change the modulations. On the one hand, failure to observe a context related change in modulations over global units leaves open the possibility that detection or recognition processes are responsible for the observed modulations. On the other hand, observing a context effect supports an explanation in terms of comprehension. Thus, contextual manipulations provide a research tool that enhances the analytical power of global measures, for they enable investigators to tease apart higher-order efferent factors.

4 Future Directions

The study of context effects promises a bright future for research on efferent factors in perception and cognition. With respect to perception, an important question concerns the extent to which lower-order efferent factors influence perceptions in various contexts. Until recently, most evidence that efferent factors affect perception came from darkroom conditions, which left open the possibility that efferent factors are ignored in more natural contexts. However, research comparing efferent factors in a darkroom context and in structured visual contexts argue against an ignoring hypothesis. Contextual manipulations suggest, in fact, that efferent factors may play a major role in shaping both the phylogeny and ontogeny of the visual system. Contextual manipulations suggest further that a key to understanding vision may be found in coming to grips with the possibility that the visual system has evolved multilevel mechanisms for using visual and efferent information in combinations that are more reliable overall than either source alone under natural conditions.

With respect to cognition, a major question concerns the extent to which higher-order efferent factors influence eye movements during complex cognitive tasks such as reading. The quest to understand higher-order efferent factors is related to a century long search for eye movement measures that reflect cognitive processes underlying reading. This search has revealed, among other things, that contextual factors are vital (Shebilske & Fisher, 1982). Superior reading is characterized by flexibility, which is the ability to adapt reading rate and approach to fit various contexts. Currently, little is done to teach flexible reading strategies because little is known about them. Although progress is being made in the use of local measures, the multilevel nature of eye movement control during cognitive tasks has impeded attempts to establish a moment-to-moment link between eye movements and cognitive processes. In the past, global measures were too imprecise to shed light on specific cognitive processes (Levy-Schoen & O'Regan, 1979; Shebilske, 1975). New global measures have been developed, however, and these can be used in conjunction with contextual manipulations to enable precise analysis of higher-order efferent factors. Specifically, the pairing of global eye movement measures with contextual manipulations creates a keen tool for analyzing comprehension processes during reading. Future research with this tool promises to reveal fundamental components of flexible reading processes and thereby get to the heart of a cardinal characteristic of superior reading. Future research may also be able to use similar contextual manipulation paradigms to articulate a multilevel model of space perception.

Reference Notes

1. Bridgeman, B. *Induced motion: Relative motion or Roelofs effect.* Paper read at the Twenty-third Annual Meeting of the Psychonomic Society, Minneapolis, Minnesota, 1982.
2. Karmiohl, C.M., & Shebilske, W.L. *Behavioral vs. formal procedures of text analysis: Discourse style, task demands, and reading strategies.* Paper submitted for publication, 1983.
3. Masse, D. Le controle des movements oculaires. Thèse, Université de Grenoble, 1976.
4. Shebilske, W.L. *Minor motor anomalies cause visual illusions that affect dart throwing.* Paper read at the Twenty-third Annual Meeting of the Psychonomic Society, Minneaplis, Minnesota, 1982.
5. Shebilske, W.L., & Karmiohl, C.M. *Toward a unified indirect perception theory.* Paper submitted for publication, 1983.

References

Abrams, S.G., & Zuber, B.L. Some temporal characteristics of information processing during reading. *Reading Research Quarterly,* 1972, *8,* 40–51.

Bouma, H., & DeVoogd, A.H. On the control of eye saccades in reading. *Vision Research,* 1974, *14,* 273–284.

Bower, T.G.R. *Development in infancy.* San Francisco: Freeman, 1975.

Bridgeman, B., Lewis, S., Heit, G., & Nagle, M. Relation between cognitive and motor-oriented systems of visual position perception. *Journal of Experimental Psychology: Human Perception and Performance,* 1979, *5,* 692–700.

Brown, E. Neuropsychological interference mechanisms in aphasia and dyslexia. In R. Rieber (Ed.), *The neuropsychology of language: Essays in memory of Eric Lenneberg.* New York: Plenum Press, 1976.

Carpenter, P.A., & Just, M.A. Reading comprehension as eyes see it. In M.A. Just & P.A. Carpenter (Eds.), *Cognitive processes in comprehension.* New York: Wiley, 1977.

Collins, C.C. The human oculomotor control system. In G. Lennerstrand & P. Bach-Y-Rita (Eds.), *Basic mechanisms of ocular motility and their clinical implications.* New York: Plenum, 1975.

Coren, S. The interaction between eye movements and visual illusions. In D.F. Fisher, R.A. Monty, & J.W. Senders (Eds.), *Eye movements: Cognition and visual perception.* Hillsdale, NJ: Erlbaum, 1981.

Craske, B., & Crawshaw, M. Spatial discordance is a sufficient condition for oculomotor adaption to prisms: Eye muscle potentiation need not be a factor. *Perception and Psychophysics,* 1978, *23,* 75–79.

Craske, B., Crawshaw, M., & Heron, P. Disturbance of the oculomotor system due to lateral fixation. *Quarterly Journal of Experimental Psychology,* 1975, *27,* 459–465.

Ebenholtz, S.M. Additivity of aftereffects of maintained head and eye rotations: An alternative to recalibration. *Perception and Psychophysics,* 1976, *19,* 113–116.

Ebenholtz, S.M. Hysteresis effects in the vergence control system: Perceptual implications. In D.F. Fisher, R.A. Monty, & J.W. Senders (Eds.), *Eye movements: Cognition and visual perception.* Hillsdale, NJ: Erlbaum, 1981.

Ebenholtz, S.M., & Wolfson, P.M. Perceptual aftereffects of sustained convergence. *Perception and Psychophysics,* 1975, *17,* 485–491.

Festinger, L., Burnham, C.A., Ono, H. & Bamber, D. Efference and the conscious experience of perception. *Journal of Experimental Psychology,* 1967, Monograph 637.

Findlay, J.M. Local and global influences on saccadic eye movements. In D.F. Fisher, R.A. Monty, & J.W. Senders (Eds.), *Eye movements: Cognition and visual perception.* Hillsdale, NJ: Erlbaum, 1981.

Fisher, D.F. Understanding the reading process through the use of transformed typography: PSG, CSG and Automaticity. In P.A. Kolers, M.E. Wrolstad, & H. Bouma (Eds.), *Processing of visible language*, (Vol. 1). New York: Plenum, 1979.

Fogelgren, L.A., & Shebilske, W.L. Central visual learning and illusory visual direction after backward head tilts. *Perception and Psychophysics,* 1979, *25,* 519–523.

Forster, K.I. Levels of processing and the structure of the language processor. In W.E. Cooper & C.T. Walker (Eds.), *Sentence processing.* New York: Wiley, 1979.

Gibson, J.J. *The perception of the visual world.* Boston: Houghton-Mifflin, 1950.

Gibson, J.J. *The senses considered as perceptual systems.* Boston: Houghton-Mifflin, 1966.

Gibson, J.J. *The ecological approach to visual perception.* Boston: Houghton-Mifflin, 1979.

Helmholtz, H. von *A treatise on physiological optics,* (Vol. 3). (J.P.C. Southall, Ed. and Trans.). New York: Dover, 1963 (originally published, 1806).

James, W. (1890) *The principles of psychology,* (Vol. 2). Holt. Reprinted, New York: Dover, 1950.

Just, M.A., & Carpenter, P.A. A theory of reading: From eye fixation to comprehension. *Psychological Review,* 1980, *87,* 329–354.

Kliegal, R., Olson, R.K., & Davidson, B.J. Eye movements in reading. In K. Rayner (Ed.), *Perceptual and linguistic aspects of reading.* New York: Academic Press, 1982.

Levelt, W.J.M. A survey of studies in sentence perception: 1970–1976. In W.J.M. Levelt & G.B. Flores d'Arcais (Eds.), *Studies in the perception of language.* New York: Wiley, 1978.

Levy, J. Autokinesis direction during and after eye turn. *Perception and Psychophysics,* 1973, *13,* 337–343.

Levy-Schoen, A., & O'Regan, K. The control of eye movements during reading. In P.A. Kolers, M.E. Wrolstad, & H. Bouma (Eds.), *Processing of visible language,* (Vol. 1). New York: Plenum Press, 1979.

Mandel, T.S. Eye movement research on the propositional structure of short texts. *Behavior Research Methods and Instrumentation,* 1979, *11,* 180–187.

Matin, L. Eye movements and perceived visual direction. In D. Jameson & L.M. Hurvich (Eds.), *Handbook of sensory physiology,* (Vol. 7). Berlin Heidelberg New York: Springer-Verlag, 1972.

Matin, L. Visual localization and eye movements. In W.A. Wagemaar, A.H. Wertheim, & H.W. Leibowitz (Eds.), *Symposium on the study of motion perception.* New York: Plenum, 1982.

Matin, L., Picoult, E., Stevens, J.K., Edwards, M.W., Young, D., & MacArthur, R. Oculoparalytic illusion: Visual-field dependent spatial mislocalization by humans paralyzed with curare. *Science,* 1982, *216,* 198–201.

McConkie, G.W. Eye movements and perception during reading. In K. Rayner (Ed.), *Perceptual and linguistic aspects of reading.* New York: Academic Press, 1982.

Newell, A. Harpy, production systems and human cognition. In R. Cole (Ed.), *Perception and production of fluent speech.* Hillsdale, NJ: Erlbaum, 1980.

Noton, D., & Stark, L. Scanpaths in saccadic eye movements while viewing and recognizing patterns. *Vision Research,* 1971, *11,* 929–942.

O'Regan, K. Saccade size control in reading: Evidence for the linguistic control hypothesis. *Perception & Psychophysics,* 1979, *17,* 578–586.

O'Regan, K. The "convenient viewing position" hypothesis. In D.F. Fisher, R.A. Monty, & J.W. Senders (Eds.), *Eye movements: Cognition and visual perception.* Hillsdale, NJ: Erlbaum, 1981.

Paap, K.R., & Ebenholtz, S.M. Perceptual consequences of potentiation in the extraocular muscles: An alternative explanation for adaptation to wedge prisms. *Journal of Experimental Psychology: Human Perception and Performance,* 1976, *2,* 457–468.

Park, J.N. Displacement of apparent straight ahead as an aftereffect of deviation of the eyes from normal position. *Perceptual and Motor Skill*, 1969, *28*, 591–597.

Rayner, K. Eye guidance in reading: Fixation locations within words. *Perception*, 1979, *8*, 21–30.

Rayner, K., & McConkie, G.W. What guides a reader's eye movements? *Vision Research*, 1976, *16*, 829–837.

Schank, R.C., & Abelson, R.P. *Scripts, plans, goals, and understanding: An inquiry into human knowledge structures.* Hillsdale, NJ: Erlbaum, 1977.

Shebilske, W.L. Reading eye movements from an information-processing point of view. In D. Massaro (Ed.), *Understanding language.* New York: Academic Press, 1975.

Shebilske, W.L. Extraretinal information in corrective saccades and inflow vs. outflow theories of visual direction constancy. *Vision Research*, 1976, *16*, 621–628.

Shebilske, W.L. Visuomotor coordination in visual direction and position constancies. In W. Epstein (Ed.), *Stability and constancy in visual perception: Mechanisms and processes.* New York: Wiley, 1977.

Shebilske, W.L. Sensory feedback during eye movements reconsidered. *Behavioral and Brain Sciences*, 1978, *1*, 160–161.

Shebilske, W.L. Visual direction illusions in everyday situations: Implications for sensorimotor and ecological theories. In D.F. Fisher, R.A. Monty, & J.W. Senders (Eds.), *Eye movements: Cognition and visual perception.* Hillsdale, NJ: Erlbaum, 1981.

Shebilske, W.L., & Fisher, D.F. Eye movements and context effects during reading of extended discourse. In K. Rayner (Ed.), *Perceptual and linguistic aspects of reading.* New York: Academic Press, 1982.

Shebilske, W.L., & Fisher, D.F. Understanding extended discourse through the eyes: How and why. In R. Groner, D.F. Fisher, R.A. Monty, & C. Menz (Eds.), *Eye movements: An international perspective.* Hillsdale, NJ: Erlbaum, 1983.

Shebilske, W.L., & Fogelgren, L.A. Eye-position aftereffects of backward head tilts. *Perception and Psychophysics*, 1977, *21*, 77–82.

Shebilske, W.L., & Karmiohl, C.M. Illusory visual direction during and after backward head tilts. *Perception and Psychophysics*, 1978, *24*, 543–545.

Shebilske, W.L., Karmiohl, C.M., & Proffitt, D.R. Induced esophoric shifts and illusory distance in reduced and structured viewing conditions. *Journal of Experimental Psychology: Human Perception and Performance*, 1983, *9*, 270–277.

Shebilske, W.L., & Reid, S.L. Reading eye movements, macro-structure and comprehension processes. In P.A. Kolers, M.E. Wrolstad, & H. Bouma (Eds.), *Processing of visible language*, (Vol. 1). New York: Plenum Press, 1979.

Shebilske, W.L., Rotondo, J.A. Typographical and spatial cues that facilitate learning from textbooks. *Visible Language*, 1981, *15*, 41–54.

Skavenski, A.A. The nature and role of extraretinal eye-position information in visual localization. In R.A. Monty & J.W. Senders (Eds.), *Eye movements and psychological processes.* Hillsdale, NJ: Erlbaum, 1976.

Stark, L., & Ellis, S.R. Scanpaths revisited; cognitive models direct active looking. In D.F. Fisher, R.A. Monty, & J.W. Senders (Eds.), *Eye movements: Cognition and visual perception.* Hillsdale, NJ: Erlbaum, 1981.

Turvey, M.T. Contrasting orientations to the theory of visual information processing. *Psychological Review*, 1977, *84*, 67–78.

Turvey, M.T. The thesis of the efference-mediation of vision cannot be rationalized. *Behavioral and Brain Sciences*, 1979, *2*, 59–94.

Winograd, T. Understanding natural language. *Cognitive Psychology*, 1972, *3*, 1–191.

Summary. The appearance of visual stimuli and the meaning of linguistic stimuli are influenced by circumstances, conditions, and objects that surround a stimulus. Such context effects are used in experiments, as opposed to correlational studies, on efferent factors. This paper reviews three kinds of experiments: (a) those that isolate motor information by measuring perception in reduced contexts; (b) those that manipulate motor information by inducing erroneous eye movements; (c) those that examine the interaction between motor and visual information. Important in this review are studies of minor motor anomalies (MMAs), which are errors in registered eye or limb position that are caused by natural conditions. These studies support the functional significance of motor information (efferent factors) during visual localization in natural tasks that are performed in a structured environment. On the one hand, this conclusion is contrary to direct perception theory as a complete account of natural event perception because ambient light information does not specify environmental events unequivocally. On the other hand, the conclusion supports the ecological approach from which direct perception theory emerged. In MMA studies, ambient light information greatly reduced illusions that would have been expected in the dark. These results raise the challenge of describing (a) real events in which MMAs occur; (b) ambient light information that is used to reduce effects of anomalous motor information; (c) the processes that pick up and combine motor information and ambient light information to determine perception of natural events. It is argued that these processes can be captured by multilevel models that are similar to those that have provided a unified account of language processing. An advantage of a multilevel space perception model is that investigators can import research techniques that have been used to articulate multilevel language models. The paper concludes with a review of some of those techniques.

8 Saccadic Eye Movements and Visual Stability: Preliminary Considerations Towards a Cognitive Approach

PETER WOLFF

Contents

1 Introduction

The images of the objects projected on the retina move with every saccade. Yet, we perceive these objects as stationary, i.e., we neither perceive the objects as moving during the saccade (dynamic component) nor do we perceive them as displaced as a result of the saccade (static component). This is the phenomenon of visual stability, which has been extensively studied since Helmholtz' (1866) early work (for overviews see Dolezal, 1982; Festinger & Cannon, 1965; Gyr, 1972; MacKay, 1973; Matin, L., 1972, 1982; McClosky, 1981; Shebilske, 1977).

As has been emphasised by Shebilske, the dynamic and static components of visual stability are independent from one another: If a target light is exposed in an otherwise dark field, a "retinal smear" caused by the fast movement of the target across the retina can be seen (e.g., Mitrani, Mateeff, & Yakimoff, 1970), although the perceived location of the target light remains invariant before and after the saccade. Thus, under dark field conditions only the static, but not the dynamic, component of visual stability is preserved.

Cognition and Motor Processes
Ed. by W. Prinz and A.F. Sanders

The dynamic component of visual stability depends on stimulus conditions, e.g., contrast ratio (Shebilske, 1977), and is thought to be caused by predominantly peripheral mechanisms, such as visual masking (Matin, E., 1974). By contrast the mechanism underlying the static component of visual stability seems to be of more central origin. How this mechanism works is still under debate.

In the present paper a cognitive approach to the static component of visual stability is outlined. In the first section I argue that, contrary to the Gibsonian position, visual stability cannot be explained without the concept of extraretinal information. The next section contains a brief discussion of the classic cancellation theories, which assume that the extraretinal information used in producing visual stability is provided by the motor system. In the subsequent sections the new theoretical viewpoint is presented, based on the assumption that the extraretinal information is not of motor, but of cognitive origin.

2 Some Definitions

The phenomenon of visual stability may be described in more precise terms by using the following definitions:

The physical object in the environment is called the *environmental object.* Its location and displacement in environmental coordinates define its *environmental location* and *environmental displacement,* respectively. The optical projection of an environmental object on the retina is called the *retinal stimulus.* Its location and displacement in retinal coordinates define the *retinal location* and *retinal displacement,* respectively. The total pattern of all retinal stimuli is called the *retinal image.* The term *exafferent retinal displacement* refers to a retinal displacement produced by an environmental displacement. A retinal displacement produced by a saccade is called *saccadic retinal displacement.* The perceived position and displacement of an object in coordinates of visual space are called *visual location* and *visual displacement,* respectively.

Using these definitions, the phenomenon of visual stability can be described as follows: Every saccade produces a retinal displacement of each retinal stimulus. These retinal displacements could also have been caused by appropriate environmental displacements with the eyes held stationary. Yet, saccadic retinal displacements do not lead to visual displacements, whereas exafferent retinal displacements do. This does not mean that saccades do not affect perception at all. What they change is *perceived gaze direction* (Gibson, 1966, 1968; Dolezal, 1982). Thus, both kinds of retinal displacements have their specific effects: Exafferent retinal displacement leads to visual displacement, saccadic retinal displacement causes a shift in perceived gaze direction.

3 The Need for Extraretinal Information

Clearly, in order to produce visual stability, the visual system must somehow be able to discriminate between saccadic and exafferent retinal displacements. Gibson (1966, 1968, 1979) has argued that this is possible on a purely visual basis, since the optical information specifies its source by itself. In our terminology this means that the exafferent retinal displacement unequivocally specifies the environmental displacement because it is distinctly different from the saccadic displacement, which in turn unequivocally specifies the eye movement.

In a fully structured ecological environment, saccadic and exafferent retinal displacements are indeed qualitatively different. A saccade causes a linear and rigid displacement of the whole retinal image: All retinal stimuli, including projections of parts of the own body (nose, orbits) are displaced in the same direction and (neglecting distortions due to optical imperfections) with the same amplitude. By contrast, environmental displacements never produce such a global, rigid transformation of the retinal image. This difference could be responsible for the differential effect of saccadic and exafferent displacements. However, a number of findings demonstrate that saccadic displacement and exafferent displacement can be distinguished by the visual system even in the absence of the qualitative differences that are critical according to Gibson's theory:

1. In many studies of eye movements, the experimental environment is restricted to a few objects in an otherwise dark field, so that parts of the body such as the nose cannot be seen. The same holds for some ecological environments, e.g., if one looks at the starry sky at night in dull weather, when only the brightest stars can be seen. Under these conditions, saccadic retinal displacements are not qualitatively different from exafferent retinal displacements which could be produced by environmental displacements. Nevertheless, visual stability is not lost as mentioned above (Section 1). Under dark field conditions saccades produce mislocations only if the stimuli are exposed briefly and if they occur shortly before, during or shortly after the saccade (Bischof & Kramer, 1968; Matin, L., 1976; Pola, 1976).

2. When the target light in a dark field is displaced (environmental displacement) while a saccade directed to this target is performed, a saccadic and an exafferent displacement are superimposed. If both are produced in the same plane, the resulting retinal displacement is identical to a saccadic displacement which would have been produced by a saccade of a different amplitude. As the field is dark, global transformations of the retinal image cannot help to differentiate between the saccadic and the exafferent components of the retinal displacement. However, if the environmental displacement exceeds a critical amplitude, it can clearly be detected (Bridgeman, Hendry, & Stark, 1975; Bridgeman, Lewis, Heit, & Nagle, 1979; Brune & Lücking, 1969).

From these results it must be concluded that visual stability is not conditional upon a global, rigid transformation of the whole retinal image. It must be assumed that additional, extraretinal information is used to produce visual stability. More direct evidence for this comes from the following observations:

3. It is known that passive movements of the globe produce visual displacements in the opposite direction (Brindley & Merton, 1960; Helmholtz, 1866). This is also observed incidentally if the retinal displacement, which is produced by the passive movement, is made similar to normal saccadic displacement, i.e., if the eye is jerked by a heavy, sudden pull. Though the retinal displacement mimics saccadic displacement, the impression of having shifted the gaze is reported only occasionally (Skavenski, Haddad, & Steinman, 1972, p. 290).

4. When the retinal displacement normally associated with a saccade does not occur, visual displacement during the saccade can be observed. Mack and Bachant (1969) found that after-images move with saccades, even in total darkness. Such movement is not observed if the eye is pushed passively (Karrer & Stevens, 1930, cited by Carpenter, 1977).

5. If the eye is forcibly held while the subject intends a saccade, a visual displacement results which has the same direction as the intended saccade (Brindley & Merton, 1960).

From these results we may conclude that *intending* a saccade is sufficient for the visual system to process the retinal information as if the corresponding saccadic retinal displacement had actually occurred, independent of whether or not it does occur. The following sections will be mainly concerned with the theoretical consequences of this conclusion. How does the intention to perform a saccade contribute to the perception of stability and diplacement? According to the "outflow" version of cancellation theory, it does so by generating a *pattern of motor commands* a copy of which is matched against the input. This explanation will be discussed in the next section. In the subsequent sections I will outline an alternative view, according to which the intention is formulated in terms of the *visual consequences* of the intended saccade.

4 Cancellation Theory

Cancellation theory assumes that the extraretinal information that guarantees visual stability stems from neural activity produced within the oculomotor system. It is assumed that this information is used by the visual system to cancel the saccadic retinal displacement. Two main versions of this theory may be distinguished. *Inflow theory* maintains that proprioceptive afferences from the spindles of the extraocular muscles which signal direction and amplitude of the *actually executed* saccade are used for cancellation (James, 1890/1950; Sherrington, 1918). Since this theory obviously cannot explain the effects of the intention to perform a saccade, it can no longer be maintained in this pure form. The currently most widely accepted version of cancellation theory is *outflow theory.* It holds that the efferent commands which specify the parameters of the saccade *to be executed* are used in cancellation (Helmholtz, 1866; v. Holst & Mittelstaedt, 1950). As L. Matin (1972, 1976, 1982) and Shebilske (1976, 1977) have argued, there are advantages in combining this idea with elements from inflow theory in a "hybrid" model.

A serious difficulty for outflow theory comes from the observation that voluntary oculomotor control is reduced in total darkness (Ditchburn, 1973). This seems to indicate that the motor signal is not sufficiently precise to cancel exactly the retinal displacement (cf. Matin, L., 1976, 1982). However, there is no indication that this lack of precision is due to the motor signal itself. The enormous precision of eye-movement control under visual guidance (Steinman, 1976) indicates that the motor signal can be very accurate indeed. Support for this conclusion comes from several studies. For example, Becker and Jürgens (1979) elicited two saccades in short succession. Their results indicate that (a) both saccades could be programmed independently, (b) the first saccade did not add to the variance of the second saccade, indicating that its amplitude was taken into account very precisely in executing the second saccade. Because of the extremely short intersaccadic interval, that accurate information about the first saccade's actual amplitude could only have been derived from its motor signal. (For further evidence on the precision of the motor signal, see Kornhuber, this volume.)

If the oculomotor system works as precisely as it apparently does, we need a non-motor explanation for the loss of precision in total darkness. One possible explanation follows immediately from the present analysis: If retinal information from a richly structured environment is used to specify the eye movement goal, such specification will be absent in total darkness. It has to be substituted by goal specification with the help of memory and imagery, which will be less precise. Thus, the precision of the *motor* signal depends on a *cognitive* event (specification of the eye movement goal) which precedes it. This cognitive event contains precise information about the intended postsaccadic position of the eye. Here we have extraretinal information, which could, in principle, serve the same purpose as the efference copy in outflow theory.

I will now explore the possibility that visual stability is indeed achieved by taking into account this kind of extraretinal information.

5 The Function of Saccadic Retinal Displacement

Cancellation theory assumes that the saccadic retinal displacement must be eliminated in order to achieve veridicality of perception. This implies that retinal displacement is a sufficient condition for visual displacement (cf. MacKay & Mittelstaedt, 1974, p. 75ff.). Because of this presupposition, cancellation theory fails to ask why an exafferent displacement *does* produce visual displacement.

Obviously cancellation theory is based on the view that the visual world is essentially an analogue projection of the retinal image and that visual processing is merely needed to rectify, complement, stabilize and "clean" the retinal image.

There are a number of arguments against this view (see, e.g., Neisser, 1976, and Turvey, 1977). One argument which is particularly relevant in the present context can be derived from the neuroanatomical structure of the retina. The analysis of local details is restricted to the tiny section of the retinal image which projects on the fovea and the macula lutea. More than 90% of the retinal image are analyzed only in terms of global features. If this kind of receptor system were designed to deliver a direct, analog representation of the environment, it would be very inadequate indeed.

One could argue, of course, that the saccades serve precisely the purpose of compensating for this anatomical inadequacy by directing the foveal receptor system to different sections of the stimulus array. But why should evolution first have invested much effort to produce an efficient motor system which then needs to be supplemented by a cancellation mechanism for the only purpose of overcoming the limitations of a poorly constructed sensory system, instead of developing a sensory system which, like the panorama eye of many fish (Gibson, 1966), does not need "compensatory" saccades? The optical equipment of the eye is relatively imperfect with respect to the quality of projection (capillarity shadows, lense distortion, chromatic aberration etc.), whereas the kinetic equipment (bearing, suspension, torque, etc.) is excellent.

This suggests a different view of the function of saccades than that implied by cancellation theory: The saccadic mobility of the eyes has not been developed as a means of compensating for the poor resolution in the non-foveal parts of the retina. It is rather that the neuroanatomic structure of the retina has been developed because it allows for an active exploration of the environment by means of voluntary saccades. According to this view, retinal displacement during a saccade does not produce useless information that has to be eliminated. It produces useful information that has to be taken into account. The critical difference between an exafferent and a saccadic retinal displacement is therefore related to intention. The latter type of displacement results from an intention directed at exploring a part of the environment, whereas the former type does not. The problem of achieving visual stability during a saccade must then be reformulated: The task for the visual system is not to cancel saccadic retinal displacements, while accepting exafferent retinal displacements as useful information. Rather it is to discriminate between retinal displacements that are caused by an intention and those that are not. Both contain useful information: the latter about the environment and the former about gaze direction.

On the basis of these considerations, a cognitive approach to the phenomenon of visual stability which contains elements of the evaluation theory of MacKay (1973, 1978, and this volume) is outlined in the following sections. In the next section I argue that an intended saccade is preceded by a attentional shift which is involved in the preparation of the saccade. Subsequently, the problem of the cognitive control of the motor saccade by visual attention is discussed. Finally, visual stability and visual displacement are explained according to the assumption that the saccades produce wanted information.

6 A Cognitive Theory of Visual Stability

6.1 Presaccadic Attention Shift

There is no doubt that eye movements and the inner process of shifting visual attention are intimately related. This is demonstrated, for example, by the fact that subjects, while exploring a visual scene under different task instructions, frequently look at those objects which are relevant for solving the task (e.g., Yarbus, 1967; Gould, 1976). Further, during read-out from iconic memory (Neisser, 1967), saccades can be observed which are directed to the environmental locations of the previously presented objects (Bryden, 1961; Hall, 1974).

The functional relationship between a shift of attention and an intended saccade can be studied in experiments which consider the detection of target stimuli at various spatial locations and at different points in time before, during or after a saccade. Data from this paradigm reported by Posner (1980) and Remington (1980) indicate that preparing a saccade is accompanied by a shift of attention away from the fixation point and towards the target's location: About 50–100 msec after the appearance of a given saccade's target detection stimuli at the target location begin to be detected faster and better than stimuli at the fixation point. This is even observed if 80% of the detection stimuli appear at the fixation point, providing a strong incentive *not* to shift attention (Posner, 1980; see also Nissen, Posner, & Snyder, Note 1). Apparently, subjects cannot avoid shifting attention to the target location while preparing the target-directed saccade.

An argument against this conclusion has been put forward by Posner (1980), based on a double stimulus study, in which the subjects had to direct their eyes to two successive targets that were separated by an invariant interval. Posner found evidence for an attention shift (despite the incentive to keep attention at the original fixation location) only prior to the first saccade step, but not prior to the second. However, because interval and targets' locations remained invariant, this may be due to the fact that the second saccade was prepared prior to the first saccade (cf. Becker & Jürgens, 1979). Thus Posner's finding does not challenge our assumption that an intended target-directed saccade is conditional on an attentional shift.

The necessary and sufficient condition for the occurrence of a presaccadic attention shift seems to be that the saccade is directed to a *specified visual target.* If there is only a cue indicating the *direction* of the saccade, the effect disappears (Remington, 1980). From this one might conclude that the appearance of the target rather than the preparation of the saccade is responsible for the attentional shift (Remington, 1980). This explanation is, however, refuted by the finding of Nissen et al. (Note 1) that the effect is not observed if the subject is instructed not to move the eyes in response to a peripheral target. On the other hand, there are conditions where attention becomes separated from eye movements. For example, under the instruction to maintain fixation, attention shifts to the target location have been found when there was a high probability that the detection stimulus would appear at that location (Posner & Cohen, Note 2).

These results suggest the following conclusion: A saccade can be performed without a preceding attention shift, and attention can be shifted without accompanying eye movements. But when a saccade is target-directed and intended, an attentional shift to the target location seems to be unavoidable.

Klein (1980) reports experimental results that seem to run counter to this conclusion. They can, however, be accommodated by the present view. In Klein's experiments the subjects had, in a given trial, *either* to perform a saccade to a peripheral target that remained identical for a block of trials, *or* to respond to a detection stimulus that could appear either at the target location or at a location opposite to the target. Klein found no difference in detection latency between the two locations of the detection stimulus and concluded from this that "shifts of visual attention do not necessarily occur when subjects get ready to move their eyes to a target location" (Klein, 1980, p. 268). This conclusion rests entirely on what is meant by "getting ready to move". Klein assumes that saccades could be prepared *prior* to the imparative stimulus because they remained unchanged during a block of trials. Based on this assumption, he argues that preparation was also present in those trials where a detection stimulus was presented instead of the imparative stimulus, and where no saccade took place. Whatever this kind of preparation may be, it is *not* the intention postulated by the present theory. The latter implies the decision to *execute* the saccade. It was therefore absent in Klein's detection trials. His argument is thus irrelevant to the present line of reasoning.

According to the present view, the shift of attention to the saccade's target is somehow involved in the preparation of the saccade. However, it obviously cannot help in setting up the motor program, since visual attention has no direct access to the motor system. I will therefore assume that the function of the presaccadic attention shift is to specify the retinal displacement, i.e., the expected shift of the retinal stimulus as the result of the intended saccade. This assumption raises two questions. First, how can retinal displacement be defined in a language that can also be used for attentional guidance? Second, how can this specification be translated into the language of motor control?

6.2 Defining the Saccadic Retinal Displacement

As already mentioned, the retinal periphery analyzes the retinal image merely in terms of global features, while only the small part of the retinal image which is projected onto the fovea and macula lutea is analyzed with respect to local details. There is evidence that the peripheral receptor system is both functionally and structurally different from the foveal receptor system (Trevarthen, 1968, 1974). I will assume that the sensory *microstructure* of the foveal area of the retinal image and the sensory *macrostructure* of the whole retinal image form two distinct representations which are used for different purposes by the visual system.

Given a stable environment, each eye position coincides with a specific combination of a sensory microstructure (containing the local information) and a sensory macrostructure (containing the global information from the retina as a whole). With each saccade this combination is changed in a specific manner. However, microstructure and macrostructure change in fundamentally different ways. While the postsaccadic microstructure is unrelated to the corresponding presaccadic pattern, the postsaccadic macrostructure can be derived from its corresponding presaccadic pattern. If the environment remains stable, the presaccadic macrostructure can be changed into the postsaccadic macrostructure by applying a particular transformation. As there is exactly one transformation for each possible saccade, each saccade can be defined in terms of the postsaccadic macrostructure.

According to this view, the presaccadic attention shift determines the intended saccade by specifying the corresponding postsaccadic macrostructure. Our view follows the general principle that movements are guided by an anticipation of their intended result (e.g., Bernstein, 1967; Greenwald, 1970). In a way the attentional shift produces the saccade's effect before it has actually been carried out. It may thus be regarded as an "inner saccade" (cf. Scheerer, Note 3).

6.3 Postsaccadic Macrostructure and Motor Control

The fact that, given a stable environment and a point of fixation, each possible saccade corresponds to one particular postsaccadic macrostructure offers a *basis* for defining the target by way of an appropriate shift of attention. It does not explain how the intended result is actually *achieved.* How does the motor apparatus know what to do in order to produce the intended macrostructure? Oculomotor control involves multiple subsystems from the brain-stem to the prefrontal cortex. The collicular component of this system may be viewed as an automatic servomechanism which produces an appropriate eye movement in response to the stimulation of a peripheral receptive field (Didday & Arbib, 1975, cited by Arbib, 1981). Correspondingly, we assume that in the case of an intended saccade the function of defining the retinal target stimulus is served by a shift of visual attention – much in the same way as it is defined by peripheral stimulation in the case of stimulus guidance.

Of course, this attentional guidance does not supersede the automatic servomechanism for stimulus guidance. What it essentially does is to provide input into the automatic mechanism which is equivalent to the external input from the stimulation of a receptive field. Thus, the intention does not control the saccade directly. Rather, it creates a condition which leads the motor apparatus to produce the desired eye movement in an automatic fashion.

6.4 Visual Stability and Instability

The findings reported in Section 3 demonstrate that, once a saccade has been intended, the visual system behaves as if that saccade has been carried out, even if the motor execution is prevented by forcibly holding the eye. According to the present theory, this is because there is no internal feedback loop from the oculomotor system to the representation of the visual world. Within the visual world there is no possibility to decide whether a saccade corresponding to the attention shift is executed: as a motor event the saccade does not exist in the visual world. The only thing which can be checked visually is whether or not the intended transformation of the sensory macrostructure has taken place.

6.4.1 Visual Stability

This check is decisive for visual stability. If the sensory macrostructure changes as anticipated by the shift of attention, that which has been intended is achieved. This is veridically reflected in perception: One perceives a change of gaze, and not a visual displacement. The basis of this perception is the identity between the intended and the actual sensory macrostructure, which *specifies* a change in gaze direction, and which does *not* specify a visual displacement. Consequently, the change in gaze direction can be perceived directly, and there is no need for cancelling anything.

This is, of course, similar to Gibson's point of view. In contrast to Gibson, however, the present theory assumes that the change in gaze direction can be perceived directly because it is *intended*, and not because the saccadic retinal displacement is global and rigid. The findings reported in Section 3 which fail to support Gibson's theory can thus be explained without the assumptions of cancellation theory.

The general principle behind the present approach is that a sensory change can be perceived as a change of gaze if and only if it was anticipated as a result of the intention to perform that particular change of gaze. This may sound trivial. However, it has the important implication that any sensory change which does *not* conform to the anticipation *cannot* be perceived as a change of gaze and must consequently be attributed to an environmental change. The importance of this implication will be shown in the next section.

Besides these perceptual consequences, the match between the intended and the actual sensory macrostructure implies that the intention has been executed. It seems natural to conclude that, as a result, it is *extinguished,* i.e., it no longer persists as a dynamic structure which guides action. Again, the implications of this will become apparent in the following section.

6.4.2 Visual Instability

Corrective Saccades. Let us now consider the case that the saccade does not succeed in reaching the target, and hence the sensory macrostructure is not transformed as specified by the presaccadic shift of attention. There can be two reasons for this: the saccade was not carried out as determined by the intention, and/or the environment has changed during the saccade. If this is the case, there will be a mismatch between the intended and the actual sensory macrostructure. Which consequence should the visual system draw from this mismatch? We may distinguish two kinds of consequences. First, how should this situation be reacted to? Second, how should it be represented in the visual world?

The answer to the first question is independent of which of the two sources has contributed to the mismatch: If the intended macrostructure has not been attained, for whatever reason, an additional saccade is necessary. According to Deubel, Wolf, and Hauske (1982) amplitude and latency of the correction saccade are independent of whether the target discrepancy is generated by an undershoot of the primary saccade (endogenous), by a displacement of the target during the saccade (exogenous), or jointly by both events. According to the present theory the mechanisms underlying correction are quite simple: If the intended transformation of the macrostructure fails to take place, the intention persists (is not extinguished), and the original target selection by the presaccadic attention shift is still in effect. Thus, a corrective movement will be executed. Its mechanism is, in principle, the same as that of the first saccade. The main difference is that no new presaccadic attention shift is necessary, since the postsaccadic macrostructure has already been specified before. Correspondingly, the corrective saccade should have a shorter latency than the main saccade. This prediction is amply supported by empirical evidence (e.g., Becker, 1976).

Finally, it follows from the present view that the corrective saccade should not lead to a visual displacement. According to the mechanism of visual stability outlined above, it should bring about the perception of having tracked the displaced target with the gaze. Note that such a mode of producing the corrective saccade does not demand any corrective mechanism over and above that which guides the main saccade itself. Its adaptive value is easy to see. Although the corrective saccade is not explicitly intended, it serves exclusively the execution of the intention. It guarantees that reality is taken into account without departing from the intended goal.

Visual Displacement. So far we have considered only the behavioral consequences of a mismatch. Let us now consider the perceptual consequences. If the visual system has no direct linkage to the oculomotor system and can only refer to the sensory structure which does not tell anything about the reason why the intended result of transformation is not achieved, the problem whether the mismatch has to be attributed to the eye movement or to a change in the environment cannot arise for this system. Two possible outcomes of a mismatch between intended and achieved sensory structure can be considered: Either the target is seen as displaced from its original location or the gaze is perceived as directed at a position other than the intended one. However, according to the present view, the observed transformation can only give rise to the perception of changing the gaze to a particular location, if this change of gaze was intended and, thus, if the result of the corresponding transformation was presaccadically anticipated. Therefore the system has no possibility to attribute a mismatch to the execution of the intention. On the other hand, if perception is to be veridical, it *must* somehow be registered that the intention is not realized as anticipated. The only possibility which remains is that the discrepancy between intended and achieved macrostructure is perceived as a change in the environment, i.e., one sees the target displaced from its original location.

An important feature of the present view is that the intended macrostructure is object-related, i.e., related to the visual location of the target. If the discrepancy between the intended and the actually resulting macrostructure gives rise to a perceptual change in the visual world which does not match this object related intention, the resulting displacement should be restricted to the target. This prediction is borne out by experiments by Brune and Lücking (1969) and Bridgeman (1981). These authors showed that if paintings are displaced during exploratory eye movements, single objects are perceived as moving, whereas the background seems to remain at rest. If the subject shifts his attention to another object, that object starts to move on the apparently stationary background.

6.4.3 Schematic Anticipation

As argued previously (Section 4) the motor signal as such seems to be capable of very high precision. However, this intrinsic precision can only be used to the degree to which the intention specifies the eye movement's target. One limitating condition is the availability of visual information from the environment, as mentioned previously. Another limiting condition will not be discussed.

Due to the poor resolution of the retinal periphery, the visual location of a target is less determined, when the target is projected in the periphery of the retina. An intended saccade to a peripheral target can, therefore, only be guided by an attentional shift to a relatively large visual area. The consequence should be that a specific transformation of the macrostructure which determines one particular saccade cannot be defined. I will assume that in this case the anticipation by attention is "schematic" in the sense defined by Neisser (1976). That is, in shifting attention to the visual area a *range* of transformations or postsaccadic macrostructures is anticipated. This range of transformations corresponds to all fixation points within this area. The particular saccade to be executed has to be selected from this group by the oculomotor system.

In this case all macrostructures which belong to the defined range are appropriate realizations of the intention. When the range of acceptable transformations is large, the postsaccadic macrostructure actually produced will tend to fall within the defined area if the target was displaced during the saccade. This explains why environmental displacements are not always detected. Bridgeman et al. (1975, 1979) found that the threshold for detecting a target displacement during a saccade increases with the amplitude of the saccade (see also Mack, 1970). Displacements within 10–20% of the saccade amplitude are not detected. Furthermore, the detection threshold is independent of whether the target is displaced opposite to the saccade or in the same direction (Bridgeman & Stark, 1979). This is not due to the fact that the discriminability of retinal displacements is generally reduced with growing amplitude, as demonstrated by the finding that exafferent retinal displacements of comparable amplitudes are precisely distinguished when produced at the stationary eye (Bridgeman, 1981). Further, the information about an undetected environmental displacement is not lost: Subjects are able to respond correctly to the target location with an unvisible pointer, even if the displacement is not detected (Bridgeman et al., 1979).

These results, which follow from the present theory, cannot be explained by outflow theory without additional assumptions. One possibility is to evoke the concept of a just noticeable difference (JND) between the efference copy and the retinal signal (Wertheim, 1981).

7 Conclusions

The central feature of the cognitive approach to the phenomenon of visual stability outlined in this chapter is the concept of intention. The intention to change the gaze direction to a visual target is a sufficient condition for an attentional shift to the visual location of that target. The function of this attention shift is to anticipate the sensory change that will be produced by the saccade which corresponds to the intended change of gaze direction. At the same time this anticipation creates the condition which causes the oculomotor system to execute the appropriate saccade automatically. In this way the intended change of gaze direction within the visual world leads to a real eye movement within the real world, with the result that the sensory structure is changed. If this sensory change corresponds to the anticipated and intended one, it specifies the intended change of gaze direction. The perception of having changed the gaze in the intended way results. Cancellation is not needed and the motor signal must not be taken into account. At the same time intention disappears. If the sensory structure does not match the anticipated structure, intention persists. The same mechanism which guided the first saccade now initiates a corrective saccade which finally realizes the intention (behavioral consequence of a mismatch). Phenomenologically the discrepancy between the intended and the actually resulting sensory structure specifies an environmental property, viz. a visual displacement. It is object-specific, i.e., only the intended object is seen as displaced since the original intention itself was object related.

This theory implies that the perception of an environmental property and the perception of the realization or failure of the intention are two sides of one coin. The mechanism suggested in this paper does not only serve to realize the intention (behavioral level) and to perceive this realization (visual level), but also to secure that, first, real barriers are considered while executing the intention (corrective saccades), and second, these barriers are perceived. In this way the visual world is continuously actualized, provided the saccades are intended.

Due to the excellent movement properties of the eye and the high precision of the oculomotor system, these barriers are, under ecological conditions, produced by changes of the environment. Thus, saccades serve to adjust continuously the lay-out of the visual world to reality. This leads us back to the statement that the retinal structure has been invented by evolution because it allows for voluntary saccades (Section 5). The poor resolution in the retinal periphery is a continuous motive to perform saccades for the purpose of precisely identifying objects forcing the visual system into a continuous interaction with objective reality.

Since this interaction does not only serve to update the visual world to environmental changes, but also to adapt the visual system permanently to reality, the present view has implications for perceptual learning. Specifically, it offers an explanation for the acquisition of local sign. This, however, is beyond the scope of the present contribution and will be treated elsewhere (Wolff, Note 4).

Acknowledgements. I thank especially Odmar Neumann for his untiring readiness for discussion and aid in presentation which helped to clarify my ideas. I thank, further, Alan Allport, Herbert Heuer, Wolfgang Prinz and Wayne Shebilske for helpful comments on earlier drafts. The assistance of Adelheid Baker and Gudrun Chafik in translation and/or preparation of the translation is greatfully acknowledged.

Reference Notes

1. Nissen, M.J., Posner, M.I., & Snyder, C.R.R. *Relationships between attention shifts and saccadic eye movements.* Paper presented to the Psychonomics Society, San Antonio, Texas, November 1978..
2. Posner, M.I., & Cohen, Y. *Consequences of visual orienting.* Paper presented to the Psychonomics Society, St. Louis, MO, November 1980.
3. Scheerer, E. *Probleme der Modellierung kognitiver Prozesse: Von der Funktionsanalyse zur genetischen Analyse.* Extended version of a paper presented to the 18. Tagung experimentell arbeitender Psychologen, Bochum 1976.
4. Wolff, P. Über die Funktion der sakkadischen Blickbewegung: Überlegungen zum Problem der Repräsentation von Umwelteigenschaften in der Wahrnehmung. (Manuscript in preparation).

References

Arbib, M.A. Perceptual structures and distributed motor control. In V.B. Brooks (Ed.), *Handbook of physiology, Sect. 1. The nervous system, Vol. II: Motor control, Part 2.* Bethesda, MD: American Physiological Society, 1981.

Becker, W. Do correction saccades depend exclusively on retinal feedback? A note on the possible role of non-retinal feedback. *Vision Research,* 1976, *16,* 425–427.

Becker, W., & Jürgens, R. An analysis of the saccadic system by means of double step stimuli. *Vision Research,* 1979, *19,* 967–983.

Bernstein, L. *The co-ordination and regulation of movement.* London: Pergamon Press, 1967.

Bischof, N., & Kramer, E. Untersuchungen und Überlegungen zur Richtungswahrnehmung bei willkürlichen sakkadischen Augenbewegungen. *Psychologische Forschung,* 1968, *32,* 185–218.

Bridgeman, B. Cognitive factors in subjective stabilization of the visual world. *Acta Psychologica,* 1981, *48,* 111–121.

Bridgeman, B., & Stark, L. Omnidirectional increase in threshold for image shifts during saccadic eye movements. *Perception & Psychophysics,* 1979, *25,* 241–243.

Bridgeman, B., Hendry, D., & Stark, L. Failure to detect displacement of the visual world during saccadic eye movements. *Vision Research,* 1975, *15,* 719–722.

Bridgeman, B., Lewis, L., Heit, G., & Nagle, M. Relation between cognitive and motor-oriented systems of visual position perception. *Journal of Experimental Psychology: Human Perception and Performance,* 1979, *5,* 692–700.

Brindley, G.S., & Merton, P.A. The absence of position sense in the human eye. *Journal of Physiology (London),* 1960, *153,* 127–130.

Brune, F., & Lücking, C.H. Oculomotorik, Bewegungswahrnehmung und Raumkonstanz der Sehdinge. *Der Nervenarzt,* 1969, *40,* 413–421.

Bryden, M.P. The role of post-exposural eye-movements in tachistoscopic perception. *Canadian Journal of Psychology,* 1961, *15,* 220–225.

Carpenter, R.H.S. *Movements of the eye.* London: Pion, 1977.

Deubel, H., Wolf, W., & Hauske, G. Corrective saccades: Effect of shifting the saccade goal. *Vision Research,* 1982, *22,* 353–364.

Ditchburn, R.W. *Eye-movements and visual perception.* Oxford: Clarendon Press, 1973.

Dolezal, H. *Living in a world transformed. Perceptual and performatory adaptation to visual distortion.* New York: Academic Press, 1982.

Festinger, L., & Canon, L.K. Information about spatial location based on knowledge about efference. *Psychological Review,* 1965, *72,* 373–384.

Gibson, J.J. *The senses considered as perceptual systems.* Boston: Houghton Mifflin, 1966.

Gibson, J.J. What gives rise to the perception of motion? *Psychological Review,* 1968, *75,* 335–346.

Gibson, J.J. *The ecological approach to visual perception.* Boston: Houghton Mifflin, 1979.

Gould, J.D. Looking at pictures. In R.A. Monty & J.W. Senders (Eds.), *Eye movements and psychological processes.* Hillsdale, NJ: Erlbaum, 1976.

Greenwald, A.G. Sensory feedback mechanisms in performance control: with special reference to the ideomotor mechanism. *Psychological Review,* 1970, *77,* 73–99.

Gyr, J.W. Is a theory of direct visual perception adequate? *Psychological Bulletin,* 1972, *77,* 246–261.

Hall, D.C. Eye movements in scanning iconic imagery. *Journal of Experimental Psychology,* 1974, *103,* 825–830.

Helmholtz, H. von *Handbuch der Physiologischen Optik.* Leipzig: Voss, 1866.

Holst, E. von, & Mittelstaedt, H. Das Reafferenzprinzip. Wechselwirkung zwischen Zentralnervensystem und Peripherie. *Naturwissenschaften,* 1950, *37,* 464–476.

James, W. *The principles of psychology,* (Vol. 2). Holt, 1890. Reprinted, New York: Dover, 1950.

Klein, R. Does oculomotor readiness mediate cognitive control of visual attention: In R.S. Nickerson (Ed.), *Attention and performance VIII.* Hillsdale, NJ: Erlbaum, 1980.

Mack, A. An investigation of the relationship between eye and retinal image movement in the perception of movement. *Perception & Psychophysics,* 1970, *8,* 291–298.

Mack, A., & Bachant, J. Perceived movement of the afterimage during eye movements. *Perception & Psychophysics,* 1969, *6,* 379–384.

MacKay, D.M. Visual stability and voluntary eye movement. In R. Jung (Ed.), *Handbook of sensory physiology,* (Vol. 7/3). Berlin Heidelberg New York: Springer, 1973.

MacKay, D.M. The dynamics of perception. In P.A. Buser & A. Rougel-Buser (Eds.), *Cerebral correlates of conscious experience.* Amsterdam: Elsevier, 1978.

MacKay, D.M., & Mittelstaedt, H. Visual stability and motor control (reafference revisited). In W.D. Keidel, W. Händler, & M. Spreng (Eds.), *Cybernetics and bionics.* Munich: Oldenbourg, 1974.

Matin, E. Saccadic suppression: A review and an analysis. *Psychological Bulletin,* 1974, *81,* 899–917.

Matin, L. Eye movements and perceived visual direction. In D. Jamesson & L. Hurvich (Eds.), *Handbook of sensory physiology,* (Vol. 7/4). Berlin Heidelberg New York: Springer, 1972.

Matin, L. Saccades and extraretinal signal for visual direction. In R.A. Monty & J.W. Senders (Eds.), *Eye movements and psychological processes.* Hillsdale, NJ: Erlbaum, 1976.

Matin, L. Visual localization and eye movements. In A.H. Wertheim, W.A. Wagenaar, & H.W. Leibowitz (Eds.), *Tutorials on motion perception.* New York: Plenum Press, 1982.

McClosky, D.I. Corollary discharges: motor commands and perception. In V.B. Brooks (Ed.), *Handbook of physiology, Sect. 1. The nervous system, Vol. II: Motor control, Part 2.* Bethesda, MD: American Physiological Society, 1981.

Mitrani, L., Mateeff, S., & Yakimoff, N. Smearing of the retinal image during voluntary saccadic eye movements. *Vision Research,* 1970, *10,* 405–409.

Neisser, U. *Cognitive psychology.* New York: Appleton-Century-Crofts, 1967.

Neisser, U. *Cognition and reality. Principles and implications of cognitive psychology.* San Francisco: Freeman, 1976.

Pola, J. Voluntary saccades, eye position and perceived visual direction. In R.A. Monty & J.W. Senders (Eds.), *Eye movements and psychological processes.* Hillsdale, NJ: Erlbaum, 1976.

Posner, M.I. Orienting of attention. *Quarterly Journal of Experimental Psychology,* 1980, *32,* 3–25.

Remington, R.W. Attention and saccadic eye movements. *Journal of Experimental Psychology: Human Perception and Performance,* 1980, *6,* 726–744.

Shebilske, W.L. Extraretinal information in corrective saccades and inflow vs. outflow theories of visual direction constancy. *Vision Research,* 1976, *16,* 621–628.

Shebilske, W.L. Visuomotor coordination in visual direction and position constancies. In W. Epstein (Ed.), *Stability and constancy in visual perception: mechanisms and processes.* New York: Wiley, 1977.

Sherrington, C.S. Observations on the sensual role of the proprioceptive nerve supply of the extrinsic ocular muscles. *Brain,* 1918, *41,* 332–343.

Skavenski, A.A., Haddad, G., & Steinman, R.M. The extraretinal signal for the visual perception of direction. *Perception & Psychophysics,* 1972, *11,* 287–290.

Steinman, R.M. Role of eye movements in maintaining a phenomenally clear and stable world. In R.A. Monty & J.W. Senders (Eds.), *Eye movements and psychological processes.* Hillsdale, NJ: Erlbaum, 1976.

Trevarthen, C.B. Two mechanisms of vision in primates. *Psychologische Forschung,* 1968, *31,* 299–337.

Trevarthen, C. Functional relations of disconnected hemispheres with the brain stem, and with each other: Monkey and man. In M. Kinsbourne (Ed.), *Hemispheric disconnection and cerebral functioning.* Springfield, IL: Thomas, 1974.

Turvey, M.T. Contrasting orientations to the theory of visual information processing. *Psychological Review,* 1977, *84,* 67–88.

Wertheim, A.H. On the relativity of perceived motion. *Acta Psychologica,* 1981, *48,* 97–110.

Yarbus, A.L. *Eye movements in vision.* New York: Plenum Press, 1967.

Summary. This chapter is concerned with the question why the objects of the environment are not perceived as displaced as a result of saccadic movement (visual stability). After a short review of traditional theories (Gibsonian view and cancellation theory) a cognitive approach to the phenomenon of visual stability is outlined. The central feature is the concept of intention. It is assumed that the intention to change the gaze direction to a visual target leads to an attentional shift to the visual target location. The function of this attentional shift is to anticipate the sensory change that will be a result of the intended saccade. This anticipation creates the condition which causes the oculomotor system to automatically execute the appropriate saccade. If the actually achieved sensory change corresponds to the anticipated change, the perception of having changed the gaze in the intended way results and the intention extinguishes. If the sensory change does not match the anticipated change, intention persists and a corrective saccade is initiated in the same way as the first saccade. The discrepancy between intended and achieved sensory change creates the perception of a displacement of the target to which directing the gaze was intended.

9 Scanning and the Distribution of Attention: The Current Status of Heron's Sensory-Motor Theory

D. J. K. MEWHORT

Contents

1 Introduction

Speech is presented to the senses as a series of phonemes extended in time. Written material is presented as a parallel spatial array, but thought, like speech, is extended in time. Reading, therefore, involves conversion from a parallel spatial representation to a temporal one. Heron (1957) addressed the problem of spatial-to-temporal conversion by postulating an attentional process that sweeps, or scans, an internal neural representation of the spatial array in an orderly way. In this chapter, I shall reconsider some of Heron's ideas about visual processing, trace the development of the problem, and, finally, re-cast his theory in the form of an information-processing account.

1.1 Heron's Scanning Theory

Heron's (1957) account derived from an empirical puzzle: suppose that subjects are presented with a row of letters tachistoscopically and are asked to report as many letters as possible. If the material is centred at fixation, the subjects will report the material in the left visual field (LVF) more accurately than that from the right visual field (RVF). If the material is divided so that half appears unpredictably in one field or the other on successive trials, the subjects will report the material in the

Cognition and Motor Processes
Ed. by W. Prinz and A.F. Sanders

RVF more accurately than that from the LVF. Thus, a LVF advantage appears for bilateral presentation, whereas for unilateral presentation the advantage goes to the material in the RVF. To obtain the LVF advantage for bilateral presentation, however, both fields must include the same class of material: if letters appear in one field while a line appears in the other, for example, the LVF advantage disappears.

Heron (1957) explained the interaction in terms of an attentional mechanism which, he speculated, was acquired as a consequence of learning to read. The mechanism distributes attention across space and was thought to develop from two conflicting eye movements used in reading. On the one hand, reading involves a sweep from left to right. On the other hand, it also involves a sweep from the end of a line to the beginning of the next, i.e., from right to left. In the tachistoscopic situation – an admittedly abnormal reading situation – Heron argued that, for a unilateral presentation, it is easier to shift attention from the central fixation point to the right than it is to shift from the fixation point to the left and then sweep from left to right across the material. Hence, for a unilateral presentation, a superiority for material in the RVF emerges. The bilateral case, however, presents a different situation: one skips to the leftmost position and then sweeps from left to right across the material. As a result, a LVF superiority emerges that reflects the left-to-right processing order.

What is meant by a shift of attention in such an exeriment: Heron's (1957) view was based on Hebb's (1949) neurophysiological theory and involved the idea that a cognitive structure can emerge from a sensory-motor interaction. Heron's idea was that when we learn to read, we start by shifting our eyes fixation-by-fixation across the letters of a word. Because such a strategy is too slow and too inefficient, as we gain proficiency, we establish an attentional mechanism which internalizes the fixation shift. Once we have acquired the mechanism, we can use it to shift "the mind's eye". For Heron, a student of Hebb, the important part of the metaphor is the idea that the attentional device involves some of the neurophysiological components of an overt eye movement re-assembled into an internalized cognitive mechanism, in much the manner Hebb had suggested for a "phase sequence".

2 The Nature of the Scanning Process

2.1 Early Evidence

Early evidence concerning the motor component of the attentional process was provided by Bryden (1961; see also Crovitz & Daves, 1962). Imagine a series of tachistoscopic trials on which a word is presented to either the LVF or the RVF in an unpredictable order. The exposure and intensity conditions have been adjusted so that the probability of reporting the material is relatively low. Can one predict on which trials subjects will report the word correctly and on which trials they will not report correctly? Bryden found that the direction of the subject's first post-exposural eye movement predicts accuracy of report from trial-to-trial: if the subject positions his eye to where the stimulus used to be some 200 msec previously, there is a good chance he will report the material correctly. If he shifts his eye in the wrong direction, however, chances are that the report will not be correct. In terms of Heron's ideas, the post-exposural shift of the eye is a motor "outflow" – a response correlated with a shift of attention. The eye's response is inhibited in normal circumstances, but it occurs in the tachistoscopic case because the motor system that controls eye movements shares a neurophysiological substrate with the attentional process.

A second line of evidence derives from experiments involving ordered report. Bryden (1960) presented a row either of letters or of common forms (such as circle, star, or diamond) and asked subjects to report from left to right or from right to left. The instructions indicating the direction of report were provided verbally following the display. When dealing with the non-alphabetic material, accuracy of report was higher for the material reported first, for both directions, i.e., accuracy was higher for material on the left when reporting from left to right, and it was better for material on the right when reporting from right to left. When dealing with the alphabetic material, however, a different pattern emerged: accuracy of report was better for material from the left regardless of the direction of report. The left side superiority associated with alphabetic material suggests left-to-right processing, and the difference in the pattern of results for the two kinds of material not only provides a link to reading habits but also suggests that left-to-right processing occurs after the class of material has been identified.

2.2 Early Criticism of Heron's Theory

Heron's (1957) theory was stated using a neurophysiological metaphor. He suggested, for example, that the attentional mechanism scans the stimulus trace. What kind of data are represented in a stimulus trace? Should we consider a "trace" to be like the pre-categorical, unprocessed, raw image that Neisser (1967) taught us to call iconic memory? Alternatively, should we consider the data scanned to be post-categorical, i.e., does scanning occur after the material has been identified? The post-categorical view seems closer to Heron's own position: to explain why the LVF advantage with bilateral presentation depends on the class of material presented to both fields, he suggested that an "abstract" representation must be involved, i.e., a level beyond the raw sensory image. Nevertheless, the bulk of the literature developed in response to Heron's ideas took the pre-categorical position, i.e., the view that the data scanned are unprocessed features and that the scanning operation is responsible, at least in part, for identifying the material.

Heron's (1957) basic evidence concerned the field superiorities, and his account offered an explanation for experiments using both bilateral and unilateral presentations. Alternative explanations have been suggested, but they usually treat the unilateral and bilateral cases separately. For the unilateral case, for example, the most popular alternative concerns neurological organization: because material presented to the two visual fields is projected contralaterally to the cortex, the RVF superiority associated with unilateral presentation is often thought to reflect an asymmetry of cerebral function. In a recent review of tachistoscopic studies involving lateralized stimuli. Bradshaw, Nettleton, and Taylor (1981) argue that "field differences are likely to reflect neural organization and that any directional scanning or processing artifacts make, at the most, a minor contribution" (p. 1). In deriving their conclusion, however, Bradshaw et al. treat scanning as a pre-categorical process, and, although they suggest that there is "little evidence of a left-to-right scan during the *recognition* of elements" (p. 4), they also acknowledge a potential post-identification role for left-to-right processing.

2.3 Criticisms of the Full-Report Procedure

For bilateral presentation, criticisms of Heron's (1957) account concern his use of ordered-report experiments. In particular, several authorities have suggested that the LVF superiority associated with bilateral displays reflects an artifact of ordered report. To test the artifact idea, one can use a partial-report task to remove the source of the artifact. Merikle, Lowe, and Coltheart (1971), for example, presented a row of letters followed by a bar marker indicating which letter to report. They argued that Heron's (1957) theory entails three propositions, namely (1) that letter features are processed in a left-to-right order, (2) that the feature representation is labile, and (3) that processing is slow enough to make the order of processing important. The three propositions predict an accuracy function decreasing from the left of the stimulus. As is typical in bar-probe studies, Merikle et al. found a symmetrical W-shaped function relating accuracy of report to stimulus position. Accepting the bar-probe task as a relatively pure measure of the accuracy of stimulus identification – at the very least, it excluded the artifact of ordered report – they took the symmetrical function as support for the artifact idea. A very similar argument was made, independently, by Smith and Ramunas (1971).

Merikle et al. (1971) treated accuracy of report in the bar-probe task as a valid measure of *identification* processes. More recent analyses of the task, however, have shown that the shape of the function relating accuracy of report to stimulus position is determined largely by localization difficulties, which occur after letter identification, not by problems of stimulus identification (e.g., Mewhort & Campbell, 1978). Further, the localization and identification components of the bar-probe task are carried out by different parts of the processing system. Mewhort, Campbell, Marchetti, & Campbell (1981), for example, showed that masking affects identification and localization separately (see also Mewhort, Marchetti, Gurnsey, & Campbell, 1983). To explain the separation, they suggest a model involving two data buffers, a pre-identification buffer (the feature buffer), and a post-identification buffer (the character buffer).

Although the recent analyses of the bar-probe task invoke different assumptions about the processing required in the task, they confirm the main argument raised by Merikle et al. (1971) concerning Heron's use of full-report experiments and bilateral presentations: it seems clear that the field superiorities Heron documented cannot reflect the order in which letters are identified. More precisely, if scanning remains a viable possibility, the material scanned cannot be a string of letter features or an image in the raw, it must be a more abstract representation, as Heron himself hinted.

3 Scanning as a Post-Identification Process

If scanning does not involve a left-to-right identification process, does the left side superiority associated with cases involving bilateral presentation reflect an artifact of ordered report? Unfortunately, proponents of the artifact idea have not been clear about what such an account entails. To escape circularity, however, the artifact idea implies that we can find an account of ordered report without invoking a scanning-like operator (cf. Bryden, 1967), and, although it is clear that scanning does not mean a left-to-right identification of features in a row of letters, it is far from clear what mechanisms are involved in ordered report – the putative artifact that the bar-probe task was used to control. From the perspective of scanning theory, there are two main possibilities. On the one hand, the artifact argument may be basically correct, i.e., ordered report in the tachistoscopic situation may have no direct or necessary relation to reading habits. If the artifact idea is correct, we should seek a general explanation for ordered report, one, perhaps, which involves ordering mechanisms related to speech processes, and we should deny any variation of a scanning mechanism. On the other hand, it may be that, even when all conceivable controls for ordered report have been satisfied, there remains a reading-related ordering function. If so, the artifact argument is wrong, and we need a more adequate understanding of the reading-related function.

3.1 Evidence Concerning the Ordering Process

Several lines of evidence suggest that a reading-related mechanism is required to explain ordering phenomena associated with tachistoscopic tasks. One line of evidence concerns the spatial nature of the organizational principle used to order report. Suppose, for example, that eight-letter pseudo-words are presented one letter at a time on a horizontal row. Each letter appears for only 5 msec, and subjects are asked to report the material in any order. Under reasonable letter size and brightness conditions, the letters themselves are easy to see, and the measure of interest is the subjects' order of report. If the inter-letter interval (ILI) separating the successive letters is greater than about 35 msec, subjects will track the input temporally and will report the letters using the temporal order in which they arrived on the screen. As the ILI is shortened, however, subjects will adopt an increasingly regular left-to-right report (Mewhort, 1974). At short ILI values, subjects ignore the temporal organization and adopt a spatial organization: they read from left to right in exactly the fashion one would expect from the scanning hypothesis. It is difficult to imagine why a general ordering mechanism would abandon temporal organization in favour of spatial organization and then translate back to a temporal organization to support sequential report. Instead, it seems reasonable to conclude that left-to-right spatial scanning is involved.

A second line of evidence concerning the scanning hypothesis derives from studies that require subjects to report in various orders. Bryden (1960) found that report of alphabetic material presented bilaterally was better from the left side regardless of the direction of report. For geometric forms, in contrast, report was better on the side reported first. Later studies showed, however, that Bryden's results depend heavily on the timing of the report cue. In his case, the instructions were given verbally (the exerimenter shouted), a technique likely to introduce considerable delay. Ayres (1966) provided instructions indicating the direction of report prior to the display. In his case, however, the material reported first was reported best, regardless of the direction of report. In particular, there was no evidence suggesting a residual LVF advantage after counterbalancing for order of report. Thus, Ayres's results, unlike Bryden's data, suggest that there is no need to invoke a reading-related mechanism. To clarify the empirical dispute, Scheerer (1972) varied the timing of the cue systematically. He presented a row of letters and, like Ayres, found that when the instructions are given in advance of the display, the material reported first was reported best, regardless of the direction of report. When the cue was delayed, however, his subjects, like Bryden's, showed a LVF superiority, even when reporting from right to left. Thus, tighter control over the timing of the direction-of-report cue provides evidence for a reading-related ordering function that operates after some processing has occurred.

A more complicated picture emerges if one not only controls the timing of the report instruction but also provides directionally biased materials. Pseudo-words can be constructed to match the characteristics of letter usage in English to various degrees, or orders of approximation. Fourth-order approximations to English, for example, are word-like strings that match the frequency in English for four-letter runs. Such materials are highly familiar word-like strings, but because the frequency constraints are directional, fourth-order pseudo-words "lose" their familiarity when spelled backwards: CRYSTEMP and SPEALTHY become PMETSYRC and YHTLAEPS, respectively. Mewhort and Cornett (1972) used such pseudo-words in both normal and reversed spellings. The direction-of-report instruction was provided immediately following the display. When subjects reported from left to right, accuracy of report was greater for the material on the left, and when they reported from right to left, accuracy of report was greater on the right. Both results are the same as Scheerer's (1972) findings for corresponding conditions. The complication concerns the subjects' ability to use the sequential constraints in the normal and reversed materials: even when subjects were required to report from right to left and showed a strong RVF advantage, their ability to exploit the sequential constraint depended on the orientation of the material, not on the match between the spelling direction and the direction of report. In short, although spelling PMETSYRC from right to left provides the same report sequence as spelling CRYSTEMP from left to right, subjects were unable to capitalize on the sequential constraint. Thus, there appear to be *two* left-to-right processes, one sensitive to sequential letter dependencies but insensitive to direction-of-report instructions and one insensitive to the sequential letter dependencies but sensitive to the direction and timing of report instructions.

3.2 The Scanning Operator as a Parser

Mewhort and Campbell (1980, 1981) explain the results in terms of a scanning operator followed by a second operation, rehearsal. The scanning mechanism is sensitive to sequential letter dependencies but not to report instructions; rehearsal, however, reflects the strategies adopted in response to report instructions. Further, while the scanning mechanism accepts data organized in space, rehearsal is associated with short-term memory and involves data extended in time. In their account, scanning occurs after letter recognition, and the data are taken from a post-categorical representation, the character buffer. The scanning mechanism is thought to use orthographic rules to parse the character string into derived units; the use of orthographic knowledge accounts for its sensitivity to sequential letter dependencies. Once the units have been formed, they are passed to short-term memory in a left-to-right order. By passing the material in order, the scanning mechanism provides rehearsal with a left-to-right organization that is not affected by instructions. The left-to-right organization provided by the scanning mechanism serves as a default. Depending on the timing of the report instruction, rehearsal may not be able to re-organize the material: with an early cue, re-organization is possible, with the result that subjects report best the material reported first. With a late cue, however, rehearsal mechanisms are forced to use the default organization and, as Scheerer (1972) noted, the material reported best comes from the LVF (see also Scheerer, 1973).

4 Conclusions

The idea of a left-to-right sweep of attention is necessary both to explain why subjects abandon temporal organization in favour of spatial organization (given rapid letter-by-letter presentation) and to explain the ordered-report results associated with instructed-recall procedures. In the discussion, I have distinguished three levels of data representation and have argued that different processing operations are associated with each level. The first level is an initial pre-categorical or feature level, and the available results make it clear that scanning does not involve data at that early level: letter identification is a prerequisite to scanning. The second level is an abstract representation, called the character buffer, which holds identified material. Presumably, it preserves at least some of the spatial attributes of the original display, and it provides input to the scanning operation. The third and final level of data is that associated with short-term memory; I assume that data at the third level are organized temporally, that the scanning operation prior to short-term memory provides a default left-to-right organization, and that rehearsal is required to preserve and, if necessary, to re-organize the material.

One controversial part of the present argument concerns the idea of different levels of data structure and of a character buffer, in particular. The idea of a post-identification data representation is not new: Smith and Spoehr (1974) acknowledged that something of the sort is required by their parsing model, but they did not explore the methodological implications of a multi-level system or offer independent evidence for the character-level construction. The major theoretical benefit offered by the character-level representation concerns the possibilities it offers for building "intelligence" into a model. Intelligence concerns the use of grammatical and/or frequency information during processing, i.e., the use of knowledge-based as opposed to stimulus-driven information. In the present context, the relevant knowledge concerns orthography, and we can distinguish two kinds of orthographic knowledge. The first, simple frequency information, is used in identifying letters (see, for example, Campbell & Mewhort, 1980). The second kind of knowledge concerns orthographic rules. Such rules involve contingent relations among letters, and there is ample evidence that subjects use orthographic rules when identifying a word (e.g., Mewhort & Beal, 1977). Scheerer-Neumann (1981), for example, has exploited subjects' use of orthographic rules to develop a very promising strategy for teaching poor readers. The use of rules, as opposed to simple frequency relations, however, requires knowledge of letters, i.e., it implies a character-level representation.

Current evidence concerning the character-level representation derives mainly from a detailed analysis of performance in partial-report tasks (e.g., Mewhort et al., 1981, 1983), but direct evidence has also been obtained in recent studies of information integration across saccades. Davidson, Fox, and Dick (1973), for example, followed a multi-letter tachistoscopic display with a mask positioned to cover one letter. The timing of the mask was arranged so that it appeared just after a voluntary eye movement had shifted the subject's gaze a few character positions. If information storage across saccades is in the form of raw information, the mask should interfere with recognition of the letter in the retinal position on which the mask was projected. The mask blanked report of the item in retinal space, i.e., it reduced accuracy of report for the item in line with the mask on the retinae. The apparent position of the mask, however, was not consistent with its effect on accuracy of report: instead, the mask appeared to be positioned over the original item, i.e., the mask's perceived location compensated for the movement of the eyes. The difference indicates a storage of additional information, and, in combination, the two effects of the mask correspond nicely to the first two levels of data, respectively, in the three-level system.

References

Ayres, J.J.B. Some artifactual causes of perceptual primacy. *Journal of Experimental Psychology,* 1966, *71,* 896–901.

Bradshaw, J.L., Nettleton, N.C., & Taylor, M.J. The use of laterally presented words in research into cerebral asymmetry: Is directional scanning likely to be a source of artifact? *Brain & Language,* 1981, *14,* 1–14.

Bryden, M.P. Tachistoscopic recognition of non-alphabetic material. *Canadian Journal of Psychology,* 1960. *14,* 78–86.

Bryden, M.P. The role of post-exposural eye movements in tachistoscopic perception. *Canadian Journal of Psychology,* 1961, *15,* 220–225.

Bryden, M.P. A model for the sequential organization of behaviour. *Canadian Journal of Psychology,* 1967, *21,* 37–56.

Campbell, A.J., & Mewhort, D.J.K. On familiarity effects in visual information processing. *Canadian Journal of Psychology,* 1980, *34,* 134–154.

Crovitz, H.F., & Daves, W. Tendencies to eye movement and perceptual accuracy. *Journal of Experimental Psychology,* 1962, *63,* 495–498.

Davidson, M.L., Fox, M., & Dick, A.O. Effect of eye movements on backward masking and perceived location. *Perception & Psychophysics,* 1973, *14,* 110–116.

Hebb, D.O. *The organization of behavior.* New York: Wiley, 1949.

Heron, W. Perception as a function of retinal locus and attention. *American Journal of Psychology,* 1957, *70,* 38–48.

Merikle, P.M., Lowe, D.G., & Coltheart, M. Familiarity and method of report as determinants of tachistoscopic performance. *Canadian Journal of Psychology,* 1971, *25,* 167–174.

Mewhort, D.J.K. Accuracy and order of report in tachistoscopic identification. *Canadian Journal of Psychology,* 1974, *28,* 383–398.

Mewhort, D.J.K., & Beal, A.L. Mechanisms of word identification. *Journal of Experimental Psychology: Human Perception & Performance,* 1977, *3,* 629–640.

Mewhort, D.J.K., & Campbell, A.J. Processing spatial information and the selective-masking effect. *Perception & Psychophysics,* 1978, *24,* 93–101.

Mewhort, D.J.K., & Campbell, A.J. The rate of word integration and the overprinting paradigm. *Memory & Cognition,* 1980, *8,* 15–25.

Mewhort, D.J.K., Campbell, A.J., Marchetti, F.M., & Campbell, J.I.D. Identification, localization, and "iconic" memory: An evaluation of the bar-probe task. *Memory & Cognition,* 1981, *9,* 50–67.

Mewhort, D.J.K., & Campbell, A.J. Toward a model of skilled reading: an analysis of performance in tachistoscopic tasks. In G.E. MacKinnon & T.G. Waller (Eds.), *Reading research: Advances in theory and practice,* (Vol. 3). New York: Academic Press, 1981.

Mewhort, D.J.K., & Cornett, S. Scanning and the familiarity effect in tachistoscopic recognition. *Canadian Journal of Psychology,* 1972, *26,* 181–189.

Mewhort, D.J.K., Marchetti, F.M., Gurnsey, R., & Campbell, A.J. Information persistence: a dual-buffer model for initial visual processing. In H. Bouman & D. Bouwhuis (Eds.), *Attention and performance 10: Control of language processes.* Hillsdale, NJ: Erlbaum, 1983.

Neisser, U. *Cognitive Psychology.* New York: Appleton, 1967.

Scheerer, E. Order of report and order of scanning in tachistoscopic recognition. *Canadian Journal of Psychology,* 1972, *26,* 382–390.

Scheerer, E. A further test of the scanning hypothesis in tachistoscopic recognition. *Canadian Journal of Psychology,* 1973, *27,* 95–102.

Scheerer-Neumann, G. The utilization of intraword structure in poor readers: Experimental evidence and a training program. *Psychological Research,* 1981, *43,* 155–178.

Smith, E.E., & Spoehr, K.T. The perception of printed English: A theoretical perspective. In B.H. Kantowitz (Ed.), *Human information processing: Tutorials in performance and cognition.* Hillsdale, NJ: Erlbaum, 1974.

Smith, M.C., & Ramunas, S. Elimination of visual field effects by use of a single report technique. *Journal of Experimental Psychology,* 1971, *87,* 23–28.

Summary. Early accounts of visual processing claimed that, when presented with a letter string placed centrally in vision, subjects scan the material from left to right. Scanning was thought to distribute attention sequentially and, thus, to provide a temporal organization for spatially parallel material. Scanning was typically thought to be part of the process used in recognizing the material. I argue in the present paper that the idea of left-to-right processing is fundamentally correct but that scanning occurs after the material has been identified.

10 The Relationship Between Motor Processes and Cognition in Tactile Vision Substitution

PAUL BACH-Y-RITA

Contents

Studies with tactile vision substitution for congenitally blind persons provide an unusual opportunity to observe the acquisition of "visual" spatial concepts in adolescents and adults. Since all aspects of the training are under the experimenter's control, the effects of each component of the process can be studied. In this paper the relationship between motor processes and cognition will be examined; specifically, the effect of placing the "eye" (television camera) under the control of the blind subject. We have noted that as long as the subject can control the movement of the camera, he can perceive in terms of the three-dimensional visual spatial world of which he is a part. It is possible to change the location and even the orientation of the tactile array (e.g., from the skin of the back to the abdomen), or the motor system controlling camera movement (either hand held, or located on spectacle frames and thus controlled by neck muscles), without compromising accurate spatial orientation.

Cognition and Motor Processes
Ed. by W. Prinz and A.F. Sanders

1 Tactile Vision Substitution System

In our sensory substitution studies, a substitute "eye" (a television camera) is placed under the motor control of a blind person. The optical signals received by the camera are transduced into stimuli that can be presented to skin receptors. A matrix of stimulators delivers images to the skin for relay to the brain. The blind person can be trained to use the information as "visual".

The instrumentation developed for the tactile sensory substitution system (TVSS) has been described (Collins & Bach-y-Rita, 1973) and our studies have been reported elsewhere (e.g., Bach-y-Rita, Collins, Saunders, White, & Scadden, 1969; Bach-y-Rita, 1972, 1982, 1983; White, Saunders, Scadden, Bach-y-Rita, & Collins, 1970).

Most of the subjects were congenitally blind college students who were paid to participate in the studies. Two of the research subjects were later included in the research team; one completed an engineering degree and became a research engineer on the project and the other became a graduate student and completed a Ph.D. while collaborating as a Research Psychologist.

Blind subjects were initially trained to control the camera, including manual control of the operation and focus and zoom. Each subject learned to direct the camera towards part of the field. Having achieved familiarity with manipulation of the camera, the person was tought to discriminate individual lines (vertical, horizontal, diagonal, or curved), subsequently shapes (circles, squares, or triangles), and solid geometric forms. When these were identified readily, a number of common objects (cup, chair, telephone) were presented, in varying positions and at different distances from the camera. As the appearance of these objects became familiar, the blind person discovered optical effects and developed visual concepts, such as shape distortion as a function of viewpoint and apparent change in size as a function of distance.

When two or more objects were presented similtaneously, the blind person learned to recognize each from minimal or partial cues. The subjects were able to describe the layout of three or four objects on a table in correct relationship even though they overlapped or were only partly visible. As training continued, techniques of visual analysis were developed. Studies with the TVSS have revealed rapid perceptual learning in spite of the poor resolution of the stimulus display. New perceptual concepts, such as the perceptual use of parallax, shadows, looming, and monocular cues of depth, were learned within surprisingly few trials, even though they had not been previously experienced by the congenitally blind persons.

Facility in directing the camera was accompanied by a change in the sensation derived from the patterned punctate stimulation of the skin. In the early stages of training (or when the camera was either immobile or under the control of another person), subjects reported experiences in terms of the sensations on the area of skin receiving the stimuli. However, when they could easily direct the camera at will, their reports were in terms of objects localized externally in space in front of them. The provision of a motor linkage (camera movement) for the sensory receptor surface on the skin produced a surrogate "perceptual organ." The receptor surface thus became part of a perceptual organ that could substitute for the normal visual perceptual organ, consisting of the eye with its receptor surface (the retina) and its motor apparatus (eye and neck muscles) (Bach-y-Rita, 1972).

We determined that the TVSS had practical educational and vocational applications; for example, we explored the presentation and recognition of forms, objects, letters, and graphic material (e.g., bar graphs), the identification of geometric projections and the ability to view objects under a microscope, such as a red blood cell or the wing of a fly. Further, instruction on spatial perception, as in the appearance of a table or a coin seen from a different angle, or the localization in depth of several objects in the field of view, or the flickering movements of a candle flame when moved by an air current, provided our blind subjects with concepts not available by any other means. A highly personal account has been published by a congenitally blind Ph.D. candidate in Philosophy (Guarniero, 1974, 1977). He later wrote his doctoral dissertation on space perception (Guarniero, Note 1).

2 The Somatosensory System: Its Relevance toTactile Sensory Substitution Studies

2.1 Tactile Pattern Stimulation

An apparently logical way to design a tactile sensory substitution system is to measure the static two-point discrimination of the target skin area and limit the display to an array of stimulators separated by at least the two-point discrimination distance. However, the central nervous system extracts information from patterns in a dynamic display. Thus, although two-point discrimination is 56 mm on the skin of the back, in our TVSS displays we were able to use vibrator separation of less than half that distance. Patterns were also successfully presented through closely spaced electrical stimulators. With both types of stimulation, the array could be moved from one part of the body to another.

It is thus apparent that the brain can adjust to considerable variation in the tactile array and its body interface. It is not necessary to use the same cutaneous receptors or even the same locus during each experimental session. Critical factors for stable perception are a reliable tactile display and motor control of the information acquisition (with the TV camera).

2.2 Information Transfer

The results obtained with tactile substitution systems reveal that the somesthetic system is capable of mediating information from an artificial "eye". The receptor matrix on the skin becomes, in essence, a relay to the brain of the information from the artificial "eye" (TV camera). In this sense its role is comparable to the dorsal column nuclei or the ventrobasal thalamus, which normally function as relay stages in the somesthetic system. This system conveys information from the cutaneous receptors centrally, whether the information is received through "normal" tactile stimulation or through the mosaic of mechanical or electrical stimulators that transmit the output of the TV camera. Undoubtedly, the information is partially processed at subcortical levels, with descending centrifugal influences and "filtering" and "funneling". The information is then relayed to perceptual regions of the brain, probably via the somesthetic cortex.

The characteristics of the cutaneous receptors allow a rapid transfer of information, with possibilities for increasing the rate of transfer. In comparison with the retina, there is a faster transmission from the skin to the brain owing to the absence of retinal delay in the somesthetic system. This is especially evident if the display is delivered to a skin region close to the brain such as the forehead to scalp. Tactile acuity is best when brief mechanical stimuli are applied repetitively and inhibition is stronger for stimuli with rapid onset, which produce a greater amount of "funneling". Indeed, a decrease in lateral spread with an increase in frequency is a universal phenomenon (von Bekesy, 1967). These factors should be reflected in the performance of a mechanical tactile sensory system. Although the only frequency of mechanical vibration we have employed has been 60 Hz (for simplicity and economy, we used the line current frequency), a higher stimulus frequency should produce increased resolution. In fact, the Linvill-Bliss Optacon studies have demonstrated that 250 Hz is the most appropriate stimulus frequency on the fingertips (Rogers, 1970). The optimum frequency may also reflect the type of cutaneous receptors activated by the skin stimulation. Future studies may employ a stimulus frequency chosen in accordance with a desired "decay rate."

2.3 Central Nervous System Factors

Electrical stimulation undoubtedly activates different receptors and different patterns of receptors than do mechanical vibrators. The short path of the electrical current between the inner and outer rings of the concentric electrodes may maximally stimulate only the superfivial free nerve endings, although the reaction times noted suggest that fast pathways are used. On the other hand, mechanical vibrators produce waves which stimulate receptors in sequence and probably also stimulate deeper receptors. The absence of subject confusion (noted in our tactile vision substitution studies) when the locus and even the type of stimulus is changed suggests strongly that the plastic changes to enable the subject to receive in the new way have occurred at a high level: certainly it is not the peripheral receptors that have changed. It is also unlikely that spinal cord mechanisms are critical, since mechanical stimulation on the back and electrical stimulation on the abdomen obviously involve different spinal cord structures at different cord levels. Our results thus lend support to the pattern theory of cutaneous innervation, but also suggest that it is the higher supracord structures that are primarily involved in subjective sensations and percepts. Further, under adequate conditions these brain processes can compensate for peripheral changes, if these changes are coupled with the movements of a subject-controlled "perceptual organ". For the perceptual regions of the brain, it is immaterial how the information enters the body as long is it is gathered by a perceptual organ with motor control and is recieved by a receptor matric that can cope with the detail of the display. So long as the display presents the information reliably, the brain can apparently be trained to use the information from the TV camera as it uses the information from any of the intact sensory systems.

2.4 "Visual" Quality of the Tactile Image

When a blind subject moves the camera across a field or an object, he obtains an image that moves across the receptors in his skin. Mechanisms similar to those in the retina (such as lateral inhibition) are available for edge enhancement. Our data suggest that, at least initially, the blind subjects obtain the "visual" information primarily by an analysis of contours (although simultaneous analysis of the information is also used), and thus artificial edge enhancement should produce improved performance.

Subjects using the TVSS learn to treat the information arriving at the skin in its proper context. Thus, at one moment the information arriving at the skin has been gathered by the TV camera, but at another it relates to the usual cutaneous information (pressure, tickle, wetness, etc.). The subject is not confused; when he scratches his back under the matrix he does not "see" anything. Even during task performance with the sensory system, the subject can perceive purely tactile sensations when he is asked to concentrate on these sensations. Further, as noted above, no relearning is necessary when the matrix is moved from one skin locus to another, provided that the camera is controlled by the blind subject. Experienced blind subjects trained with the mechanovibratory matrix on the back immediately adapted to the electrical stimulus matrix on the abdomen. The vibrators produce waves of skin movement, which travel across the skin. (Von Bekesy, 1967, has reported skin waves on the arm of the order of 2 cm for vibrations of 50 Hz. An increase in vibrator frequency up to 150 Hz reduces the wavelength to 0.6 cm.) In contrast, the effect of electrical stimulation is limited to a small region between the inner and outer rings of the concentric electrodes.

Blind subjects apparently learn how to interpret the information relayed through the skin stimulators in terms of "visual" images. The learning process may be similar to that which takes place in children with normal sensory and motor systems, or in adults learning a foreign language or Morse code, or in deaf persons learning manual communication. Blind subjects who have trained with the TVSS demonstrate perceptual equivalence between and across modalities. However, this is also frequently noted under other circumstances: Gibson (1966) noted that "fire" is the same whether the information has been obtained by hearing, feeling, looking, or smelling. There is a common aspect of perceptual activity that permits one to utilize information from several channels in such a way that invariant properties of objects are extracted.

As learning progresses, the information extraction processes become more and more automatic and unconscious, and the "chunking" phenomena discussed by Miller (1956) allow the number of bits per chunk to increase. For example, a blind subject "looking" at a display of objects must initially consciously perceive each of the relative factors such as the perspective of the table, the precise contour of eac each object, the size and orientation of each object, and the relative position of parts of each object to others nearby. With experience, information regarding several of these factors is simultaneously gathered and evaluated. Thus concepts of "chunking" appear to apply to the development of increased information transfer through a sensory substitution system. This highly complex "visual" work can thus be reduced, by selective processes, to manageable proportions, allowing the input to be mediated by the somesthetic system or, in Gibsonian (1966) terms, the subject learns to extract the relevant information. Since the channel capacity of the somesthetic system does not differ markedly from that of the visual system (Miller, 1956), it is possible that with a high-resolution sensory substitution system, the information transfer rate may be comparable to that in a normal visual system, if the chunking processes can be developed.

2.5 Overload

Normal sensory systems do not usually overload, since the central nervous system is able selectively to inhibit information not needed for any particular perceptual task. We have discussed this elsewhere: "Many efforts at creating sensory aids set out to provide a set of maximally disciminable sensations. With this approach, one almost immediately encounters the problem of overload – a sharp limitation in the rate at which the person can cope with the incoming information. It is the difference between landing an aircraft on the basis of a number of dials and pointers that provide readings on such things as airspeed, pitch, yaw, and roll, and landing a plane with a contact analog display. . . Visual perception thrives when it is flooded with information, when there is a whole page of prose before the eye, or a whole image of the environment; it falters when the input is diminished, when it is forced to read one word at a time, or when it must look at the world through a mailing tube. It would be rash to predict that the skin will be able to see all the things the eye can behold, but we would never have been able to say that it was possible to determine the identity and layout in three dimensions of a group of familiar objects if this system had been designed to deliver 400 maximally discriminable sensations to the skin. The perceptual systems of living organisms are the most remarkable information-reduction machines known. They are not seriously embarrassed in situations where an enormous proportion of the input must be filtered out or ignored, but they are invariably handicapped when the input is drastically curtailed or artificially encoded.

Some of the controversy about the necessity of preprocessing sensory information stems from disappointment in the rates at which human beings can cope with discrete sensory events. It is possible that such evidence of overload reflects more an inappropriate display than a limitation of the perceiver. Certainly the limitations of this system are as yet more attributable to the poverty of the display than to taxing the information-handling capacities of the epidermis" (White et al., 1970).

2.6 Transduction into the Mode of Information Handling of the Substitute Sensory System

In one of the most successful sensory substitution systems, American Sign Language (ASL) for the deaf, information usually presented to the auditory system (which is capable of high frequency analysis, but receives relatively little information simultaneously) is presented to the visual system (which has a poor frequency response, but can receive a large amount of information simultaneously). Although the hand movements of the deaf ASL communicators are slow in relation to the auditory system frequency capacities, the hand movements and associated facial and body movements transmit a great amount of information simultaneously and allow the deaf persons to "speak" in real-time. Similarly, to transmit visual information to the somatosensory system (which is capable of a frequency resolution an order of magnitude greater than the eye but much less than the auditory system, but handles more simultaneous information than the ear although less than the visual system), the visual information must be transduced to allow the somatosensory system to operate most efficiently. This is discussed elsewhere (Bach-y-Rita, 1983).

2.7 Somatosensory Cortical Evoked Potentials in Blind and Sighted Subjects

To evaluate and identify further brain plasticity mechanisms demonstrated by the sensory substitution studies, we have evaluated differences in the cortical potentials evoked by patterned fingertip stimulation between sighted subjects and blind persons with extensive tactile training. The early components of the evoked potential are the same, but later components reveal shorter latencies in the blind subjects, with the N_3*3* latency being 13% shorter (Feinsod, Bach-y-Rita, & Madey, 1973). These data may reflect plastic changes due to training.

3 Motor Control and Cognition

3.1 TVSS Motor Control and Cognition Studies

Systematic studies of the motor control aspects of tactile vision substitution have not been undertaken. However, some of the anecdotal reports relating to motor control will be discussed briefly.

To differentiate between camera movement and object movement, both proprioceptive input and motor command [or in terms discussed by von Holst and Mittelstaedt (1950), the distinction between afferent stimulation (environment) and reafferent stimulation (active motion of the perceiver)] may play a role. This is comparable to the adaptation to prism and reversal lenses in vision studies, which requires some orderly relation between sensation and the sensory consequences of self-produced motion (Teuber, 1960). After training with the TVSS, Guarniero (1977) noted that "watching an image move in the direction opposite to the one in which the camera was moving was at first a most unsettling phenomenon, but eventually I learned to make the necessary adaptation so that objects appeared stationary as I scanned them. This was especially difficult for me to become accustomed to, because nothing in my tactile experience had prepared me for it." A further adaptation had to be made when he was given motor control of zoom and lens aperture: "After an hour or so of using the new camera, objects started to regain their familiar sizes." Guarniero (Note 1, p. 137) points out that only touch, vision, and the TVSS require movements of the receptors to explore the environment.

In vision, the muscles (oculorotary) and the receptor matrix (retina) controlled by them are in close proximity. In the TVSS studies muscles could be distant from the sensory input, and muscle control mechanisms could be interchanged. Thus, with the skin stimulation matrix kept in one area (e.g., the skin of the abdomen), the camera could be head mounted (controlled by neck muscles) or changed to hand held, with no noticeable effect on perception. Furthermore, the motor control can be complex: at the earliest stage of the TVSS project the TV camera was mounted on a tripod; vertical and horizontal movements were controlled by separate hand cranks, yet the blind subjects easily adapted to the awkward movement control system.

The change from the perception of a tactile stimulation to a spatial three-dimensional perception occurred in most TVSS subjects after 5–10 h of training. Subjects would begin to report perceptions in terms of space and distance ("out there") instead of merely describing the shape of the object. This important change was not easily recognized by the subjects; in fact Guarniero (1977), a highly analytical subject, was not convinced that he perceived spatially. However, in addition to reports of spatial percepts, incidents occurred that re-inforced this observation. For example, one subject was observing changes in perspective by viewing a large cardboard checkerboard; the cardboard was tilted in different directions and the subject was asked to judge the directions of the tilt. At one point, the cardboard slipped out of the hands of the instructor and fell against the subject. A few days later, the instructor moved the zoom control of the camera as the subject viewed the checkerboard. He had a clear defensive reaction, moving backward to avoid what he perceived to be the cardboard checkerboard falling on him. Thus, the increase in size of the checkerboard was perceived by him to be a looming object and the movement was perceived to be toward his face, even though the tactile array was placed against the skin of his back. The subject was controlling the camera movement, and thus perceived the spatial characteristics of the stimulus.

A number of the tasks performed by blind TVSS subjects can be interpreted as demonstrating hand-"eye" coordination, and thus requiring three-dimensional spatial (cognitive) and motor interaction. Among these tasks are the following:

(a) Assembly and Inspection of Miniatur Electronic Components. In cooperation with an electronics manufacturing firm, a highly trained TVSS subject (who was totally blind from the age of 2 months) learned to perform complex miniature diode assembly and inspection tasks requiring a considerable degree of hand-"eye" coordination while working on the assembly line. A small television camera was placed in the ocular of a dissecting microscope, and the subject received the image on the skin of the abdomen through a vibrotactile array fixed to the workbench. He attained a high level of performance in both the inspection and the assembly components of the job (described elsewhere; e.g., Bach-y-Rita, 1982).

(b) Tactually Guided Batting. The aim of this investigation was to study to what extent it is possible to perform a predictive task such as catching a ball guided only by information presented tactually. The tactile display was a 20 × 20 matrix of vibrators presented to the back of two well-trained blind subjects. The perceptual-motor task consisted of "batting" a ball which was rolled towards the subject with tactual information about its path. The results demonstrated that it is possible to pick up tactual information in this form, organize the response and perform appropriate movements within the time available (Jansson & Brabyn, Note 2).

(c) Handshaking. Congenitally blind TVSS subjects learned to perceive an outstretched hand and reach out to clasp it accurately with their own hand, monitoring the relative movement of the two hands (e.g., Guarniero, 1977).

TVSS subjects interviewed 10 years after completing their participation in the project emphasize the importance the training has had in developing visual concepts. As one subject stated, "I know what sighted people are seeing as they walk around a desk." The changes in perspective with self-generated motion while maintaining a unitary percept (the desk did not change shape) was thus particularly noteworthy to him.

3.2 Some Theoretical Questions

The theoretical questions that may be studied with tactile vision substitution are intriguing: Is the visual cortex necessary for tactile vision substitution? To date we have not trained cortically blind persons. In any case, it is by no means certain that the visual cortex serves only visual functions. We showed many years ago that cat primary visual cortex cells received inputs from skin and auditory (we did not test for other modalities) receptors (Murata, Cramer, & Bach-y-Rita, 1965), and Rosenzweig, Krech, Bennett, and Diamond (1962) have shown that in rats blinded at birth, the principal cortical changes produced by an enriched environment were in the occipital cortex, even though the rats had developed without any visual input. There is considerable interest in the possibility that the occipital cortex is an area of spatial function (e.g., Thompson, 1982; Doty, 1982). However, non-visual functions of the visual cortex had earlier been demonstrated by Lashley (1943). Tactile vision substitution studies may help to clarify the spatial orientation role of the occipital cortex.

Among questions that may be answered by further study are the following:

(a) How is the experience of a continuous visual world developed? A puzzle in the psychology of perception has been the appearance of the visual world as a coherent whole despite our viewing it through a temporally discontinuous series of eye fixations. Jonides, Irwin, and Yantes (1982) discuss the importance of a briefly lasting memory in which temporally separate glimpses of a display are stored simultaneously and are spatially reconciled with one another, thus allowing a coherent view of a display that is constructed from the individual glimpses of which it is made. They further discuss the effect of saccades and saccadic suppression, and they hypothesize that the integration of information indicated by the saccade condition requires the use of a special memory.

Blind TVSS subjects have also experienced a continuous stable "visual" world. It apparently must be learned: for example, Guarniero (1977) described his initial impression, while sweeping across a field with the TV camera, that the field was moving in the opposite direction. Tactile vision substitution studies, in which all factors can be controlled (including the ability to provide discontinuous input and possibly to create "saccades") may help to clarify these questions.

(b) Visual illusions. The fact that we have demonstrated "visual" illusions (e.g., the waterfall effect) with blind TVSS subjects strongly suggests that central, rather than retinal receptor mechanisms, underlie these phenomena.

A number of other theoretical questions, such as the cortical representation of functions including motor localization, may be appropriately studied with sensory substitution models. In particular, the development of a primate tactile vision substitution model would allow neuronal mechanisms to be evaluated. However, a discussion of these topics is beyond the scope of this chapter. Results to date with the TVSS have demonstrated, however, that sensory substitution models are of considerable value in the study of perception, cognition, and motor control.

References Notes

1. Guarniero, G. *The senses and the perception of space.* New York University Department of Philosophy, Ph.D. Thesis. University Microfilms (Nos. 78-3097), Ann Arbor, MI, 1977.
2. Jannson, G., & Brabyn, L. Tactually guided batting, Uppsala Psychological Reports no. 304, Uppsala, Sweden, 1981.

References

Bach-y-Rita, P. *Brain mechanisms in sensory substitution,* p. 192. New York: Academic Press, 1972.

Bach-y-Rita, P. Sensory substitution in rehabilitation. In L. Illis, M. Sedgwick, & H. Granville (Eds.), *Rehabilitation of the neurological patient,* pp. 361–383. Oxford: Blackwell Press, 1982.

Bach-y-Rita, P. Tactile vision substitution: Past and future. *International Journal of Neuroscience,* 1983, *19,* 29–35.

Bach-y-Rita, P., Collins, C.C., Saunders, F., White, B., & Scadden, L. Vision substitution by tactile image projection. *Nature,* 1969, *221,* 963–964.

Collins, C.C., & Bach-y-Rita, P. Transimission of pictorial information through the skin. *Advances in Biological and Medical Physics,* 1973, *14,* 285–315.

Doty, R.W. Lashley as iconoclast in the temple of neuroscience: Some thrusts he would have enjoyed today. In J. Orbach (Ed.), *Neuropsychology after Lashley,* pp. 229–246. Hillsdale, NJ: Erlbaum, 1982.

Feinsod, M., Bach-y-Rita, P., & Madey, J.M. Somatosensory evoked responses – Latency differences in blind and sighted persons. *Brain Research,* 1973, *60,* 219–223.

Gibson, J.J. *The senses considered as perceptual systems.* Boston: Houghton, 1966.

Guarniero, G. Experience of tactile vision. *Perception,* 1974, *3,* 101–104.

Guarniero, G. Tactile vision: A personal view. *Journal of Blindness and Visual Impairment,* 1977, *71,* 125–130.

Jonides, J., Irwin, D.E., & Yantes, S. Integrating visual information from successive fixations. *Science,* 1982, *215,* 192–194.

Lashley, K.S. Studies of cerebral function in learning. XII. Loss of the maze habit after lesions in blind rats. *Journal of Comparative Neurology,* 1943, *79,* 431–462.

Miller, G.A. The magical number seven, plus or minus two: Some limits on our capacity for processing information. *Psychological Review,* 1956, *63,* 81–97.

Murata, K., Cramer, H., & Bach-y-Rita, P. Neuronal convergence of nocious, acoustic and visual stimuli in the visual cortex of the cat. *Journal of Neurophysiology,* 1965, *28,* 1223–1239.

Rogers, C.H. Choice of stimulator frequency for tactile arrays. IEEE Trans. *Man-Machine System,* 1970, *11,* 5–11.

Rosenzweig, M.R., Krech, D., Bennett, E.L., & Diamond, M.C. Effects of environmental complexity and training on brain chemistry and anatomy: a replication and extension. *Journal of Comparative and Physiological Psychology,* 1962, *55,* 429–437.

Teuber, H.-L. Perception. In J. Field (Ed.), *Handbook of physiology,* (Vol 3, pp. 1595–1668). Baltimore: Williams and Wilkins, 1960.

Thompson, R. Functional organization of the rat lesion. In J. Orbach (Ed.), *Neuropsychology after Lashley,* (pp. 207–208). Hillsdale, NJ: Erlbaum, 1982.

von Bekesy, G. *Sensory inhibition.* Princeton, NJ: Princeton University Press, 1967.
von Holst, E., & Mittelstaedt, H. Das Reafferenzprinzip (Wechselwirkungen zwischen Zentralnervensystem und Peripherie). *Naturwissenschaften,* 1950, *37,* 464–476.
White, B.W., Saunders, F.A., Scadden, L., Bach-y-Rita, P., & Collins, C.C. Seeing with the skin. *Perception and Psychophysics,* 1970, *7,* 23–27.

Summary. A vision substitution system, in which "visual" information is acquired by congenitally blind persons by means of a tactile display of the image captured by a television camera, offers an opportunity to study the relationship between motor processes and cognition. As the blind subjects learn to use the system, spatial perception is related to the camera if its movement is under the subject's control. This occurs whether he is controlling camera movement with head or hand movements; the two control modes can be interchanged at will. To differentiate between camera movement and object movement, both proprioceptive input and motor command may play a role. A number of other practical and theoretical questions can be answered by using the tactile vision substitution system, and the development of an animal model would allow the study of underlying neural mechanisms.

III Mediating Structures and Operations Between Cognition and Action

11 Mechanisms of Voluntary Movement

H. H. KORNHUBER

Contents

1 Introduction

Movements are usually parts of meaningful actions guided on the one hand by motivation and on the other hand by cognition. Since its discovery by Fritsch and Hitzig in 1870 there has traditionally been an overestimation of the role of the motor cortex in voluntary movement. Therefore, a lively discussion started, when the author in 1970 replaced the omnipotency of the motor cortex by a more realistic theory of voluntary movement (Kornhuber, 1971a). In recent years, the hypotheses of 1970 have been supported by new findings while some details have been modified by recent research so that it seems timely to give an updated summary.

To use a comparison from politics: the traditional picture of the monarchy of the motor cortex has been changed into a republican cooperation of numerous centers and functions in a *distributed system.* The distribution is partly the consequence of the evolution of more advanced systems in the course of phylogenesis, without abolishing the older systems. The result is a multiple representation of motor functions with partial division of labor, which is an advantage for compensation of lesions by learning. For instance, a bilateral lesion of the superior colliculus in the monkey causes very little functional disturbance; similarly, bilateral destruction of the frontal eye fields results in little deficit. A combined lesion, however, of the frontal eye fields and the superior colliculus causes a severe and long lasting paralysis of all voluntary and visually guided eye movements (Schiller, True, & Conway, 1979).

Cognition and Motor Processes
Ed. by W. Prinz and A.F. Sanders

The fact, however, that the frontal eye field may compensate for the destruction of the midbrain optic tectum does not mean that both structures have the same function. Normally the visually guided eye movements are mediated by the superior colliculus (with the help of the forebrain visual system). This may be concluded from the following facts: (1) After destruction of the superior colliculus the saccadic eye movements in response to a visual stimulus have a longer latency; (2) there is no *Bereitschaftspotential* or motor potential over the frontal eye field preceding saccadic eye movements (Becker, Hoehne, Iwase, & Kornhuber, 1972); (3) in the monkey the neurons of the frontal eye field do not fire preceding saccadic eye movements (Bizzi, 1968). It has been suggested that one function of the frontal eye field may be to represent at the cortical level the actual position of the eyes; this information is important for the coordination with goal-directed hand movements (Kornhuber, 1974b).

Thus, the superior colliculus of the midbrain seems normally to be something like the sensorimotor cortex for eye movements. The sensorimotor cortex for speech, however, is probably Wernicke's speech area in the temporal lobe. This hypothesis is based on the following facts: (1) Lesions of the precentral motor cortex do not cause aphasia. (2) For the development of speech in the child, sufficient hearing is a necessary condition. The auditory cortex is juxtaposed to Wernicke's area in the temporal lobe. (3) Besides auditory feedback, verbal memory is necessary for language. This is obviously also localized in and near Wernicke's area, as shown by the amnestic aphasia in the course of left temporo-parietal lesions. (4) Lesions of Broca's frontal speech area cause only transient aphasic symptoms; lesions of Wernicke's area, however, cause longer lasting aphasia. (5) The most severe kinds of aphasia, however, result from combined lesions of Wernicke's cortex and the basal ganglia (Brunner, Kornhuber, Seemüller, Suger, & Wallesch, 1982). Thus, a motor structure, the basal ganglia, is able at least partially to compensate for lesions of Wernicke's speech area.

2 Motivation and Planning

After these examples to demonstrate the distributed nature of the motor system (Kornhuber, 1980b), let us treat the motor system in a more systematic way, starting with *motivation.* The analysis of the cerebral potentials preceding voluntary movement (Kornhuber & Deecke, 1965) shows that the earliest and largest *Bereitschaftspotential* preceding movements of the finger or hand (Deecke, Grözinger, & Kornhuber, 1976), of the eyes (Becker et al., 1972), or preceding speech production (Grözinger, Kornhuber, & Kriebel, 1979) is over the anterior midline of the forebrain, i.e., over the so-called supplementary motor area. This midline maximum is not due to summation from potentials of more laterally localized motor fields such as the precentral hand area, because there are cases in which the *Bereitschaftspotential* completely disappears over the lateral motor cortex whereas it is fully preserved over the midline (Deecke & Kornhuber, 1978). Furthermore, the temporal course of the cerebral potentials over different parts of the forebrain shows that the maximum of the *Bereitschaftspotential* over the mesial cortex is not an artifact due to volume conduction. There are, for instance, experimental situations in which the large (surface negative) *Bereitschaftspotential* over the supplementary motor area disappears 500 msec prior to another large (surface negative) potential, the directed attention potential, which has its maximum over the lateral parietal area (Deecke, Heise, Kornhuber, Lang, & Lang, 1983).

It has been concluded that "the juxtalimbic supplementary motor area obviously is one of the ways that channel drives, will and planning into action" (Kornhuber, 1980a). The anticipatory behavior of the *Bereitschaftspotential* over the supplementary motor area (Deecke et al., 1983) supports this conclusion. Preceding simple, self-paced finger or hand movements the *Bereitschaftspotential* starts to decline about 90 msec prior to the onset of movement in the electromyogram (Deecke et al., 1976). In a stimulus situation, however, with a fixed time program that allows temporal anticipation, the *Bereitschaftspotential* over the supplementary motor area starts to decline 300 msec prior to the change in movement (Deecke et al., 1983).

3 Classic Theory in Need of Revision

In Liepmann's (1900) classic model for the organization of voluntary movements (which has dominated the textbooks until now) there was no *motivation* at all. The basis for voluntary movement was thought to consist of fiber connections from the posterior sensory areas to the precentral motor cortex. It was an idea of a higher reflex mechanism: the subject was supposed to be guided exclusively by his perception of the external world. Thus, the introduction of motivation (drive or will) and corresponding planning is one of the major points of the modern theory of voluntary movement (Kornhuber, 1974a, 1980a,b).

Another error of the classic theory was the *centralization* of all different kinds of voluntary movement into the precentral motor cortex – with the consequence that even the eye movements were thought to be localized in the motor cortex, although all the available evidence shows that the eye movements are not represented in the motor cortex. It was forgotten that on electrical stimulation not only the motor cortex gives rise to movements but many other cortical areas as well (Vogt & Vogt, 1919).

A third point of the older theory in need of correction is the assumption that the precentral motor cortex is able to initiate and program voluntary movements by itself. The fact is that the motor cortex needs afferents from subcortical ganglia, especially from the basal ganglia and/or the cerebellum (via the thalamus) in addition to motivational and sensory information in order to initiate movements (Kornhuber, 1971a,b, 1974a). Lamarre, Spidalieri, Busby, and Lund (1980) found that preceding a conditioned arm movement neurons of the monkey motor cortex show an early sensory response (of about 25 msec latency to auditory stimuli and about 35 msec latency to somatosensory stimuli), but this early sensory response was temporally not related to the onset of movement. By contrast, there was a later motor discharge of the same neurons preceding the movement onset and correlated with it in time which was temporally unrelated to the sensory discharge of the same neuron.

The older theory of the motor system assumed that the activity of subcortical ganglia (especially the basal ganglia and the cerebellum) followed the motor cortex in time during a movement, making it more smooth. The conclusion that the cerebellar discharge precedes the discharge of the motor cortex was first drawn from the dysmetria following cerebellar lesions (Kornhuber, 1968, 1971a,b). There is now additional evidence from neuronal data: Preceding a forearm movement the neurons of the cerebellar dentate nucleus on average start to discharge 20 msec prior to the neurons of the motor cortex hand area (Thach, 1975). Furthermore, after cerebellar lesions in the monkey the arm movements to a sensory stimulus start 150 msec later than in the intact animal, and the neurons of the motor cortex are also delayed by 150 msec after the cerebellar lesions (Lamarre & Jacks, 1978).

Voluntary actions are usually preceded by *planning*, which consists of two aspects: strategy and tactics. The strategical aspects of planning belong to motivation and are localized at the cortical level in the mesial (limbic and paralimbic) and frontal cortex. The tactics, however, adjust the actions to the environment, which is analyzed by the teleceptive mechanisms of the posterior and lateral parts of the forebrain. It is this aspect of planning that is disturbed when neurologists diagnose apraxia.

Apraxia is the inability to act in the absence of paralysis. Both the apraxia for simple gestures (ideomotor apraxia) and for complex actions (ideational apraxia) are usually due to left parietal lesions (for review see Hécaen, 1981). Some neurophysiological data support this localization: first, cooling the parietal area 7 in the monkey disturbs aimed arm movements towards the contralateral side without causing paresis (Stein, 1978). Second, there is a wealth of neuronal data from Mountcastle's investigations (Mountcastle, Lynch, Georgopoulos, Sakata, & Acuna, 1975; Mountcastle, 1976; Lynch, Mountcastle, & Talbot, 1977) that demonstrate attention and command functions in the monkey parietal lobe. Third, there is a directed attention potential in man over the parietal area, when a human pays attention to sensory stimuli in the contralateral visual half field or on the contralateral hand (Deecke et al., 1983). Attention is necessary for planning. There are other kinds of apraxia (such as constructional apraxia and dressing apraxia) which may also be found following lesions of the posterior right hemisphere.

4 Interaction Between Cortical and Subcortical Control

4.1 Voluntary Movements

Regarding the programs for voluntary movements, I believe that they largely generate in subcortical ganglia (Kornhuber, 1971a,b, 1974a, 1980b). Akinesia, a symptom in basal ganglia disease, is the defective generation of movements with voluntary pattern and speed which are independent of external stimuli. The same movements, however, that cannot be carried out spontaneously by the patient, can often be evoked by external stimuli. With tactile stimuli there are magnet responses of the head in spastic torticollis or torsion dystonia, or magnet responses of the hand or foot in athetosis (Aschoff & Kornhuber, 1975). This difference in motor performance with and without guidance by external stimuli has also been found in experiments of basal ganglia cooling in monkeys (Hore, Meyer-Lohmann, & Brooks, 1977). Because of the difficulty in generating varing motor programs by himself, the basal ganglia patient tends to repeat existing programs, resulting in echopraxia, palipraxia, palilalia, recurring utterances. Global aphasia is due to combined lesions of Wernicke's perisylvian speech cortex and basal ganglia; in contrast to aphasia resulting from cortical lesions it is characterized by speech automatisms and recurring utterances (Kornhuber, 1977; Brunner et al., 1982).

In technical machines function generators and program generators are required for the production of self-generated movement. One of these function generators in our brain is obviously in the basal ganglia. After a stroke the hemiplegia (as well as the aphasia in lesions of the left hemisphere) has a much longer duration if the basal ganglia are involved in addition to the cortex.

Let us now return to the *precentral motor cortex.* In humans, there are three cerebral potentials preceding a voluntary finger or hand movement (Deecke, Scheid, & Kornhuber, 1969). First, there is the *Bereitschaftspotential* (already mentioned) with its maximum over the midline, starting about 800 msec prior to movement. Second, there is the pre-motion positivity that starts about 90 msec before the firing in the forearm electromyogram (EMG). Third, there is a small potential that starts about 50 msec prior to the EMG activity; it has been called motor potential and is localized over the precentral motor cortex hand area. It corresponds to the pyramidal cell firing of area 4. This discharge is, however, a late event in the course of information processing preceding movement. Having relieved the precentral motor cortex from some duties, what remains as its function? The motor cortex is juxtaposed to the somatosensory cortex. From the sensory point of view, the motor cortex is a somatic and proprioceptive-vestibular association area. The vestibular cortex belongs to the somatosensory area (Fredrickson, Kornhuber, & Schwarz, 1974). Most neurons of the cat's motor cortex respond to somatic stimuli and many to vestibular stimuli (Kornhuber & Aschoff, 1964; Aschoff & Kornhuber, 1975). The tactile placing reaction, the proprioceptive hopping reaction (Bard, 1938), the tactile grasping response (Denny-Brown, 1960) and the vestibular placing reaction (Bard & Orias, 1933) are lost after lesions of the motor cortex, while the visual placing reaction is preserved (Bard & Orias, 1933). These facts suggest that the function of the motor cortex is advanced tactile and proprioceptive adjustment of those movements that need this type of control: the finger, toe, lip and tongue movements. However, not all movements of these limbs are involved; for instance, lip and tongue movements for speech are guided by Wernicke's perisylvian speech area. While in rodents mainly the lips and the tongue are represented in the motor cortex, the movements for which the motor cortex is most indispensable in primates are the finger movements. It is probably because of this that in the monkey the visual signals for regulation of independent finger movements were also conveyed to the motor cortex (Haaxma & Kuypers, 1975). The evolution of the motor cortex in primates was an adjustment for climbing trees, which called for sophisticated finger, hand, and toe movements with tactile and proprioceptive beedbacks.

4.2 Saccadic Eye Movements

The evolution of eye movements was comparable to that of finger movements. Our subprimate ancestors had lateral eyes with panoramic vision and poor depth perception, while primates have front eyes with high foveal acuity, precise saccades, vergence, and a good depth perception which was necessary for jumping from tree to tree. The eye movements, however, did not enter the motor cortex but stayed in the midbrain tectum with analysis of foveal and perifoveal details by the visual, paravisual and inferotemporal cortex of the forebrain.

The regulation of eye movements uses the following afferents: vision, proprioceptive discharge from the extraocular muscle spindles, vestibular signals, and local feedback from the oculomotor system itself. For visual size perception, only visual information is used (Bechinger, Kongehl, Kornhuber, & Walther, 1974). While previously it was thought that the visual information used for programming saccadic eye movements is gathered discontinuously at intervals of about 200 msec, it is now clear that the amplitude information in the oculomotor system is continuously updated. There is a window of the order of 100 msec duration over which the spatial stimulus information is averaged by the brain (Becker & Jürgens, 1979). This moving average window is continuously shifted over the stimulus situation. The speed of our saccadic eye movements depends on the amplitude of the saccade and cannot be changed at will. With large target displacements, the saccadic system uses a double step approach: it generates first a large saccade that covers about 90% of the distance, which is followed by a small correction saccade of up to 5–6° amplitude (Becker, 1972). This is related to the fact that (in the monkey) the peripheral visual field projects in a retinotopic manner to the superior colliculus while the fovea has no direct connections to the midbrain. There is, however, a dense projection to the superior colliculus from that part of the monkey's striate cortex which serves the central 5° of the visual field. Experiments with double step stimuli (Becker & Jürgens, 1979) show that the preparatory processes for the main and for the small correction saccade partially overlap in time.

The tranformation of the retinal coordinates into head or orbita coordinates is not possible without taking into consideration the initial position of the eye within the orbit. The oculomotor system has to take into account that eye movements are limited. Furthermore, the viscoelastic properties of the eye muscles must be accounted for; they result in faster saccades from lateral to midposition than from center to lateral position. In clinical cases dysmetria of saccadic eye movements has been found resulting from cerebellar lesions (Kornhuber, 1968), and the eye muscle afferents were therefore studied in the cat using adequate stimuli. They were found in the middle part of the cerebellum (Fuchs & Konrhuber, 1969), which on electrical stimulation gives rise to saccadic eye movements (Hampson, Harrison, & Woolsey, 1952). These results have been confirmed and extended (Ritchie, 1976) by experiments in the monkey which showed that cerebellar dysmetria of saccadic eye movements is position-dependent: there is undershoot of large saccades from central to lateral positions and overshoot of saccades from lateral to central positions after middle cerebellar lesions. These results are exactly what one would expect if the system is not able to utilize orbital position information from the eye muscle receptors.

Nevertheless, it was not clear whether the saccade follows an open loop ballistic program (Robinson, 1973) or is continuously regulated by some fast feedback (as proposed by Vossius, 1960). Feedback from the eye muscle spindles was postulated since it was clear that visual feedback is too slow to regulate continuously the saccadic eye movements. Eventually a crucial experiment was performed (Jürgens, Becker, & Kornhuber, 1981): While the human subject performed aimed saccades to random step stimuli, diazepam was injected intravenously to slow the velocity of the saccades. The velocity dropped as much as 60%. There was, however, an immediate compensatory increase of saccadic duration, so that the amplitude of saccades remained unchanged. This could not be explained by learning since the variability of saccadic velocity increased greatly. This was conclusive evidence for the existence of a fast internal, non-visual feedback which helps to carry out precisely the commands from the visual system – despite the varying conditions of the motor system (e.g., slow saccadic speed in drowsiness).

There is additional evidence for fast non-visual feedback in the saccade by microstimulation of the pauser neurons in the raphe pontis, i.e., in the midline of the brain stem below the abducens nucleus of trained monkeys (Becker, King, Fuchs, Jürgens, Johanson, & Kornhuber, 1981). These neurons discharge continuously and inhibit the short lead burst neurons involved in the generation of the saccadic pulse. Stimulation of the pauser neurons therefore interrupts the saccades in mid-flight. If the concept of internal feedback is correct, a short interruption of a saccade should increase the duration of the saccade but should not affect the accuracy of the amplitude. Precisely this was the result. The most likely source of the internal feedback are lower centers of the oculomotor system itself such as brain stem oculomotor neurons.

5 Conclusion

In the advanced motor system of a freely moving animal, several feedbacks have to cooperate and a large amount of simultaneous (parallel) information processing is necessary in motivation, planning, cognition, programming, and feedback regulation. The cerebral cortex is an ensemble of a great number of cooperating parallel computers, the basic computations of which are probably performed in a way similar to an analog computer. To understand the coordination of the simultaneous activity of a large number of analog computers is a task for to be undertaken by neuroscientists in the coming years.

The distributed system theory of voluntary movement, which has not been the prevailing theory so far in physiology or neurology, is stressed in this paper in order to encourage psychologists, physicians and teachers to believe in the possibility of finding roundabout ways by learning and to prevent them from giving up too early in the training of patients with lesions of the central nervous system. The use of the motor system in primates is largely the result of learning; strategies once learned, however, are often pursued after lesions when they are no longer practical. Training with a new approach may therefore succeed even after years, for instance in cases of aphasia with total muteness.

References

Aschoff, J.C., & Kornhuber, H.H. Functional interpretation of somatic afferents in cerebellum, basal ganglia, and motor cortex. In H.H. Kornhuber (Ed.), *The somatosensory system,* (p. 145–157). Stuttgart: Thieme, 1975.

Bard, P. Studies in the cortical representation of somatic sensibility. *Harvey Lectures,* 1938, *33,* 143–169.

Bard, P., & Orias, O. Localized cortical management of visual and labyrinthine placing reactions. *American Journal of Physiology,* 1933, *105,* 2–3.

Bechinger, D., Kongehl, G., Kornhuber, H.H., & Walther, C. What do the eye movements contribute to the visual perception of length? *Pflügers Archives of Physiology,* 1974, *347,* Suppl., R 63.

Becker, W. The control of eye movements in the saccadic system. *Bibliotheca Ophthalmologica,* 1972, *82,* 233–243.

Becker, W., Hoehne, U., Iwase, K., & Kornhuber, H.H. Bereitschaftspotential, prämotorische Positivierung und andere Hirnpotentiale bei saccadischen Augenbewegungen. *Vision Research,* 1972, *12,* 421–436.

Becker, W., & Jürgens, R. An analysis of the saccadic system by means of double step stimuli. *Vision Research,* 1979, *99,* 967–983.

Becker, W., King, M., Fuchs, A.F., Jürgens, R., Johanson, G., & Kornhuber, H.H. Accuracy of goal-directed saccades and mechanisms of error correction. In A.F. Fuchs & W. Becker (Eds.), *Progress in oculomotor research,* (pp. 29–37). New York: Elsevier, 1981.

Bizzi, E. Discharge of frontal eye field neurones during saccadic and following eye movements in unanesthetized monkeys. *Experimental Brain Research,* 1968, *10,* 69–80.

Brunner, R.J., Kornhuber, H.H., Seemüller, E., Suger, G., & Wallesch, C.W. Basal ganglia participation in language pathology. *Brain and Language,* 1982, *16,* 281–299.

Deecke, L., Scheid, P., & Kornhuber, H.H. Distribution of readiness potential, pre-motion positivity and motor potential of the human cerebral cortex preceding voluntary finger movements. *Experimental Brain Research,* 1969, *7,* 158–168.

Deecke, L., Grözinger, B., Kornhuber, H.H. Voluntary finger movement in man: cerebral potentials and theory. *Biological Cybernetics,* 1976, *23,* 99–119.

Deecke, L., & Kornhuber, H.H. An electrical sign of participation of the mesial "supplementary" motor cortex in human voluntary finger movement. *Brain Research,* 1978, *159,* 473–476.

Deecke, L., Heise, B., Kornhuber, H.H., Lang, M., & Lang, W. Brain potentials associated with voluntary manual tracking: Bereitschaftspotential, conditioned premotion positivity (cPMP) directed attention potential (DAP) and relaxation potential (RXP); anticipatory activity of the limbic and frontal cortex. *Annals of the New York Academy of Sciences,* 1983.

Denny-Brown, D. The general principles of motor integration. In J. Field, H.W. Magoun, & V.E. Hall (Eds.), *Handbook of physiology,* Section I (Neurophysiology). Washington D.C.: American Physiological Society, 1960.

Fredrickson, J.M., Kornhuber, H.H., & Schwarz D. Cortical projections of the vestibular nerve. In H.H. Kornhuber (Ed.), *Vestibular system, Handbook of sensory physiology,* (Vol. 6, Part 1, pp. 565–582). Berlin, Heidelberg, New York: Springer, 1974.

Fritsch, G.T., & Hitzig, E. Über die elektrische Erregbarkeit des Großhirns. *Archiv Anat. Physiol. Wiss. Medizin* (Leipzig), 1870, 300–332.

Fuchs, A.F., & Kornhuber, H.H. Extraocular muscle afferents to the cerebellum of the cat. *Journal of Physiology* (London), 1969, *200,* 713–722.

Grözinger, B., Kornhuber, H.H., & Kriebel, J. Participation of mesial cortex in speech: evidence from cerebral potentials preceding speech production in man. In O. Creutzfeldt et al. (Eds.), *Hearing mechanisms and speech,* (pp. 189–192). Berlin, Heidelberg, New York: Springer, 1979.

Haaxma, R., & Kuypers, H.G.J.M. Intrahemispheric cortical connexions and visual guidance of hand and finger movements in the Rhesus monkey. *Brain,* 1975, *98,* 239–260.

Hampson, J., Harrison, C., & Woolsey, C. Cerebro-cerebellar projections and the somatotopic localization of motor function in the cerebellum. *Res. Publ. Ass. North Mental Diseases,* 1952, *30,* 299–316.

Hécaen, H. Apraxias. In S.B. Filskov & T.J. Boll (Eds.), *Handbook of clinical neuropsychology,* (pp. 257–286). New York: Wiley, 1981.

Hore, J., Meyer-Lohmann, J., & Brooks, V.B. Basal ganglia cooling disables learned arm movements of monkeys in the absence of visual guidance. *Science,* 1977, *195,* 584–586.

Jürgens, R., Becker, W., & Kornhuber, H.H. Natural and drug-induced variations of velocity and duration of human saccadic eye movements: evidence for a control of the neural pulse generator by local feedback. *Biological Cybernetics,* 1981, *39,* 87–96.

Kornhuber, H.H. Neurologie des Kleinhirns. *Zentralblatt für die gesamte Neurologie Psychiatrie,* 1968, *191,* 13.

Kornhuber, H.H. Das vestibuläre System, mit Exkursen über die motorischen Funktionen der Formatio reticularis, des Kleinhirns, der Stammganglien und des motorischen Cortex sowie über die Raumkonstanz der Sehdinge. In W.D. Keidel & K.H. Plattig (Eds.), *Vorträge der Erlanger Physiologentagung 1970.* Berlin, Heidelberg, New York: Springer, 1971. (a)

Kornhuber, H.H. Motor functions of cerebellum and basal ganglia. *Kybernetik,* 1971, *8,* 157–162. (b)

Kornhuber, H.H. Cerebral cortex, cerebellum and basal ganglia: an introduction to their motor functions. In F.O. Schmitt and F.G. Worden (Eds.), *The neurosciences, third study program,* (pp. 267–680). Cambridge, London: MIT Press, 1974. (a)

Kornhuber, H.H. The vestibular system and the general motor system. In H.H. Kornhuber (Ed.), *Vestibular system, part 2, Handbook of sensory physiology,* (Vol. 6, pp. 581–620). Berlin, Heidelberg, New York: Springer, 1974. (b)

Kornhuber, H.H. A reconsideration of the cortical and subcortical mechanisms involved in speech and aphasia. In J.E. Desmedt (Ed.), *Language and hemispheric specialization in man: cerebral ERP's. Progress in Clinical neurophysiology,* (Vol. 3, pp. 28–35). Basel, Karger, 1977.

Kornhuber, H.H. Introduction. In H.H. Kornhuber & L. Deecke (Eds.), *Motivation, motor and sensory processes of the brain: electrical potentials, behaviour and clinical use. Progress in Brain Research,* (Vol. 54, pp. IX–XII). Amsterdam: Elsevier, 1980. (a)

Kornhuber, H.H. Physiologie und Pathophysiologie der cortikalen und subcortikalen Bewegungssteuerung. In H.G. Mertens & H. Przuntek (Eds.), *Pathologische Erregbarkeit des Nervensystems und ihre Behandlung. Verhandlungen der deutschen Gesellschaft für Neurologie,* (Vol. 1, pp. 17–32). Berlin, Heidelberg, New York: Springer, 1980. (b)

Kornhuber, H.H., & Aschoff, J.C. Somatisch-vestimuläre Integration an Neuronen des motorischen Cortex. *Naturwissenschaften,* 1964, *51,* 62–63.

Kornhuber, H.H., & Deecke, L. Hirnpotentialänderungen bei Willkürbewegungen und passiven Bewegungen des Menschen: Bereitschaftspotential und reafferente Potentiale. *Pflügers Archives of Physiology,* 1965, *248,* 1–17.

Lamarre, Y., & Jacks, B. Involvement of the cerebellum in the initiation of fast ballistic movement in the monkey. In W.A. Cobb & H. van Duijn (Eds.), *Contemporary clinical neurophysiology. EEG and Clinical Neurophysiology,* (Suppl. No. 34, pp. 441–447). Amsterdam: Elsevier, 1978.

Lamarre, Y., Spidalieri, G., Busby, L., & Lund, J.P. Programming of initiation and execution of ballistic arm movements in the monkey. In H.H. Kornhuber & L. Deecke (Eds.), *Motivation, motor and sensory processes of the brain: electrical potentials, behaviour and clinical use. Progress Brain Research 53.* Amsterdam: Elsevier, 1980.

Liepmann, H. Das Krankheitsbild der Apraxie. *Monatsschrift für Psychiatrie,* 1900, *8,* 15, 102, 182.

Lynch, J.C., Mountcastle, V.B., Talbot, W.H. et al. Parietal lobe mechanisms for directed visual attention. *Journal of Neurophysiology,* 1977, *40,* 362–389.

Mountcastle, V.B. The world around us: neural command functions for selective attention. *Neuroscience Research Program Bulletin* (Suppl.), 1976, *14,* 1–47.

Mountcastle, V.B., Lynch, J.C., Georgopoulos, A., Sakata, H., & Acuna, C. Posterior parietal association cortex of the monkey: command functions for operations within extrapersonal space. *Journal of Neurophysiology,* 1975, *38,* 871–908.

Ritchie, L. Effects of cerebellar lesions on saccadic eye movements. *Journal of Neurophysiology,* 1976, *39,* 1246–1256.

Robinson, D.A. Models of the saccadic eye movement control system. *Kybernetik,* 1973, *14,* 71–83.

Schiller, P.H., True, S.D., & Conway, J.L. Effects of frontal eye field and superior colliculus ablations on eye movements. *Science,* 1979, *206,* 590–592.

Stein, J. Long loop motor control in monkeys. The effects of transient cooling of parietal cortex and of cerebellar nuclei during tracking tasks. In J.E. Desmedt (Ed.), *Cerebral motor control in man: long loop mechanisms. Progress Clinical Neurophysiology,* (Vol. 4, pp. 107–122). Basel: Karger, 1978.

Thach, W.T. Timing of activity in cerebellar dentate nucleus and cerebral motor cortex during prompt volitional movement. *Brain Research,* 1975, *88,* 233–241.

Vogt, C., & Vogt, O. Allgemeinere Ergebnisse unserer Hirnforschung IV: Die physiologische Bedeutung der architektonischen Rindenfelderung auf Grund neuer Rindenreizungen. *Journal für Psychologie und Neurologie* (Leipzig), 1919, *25,* 339.

Vossius, G. Das System der Augenbewegung. *Zeitschrift für Biologie,* 1960, *112,* 27–57.

Summary. Voluntary movement is usually guided both by cognitive mechanisms directed towards the external world and by motivation derived from internal stimuli, inborn information and experience stored in memory. In the case of simple movements, motivation seems to be channelled into action by way of the limbic system and supplementary motor area, localized anterior-medially in the forebrain. On the other hand, the mechanisms of perception and attention directed toward the external world are localized over the posterior and lateral parts of the hemisphere. The spatiotemporal functions and programs for voluntary movement need subcortical mechanisms of the basal ganglia, brain stem and cerebellum. The precentral motor cortex alone is unable to generate voluntary movements. The function of the motor cortex is probably advanced tactile and proprioceptive adjustment of the finger, hand, toe, lip and tongue movements. The same tongue movement, however, which during chewing is regulated by the motor cortex, is guided during speech production by Wernicke's cortex of the temporal lobe. Although eye movements had an evolution comparable to finger movements, they did not enter the motor cortex, but stayed in the midbrain tectum, with help from the visual areas and motivational mechanisms of the forebrain. A relatively simple movement such as a saccadic eye movement has a multiple feedback regulation by vision, extraocular muscle proprioception and fast internal feedback (by efference copy from the brain stem oculomotor system). The system of voluntary movement is a distributed system with alternative pathways and with possibilities to find roundabout ways after lesions.

12 Evaluation:
The Missing Link Between Cognition and Action

D. M. MacKay

Contents

1 The Information-Engineering Approach

Information engineering has grown out of the need to understand – and to contrive – systems concerned primarily with the determination of *form*. In processes of communication and control such as printing, telephony or automation, what matters is how the form of one activity or state of affairs determines the form of another, without explicit regard to the energy involved. Questions of energetics – the determination of *force* by *force* – are the complementary province of physics. Some physical balance-sheet must apply to any information-engineering transaction; but the information engineer normally takes this for granted, having assured himself that the energy supply is more than adequate for his purposes (and/or having restricted his purposes to respect the limitations imposed by energetic considerations).

Except in the simplest cases (for example in linear amplification) the process of form determination involves *computation*. It is important to recognize that the concept of computation (or information processing) is not restricted to digital operations such as counting and binary logic. It also covers such diverse processes as

(a) setting up interactions between physical quantities representing informational variables, and allowing the physical magnitude of the result to determine the form of the outcome (as in "analog" computing);

Cognition and Motor Processes
Ed. by W. Prinz and A.F. Sanders

(b) detection of those signals that coincide or co-vary in a large population;
(c) modulation of stochastic processes, in which information may be represented for example by a conditional "transition probability matrix" determining only the relative probabilities of alternative forms of outcome in different circumstances, and so on.

In all of these, what matters is how the form of one set of variables determines (uniquely or statistically) the form of another set. If (as in analog computing) considerations of physical energetics are relevant, they are taken into account as *determinants of form.*

2 Evaluation

One of the key concepts of information engineering is *evaluation.* As use of this technical term is often mistaken by laymen to imply the attribution *ipso facto* of conscious mental activity, it may be as well to introduce it by way of a down-to-earth example. Figure 1 shows the basic information-flow diagram of a simple engineering servo-mechanism, as in a car with "power steering". It has essentially four components:

(i) An effector (motor) system (E_s) determines the angle of inclination (θ) of the road wheels.
(ii) A receptor system (R_s) provides an indication θ_I of the current value of θ.
(iii) An evaluator (C_s) compares the indicated value θ_I against the angle θ_G prescribed by the manually operated steering wheel, and generates a signal (normally called the "mismatch signal") which indicates the extent and nature of any deviation of θ_I from θ_G.
(iv) A computational organizing system (O_s) determines the form of any corrective action required from E_s, according to the information content of the mismatch signal (including in general its recent history).

"Evaluation" in this case has obviously no necessary "cognitivist" implications. The information engineer uses it to refer to a form-determining function that *could* in principle be performed by a conscious cognitive agent using his mental capacities (hence the "mentalistic" flavour of the term), but a function that can here be taken over by a mindless automaton.

3 Setting of Evaluative Criteria

The flow-pattern of informational cause-and-effect in Figure 1 is of course incomplete. The *criterion* of evaluation θ_G is taken as given, and its source is not shown.

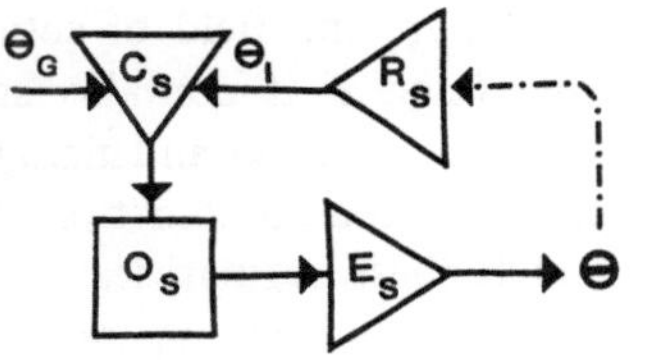

Figure 1. Information-flow in a simple steering servo. Evaluator C_s compares indicated wheel angle θ_I, derived from receptor system R_s, against goal criterion θ_G, and generates a mismatch signal from which O_s calculates the action by effector system E_s that should bring θ_I to match θ_G.

As a first step forward, we can elaborate our diagram as in Figure 2, where the setting of θ_G is performed by the effector system (E) of the human car driver. In relation to the steering servo his goal-setting function is *supervisory*.

The driver's CNS in turn has a receptor system (R), providing a (multi-channel) indication (I_F) of the state of his field of action (F), and particularly of the course of the car relative to the road. In information engineering terms I_F is presumably evaluated in turn against some current goal-criteria (distance of car from road-centre, or whatever), which we represent in Figure 2 by I_G. Once again, we depict evaluation formally as a comparison process, in an information-processing stage (C) which may now have to embody quite complex associative computations.

The output of the evaluative process will in general have many degrees of freedom. For example, when controlling the motion of an object in three dimensions we may need to evaluate three orthogonal position coordinates and three components of velocity for its centre of gravity alone, together with other variables governing its orientation in space. The evaluative assessment with respect to each degree of freedom has to be supplied separately to an organizing system (O), whose function is to calculate the form of an action most likely on average to maximize positive evaluation, or minimize negative evaluation, for each degree of freedom.

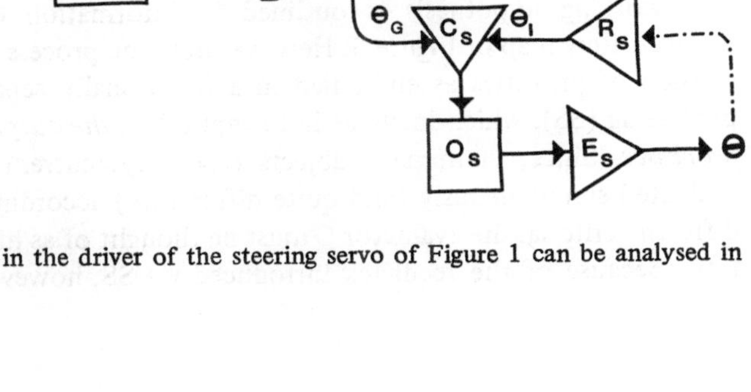

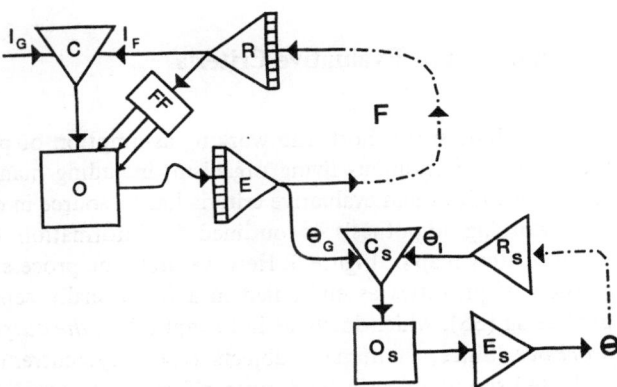

Figure 2. Information-flow in the driver of the steering servo of Figure 1 can be analysed in similar terms (see text).

Where the field of action (F) manifests some degree of regularity in space or time (e.g., if it contains stable objects like chairs and tables, or shows regular sequences like day and night) it will pay the organizing system to set up subroutines – prefabricated combinations of selective commands or conditional instructions – to match these regularities in the sense of equipping it to take adaptive account of them in the planning or execution of action relating to them. To take a simple example I have used before (MacKay, 1956), if a road has a regularly alternating sequence of left- and right-hand bends, it would help an automaton to find its way along it if it were equipped automatically to try turning left (L) at the next corner after a right-hand turn (R), and vice versa. If the shape were to change to LLRRLL. . ., a new subroutine would have to be developed to match it. Each of these subroutines would in an operational sense *represent* the abstract feature of the road to which they were adaptive.

To help it "home in" on appropriate actions in real time with a minimum of trial-and-error 0 will need "feed-forward", filtered by FF in suitable categories from the array of receptor signals, as shown in Figure 2. In general, then, as I have argued elsewhere, the resulting "state of conditional readiness" (SCR) set up in 0 to reckon with the field of action (F) can be thought of as implicitly representing the *structure* of that field (MacKay, 1951, 1956, 1970).

So far, so good. Information-flow analysis in these terms has found plenty of fruitful applications to the functioning of motor control in living organisms, including man. (See, for example, Dichgans & Bizzi, 1972.) But in our search for a link between cognitive and motor processes, Figure 2 seems hardly to offer any more help than Figure 1. For all its multidimensionality, our flow diagram representing goal pursuit by the driver is, after all, indistinguishable in essentials from that for the power-steering mechanism. Once again, it has a flow line ending in thin air – the input I_G of the higher-level criteria by which the driving performance as a whole is evaluated and its goals determined.

4 Evaluation of Evaluative Criteria

To cut a long story short, the working assumption of physiological psychology is that in the autonomous living organism, including man, the changing pattern of relative priorities and evaluative criteria has its source in other activities in the CNS. This working hypothesis is outlined in information engineering terms by the skeleton flow map of Figure 3. Here we show the process of ongoing evaluation and revision of priorities as embodied in a functionally separate level of information processing (SS), which includes in its input data *the output from current evaluative processes.* Since, in human subjects especially, current states of affairs may be evaluated simultaneously (and quite differently) according to a whole hierarchy of different criteria, the evaluator C must be thought of as hierarchic in basic organization. Because of the feedback introduced via SS, however, the system as a whole

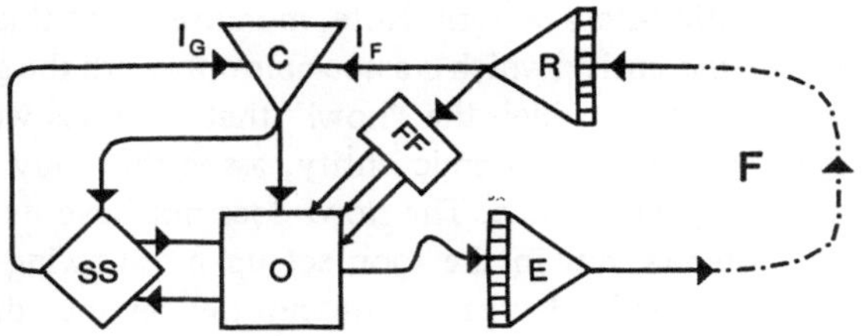

Figure 3. Basic information-flow map of autonomous evaluative agency (see text). The supervisory system SS determines goal criteria and relative priorities and keeps up-to-date the "state of conditional readiness" (SCR) of the organizing system O, in matching response to sensory feedforward from FF.

has the possibility of functioning heterarchically. Indeed, in an important sense the closing of this loop has made the system as a whole *its own programmer.* In informational terms, it has become both the selector and the evaluator of its own purposes.

The supervisory system SS has a second major function. We noted earlier that the organizing system O needs feedforward (from FF) to keep it up to date in matching the changing state of the field of action F. How is this to be achieved? To some extent we could expect passive associative networks to do the job; but where there is a gross excess of sensory information (as in visual search, or with ambiguous figures, for example) some "active filtering" and internal experimentation may again be required in order that the appropriate matching SCR may be set up with minimal delay. If the evaluative supervisory system SS were equipped to perform internal experiments on these lines, taking the organizing system O as its field of action in something of the sense in which O takes the external field F, then it could not only steer the ongoing process of keeping the conditional probabilities in O matched to the sensory demands from FF, but could also fulfil the need (first envisaged explicitly by Craik, 1943) for internalized trials of possible scenarios, with evaluations of their envisaged outcomes, leading to the selection or modification of goals and states of readiness on that basis.

Here, in this supervisory activity, it seems to me that we have an information-flow structure with some of the necessary features to serve as a direct correlate of our conscious experience (MacKay, 1951, 1966, 1981). The most characteristic aspect of being conscious, as we all know at first hand, is that we both evaluate the ongoing state of affairs and determine or revise "at will" our goal-priorities and criteria of evaluation. To "give one's mind to" some current activity means primarily to evaluate it and its foreseen implications. To fail to do so does not necessarily mean that we fail to react to stimuli, but rather that we fail to evaluate their implications for our own SCR.

If I am right, then the search for an "anatomical substrate of consciousness" is not meaningless; but the direct correlate of a specific experience would not be the activity of specific neurons, but rather the presence of a correspondingly specific information-flow pattern. Note, moreover, that this pattern of supervisory agency is a *systemic* entity, which is associable only with the supervisory structure as a whole – in the sense in which the "howl" that develops when a microphone is too near its loudspeaker is a systemic entity, associable only with the resulting positive feedback loop as a whole. The howl does not have its origin in any one of the chain of components, but in the loop set up by allowing the amplifier to supply its own input. Similarly, I suggest, conscious experience does not have its origin in any one of the participating brain nuclei, but in the positive-feedback chain-mesh that is set up when the evaluative system *becomes its own evaluator.*

It is no part of my purpose to claim that this is a *sufficient* condition for consciousness – only that it offers a hypothetically necessary cerebral correlate, in hardnosed information-engineering terms, for any conscious experiencing that does occur.

5 Implications

Let me conclude by mentioning briefly some practical scientific implications of the foregoing suggestions.

5.1 Motivation

In these terms, the concept of "motivation" emerges as a perfectly respectable scientific concept, provided that we see it as a compound of several ingredients. First, action may be stirred by either positive or negative evaluation; there can be pursuit of "match" as well as avoidance of "mismatch". But the "match" or "mismatch" signal itself cannot be the correlate of motivation, since it may fluctuate or reverse as action proceeds. The most direct correlate might seem to be the current evaluative criterion (I_G); but this in turn is a function of the supervisory evaluative action of the normative system SS. Thus the correlate of motivation (in its usual psychological sense) turns out to be another systemic property, subject in principle to modulation by many associative processes within the supervisory system. Certainly on these grounds it would seem reasonable, as suggested by Kornhuber (this volume), to include the association cortex within the "motivational system".

5.2 Perceptual Stability

From the information-engineering standpoint, what matters about the field of action is the constraints and enablements it imposes on action and the planning of action (MacKay, 1954, 1956). The *updating of the SCR* by the supervisory system, in matching response to sensory demands, has precisely the informational function of *perception.* In information-flow terms, then, the internal representation of the world-as-perceived is not the SCR, but the supervisory traffic involved in its updating by the sensory input.

This has a particular practical implication for theories of perceptual stability during exploratory movements. To perceive instability, on this view, requires *information of mismatch* between the current SCR and sensory evidence, significant enough to demand a change in the SCR. Now an exploratory eye movement is (by definition) an action calculated to change the region of the visual field being sampled by the fovea. If all goes well, the sensory changes produced will thus be evaluated as positive evidence of the correctness of whatever SCR determined the action. The stability of the world-as-perceived is confirmed, rather than disturbed, by the sensory consequences, provided that they co-vary appropriately with the exploratory movement (MacKay, 1957, 1973, 1978).

Note again that "evaluation" here does *not* imply conscious detection and discounting of the sensory changes. No activity of the cognitive supervisory level need be aroused unless the sensory changes fall outside the range of tolerance of the lower-level evaluatory criteria that assess the exploratory movement.

On this quantitative point, it is interesting to note that when the eyeball is rotated by external means (e.g., by finger pressure on the canthus) the *extent* of illusory perceived world-movement is much greater in the foveal than in the peripheral field, reflecting the greater weight of discrepant evidence supplied from the foveal system (MacKay, 1958a, 1973).

5.3 Interpersonal Cognition

The joint information-flow structure formed when two individuals are in dialogue offers a qualitatively new possibility. If the interaction is fully reciprocal, each system participates in a closed-loop situation in which each receives feedback by way of the evaluative process of the other. The interplay of the two supervisory evaluators can now give rise to a distinctive mode of cooperativity, which offers itself as the correlate of *interpersonal* cognition. Without going into detail, it is significant that in such reciprocal dialogue each system becomes essentially *indeterminate* to both participants (MacKay, 1958b, 1964, 1967).

5.4 Split Brains: How Many Minds?

As a final example, from this standpoint the question whether or not commissurotomized patients have "two free wills in the one cranial vault" (as suggested by Sperry, 1966) becomes quite a down-to-earth empirical question in engineering terms. The classic evidence on which Sperry based his claim was that the left- and right-hand systems of such patients can engage in physically conflicting actions and show incompatible goal preferences. In terms of Figure 3 this would certainly indicate that each "half" had separate evaluators at an executive level, embodying (in these circumstances) conflicting criteria of evaluation.

The question made explicit by Figure 3, however, is whether cutting the corpus callosum has also separated the *normative supervisory* system into two independent subsystems, each capable of independently evaluating and re-ordering priorities and goal criteria for its own executive evaluators. To this question none of the evidence so far published appears to justify an affirmative answer.

One way of probing the situation would be to see to what extent and at what level interpersonal cognition can be sustained between the two "halves". In a series of recent experiments to this end, my wife and I have found ample evidence of the capacity of each half to struggle against (or more often to cooperate with) the other at an "executive" level, and even to exchange information with the other by way of questions and answers addressed to (and elicited by) the experimenter as an intermediary. So far, however, we have found no evidence of independent centres of motivation, or of independent capacities for normative supervisory control of priorities on the two sides (MacKay, 1981; MacKay & MacKay, 1982).

My purpose in mentioning these results is not to stress the negative outcome (which is open to a number of interpretations), but rather to illustrate the way in which information-engineering analysis can help to crystallize questions and stimulate the testing of operational hypotheses.

6 Conclusion

The argument of this paper has been that for the scientific understanding of cognitive agency we need to organize our explanatory modelling around the concept of evaluation as its essential core. Except in the most primitive cases (where they can serve well enough), "stimulus" and "response" are well-defined only in relation to the criteria of evaluation of the cognitive agent concerned. The fact that the activities involved in setting and applying criteria of evaluation are embodied in structures deep within the CNS means, not that psychologists should or can pretend that they do not exist, but rather that closer collaboration is needed between psychologists and physiologists, if progress is to be made in elucidating them and so bridging the gap between cognition and action. In this collaborative enterprise information-engineering analysis can serve as a useful conceptual intermediary, enabling theories and data from each side to suggest experimental questions and interpretations of results on the other, and (sometimes at least) saving wasted effort on wrongly formulated problems.

References

Craik, K.J.W. *The nature of explanation.* Cambridge: Cambridge University Press, 1943.

Dichgans, J., & Bizzi, E. (Eds.) *Cerebral control of eye movements and perception of motion in space.* Basel: Karger, 1972.

MacKay, D.M. Mindlike behaviour in artefacts. *British Journal of the Philosophy of Science,* 1951, *II,* 105–121.

MacKay, D.M. Operational aspects of some fundamental concepts of human communication. *Synthese,* 1954, *9,* 182–198. Reprinted in MacKay, D.M. *Information, mechanism and meaning.* Cambridge, MA; MIT Press, 1969.

MacKay, D.M. Towards an information-flow model of human behaviour. *British Journal of Psychology,* 1956, *47,* 30–43. Reprinted in W. Buckley (Ed.), *Modern systems research for the behavioral scientist.* Chicago: Aldine, 1968.

MacKay, D.M. The stabilization of perception during voluntary activity. *Proceedings of the 15th International Congress of Psychology,* (pp. 284–285). Amsterdam: North-Holland, 1957.

MacKay, D.M. Perceptual stability of a stroboscopically lit visual field containing self-luminous objects. *Nature,* 1958, *181,* 507–508. (a)

MacKay, D.M. On the logical indeterminacy of a free choice. *Proceedings of the XIIth International Congress of Philosophy,* Venice, (Vol. 3, pp. 249–256). Florence: Sansoni, 1958. Reprinted in a expanded version in *Mind,* 1960, *69,* 31–40. (b)

MacKay, D.M. Communication and meaning – A functional approach. In H. Livingston (Ed.), *Cross-cultural understanding: epistemology in anthropology,* (pp. 162–179). New York: Harper & Row, 1964.

MacKay, D.M. Cerebral organization and the conscious control of action. In J.C. Eccles (Ed.), *Brain and conscious experience,* (pp. 422–445). New York: Springer, 1966.

MacKay, D.M. The mechanization of normative behaviour. In L. Thayer (Ed.), *Communication theory and research,* (pp. 228–245). Springfield, IL: Thomas, 1967.

MacKay, D.M. Perception and brain function. In F.O. Schmitt (Ed.), *The neurosciences: Second study program,* (pp. 303–316). Rockefeller University Press, 1970.

MacKay, D.M. Visual stability and voluntary eye movement. In R. Jung (Ed.), *Handbook of sensory physiology,* (Vol. 7, Part 3A, pp. 307–331). Berlin, Heidelberg, New York: Springer, 1973.

MacKay, D.M. The dynamics of perception. In P.A. Buser & A. Rougeul-Buser (Eds.), *Cerebral correlates of conscious experience,* (pp. 53–68). Amsterdam: Elsevier, 1978.

MacKay, D.M. Neural basis of cognitive experience. In G. Szekely, E. Labos, & S. Damjanovich (Eds.), *Advances in physiological sciences,* (Vol. 30, pp. 315–332). Oxford, Budapest: Pergamon and Akademiai Kiado, 1981.

MacKay, D.M., & MacKay, V. Explicit dialogue between left and right half systems of split brains. *Nature,* 1982, *295,* 690–691.

Sperry, R.W. Brain bisection and mechanisms of consciousness. In J.C. Eccles (Ed.), *Brain and conscious experience,* (pp. 298–313). New York: Springer, 1966.

Summary. (1) Not all motor activities are conscious actions.

(2) Not all sensory information-processing mediates conscious experience.

(3) Conscious experience (thinking, planning, longing, suffering) can occur even in the absence of overt action and sensory input.

This paper extends the argument advanced elsewhere that the core of conscious cognitive agency is *evaluation* and that an appropriate information-flow model of cognitive agency must have as its core a hierarchic evaluative or normative system which becomes heterarchic by including its own activity among its data. The information-flow traffic in such a system would seem to offer an appropriate correlate of our *conscious experience* in cognitive agency.

Possible anatomical implications are discussed with particular reference to perceptual stability during motoric agency and to recent investigations of "split-brain" patients.

13 Modes of Linkage Between Perception and Action

WOLFGANG PRINZ

Contents

The problem of understanding the nature of the linkage between perception and action can be raised at two different levels: execution and initiation. At the level of execution the linkage problem is mainly related to the conditions and mechanisms of interaction between efferent and afferent information during the execution of a given movement or response. Relevant issues such as open vs. closed loop control or feed backward vs. feed forward have recently attracted considerable interest in the field of motor learning and motor control (e.g., Adams, 1971; Schmidt, 1975; Keele, 1981; see also Shebilske and Wolff, this volume).

At the level of initiation the linkage problem refers to the issue how stimulus-related information and action-related information can be brought into contact with each other in a meaningful way. In the traditional bottom-up version the problem reads: How is a particular response selected and triggered by a given stimulus? Or, alternatively, in the top-down version: How is a particular stimulus selected and used to initiate an intended action? In both versions, the basic problem arises as to how representations of stimuli and representations of actions can be linked to each other at all. As it is usually assumed that "stimuli" and "actions" are represented at different and independent levels of coding it must be concluded that they cannot be compared with each other in a direct way.

On the one hand, the problems at the execution level and the initiation level are similar in that they are both concerned with mechanisms which allow incomparable codes to be linked to each other. Yet, on the other hand, they are clearly different and must be kept separate. At the level of execution, the linkage problem refers to the relationship between efferent and afferent information. At the level of initiation, it refers to the gap between afferent information and the stored representations of the conditions for producing efferent information.

Cognition and Motor Processes
Ed. by W. Prinz and A.F. Sanders

This chapter is solely concerned with the perception-action linkage at the initiation level. It attemps to provide a brief account of some ideas that have been discussed so far to bridge this gap. The first section deals with the old idea that stimuli and responses are linked to each other by stored connections – either learned or built in. This is the traditional connectionistic view. The second section is concerned with the more recent idea that perception and action are brought into contact with each other by matching information from the stimulus against information required for the initiation of an action. In the third section it is argued that, to give a satisfactory account of the linkage between perception and action, elements of the connectionist and the matching views should be combined. In conclusion three different ways of combining these two views are discussed.

1 Connections

The main principle underlying the connectionist view has been described in terms of either observable events or of inferred internal representations of these events (Figure 1). In its classic behaviorist version connectionism is only concerned with relationships between observable stimuli and responses and not with what is going on inside the black box. In more recent versions connectionism is concerned with connections between internal representations of stimuli and responses within the black box. These versions can be either phrased in neobehavioristic language or in simple cognitivistic terminology.

The simple scheme in Figure 1 suggests that the stimulus (or its representation) triggers the response (or its program), i.e., that the occurrence of the stimulus is both a necessary and sufficient condition for the elicitation of the response. In more elaborated versions of this view two types of additional assumptions have been introduced: first, that the connections between stimuli and responses are mediated by various internal instances (such as mediating responses, internal stimuli, concepts, categories, etc.) and second, that a given stimulus is not only connected to a single response but rather to a hierarchy of responses and that at the same time a given response is connected to (can be triggered by) a hierarchy of stimuli.

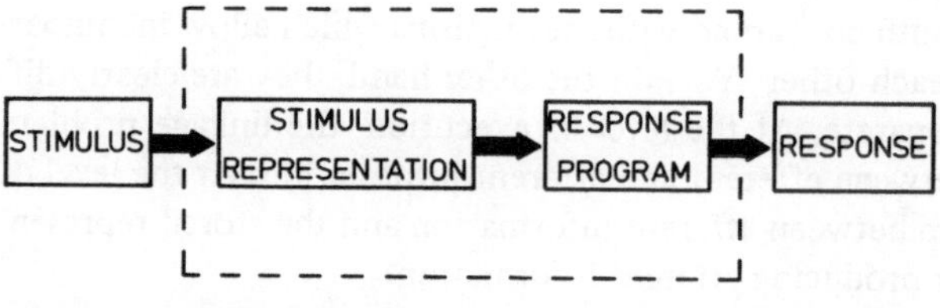

Figure 1. Connectionistic views regarding the linkage between perception and action. In the classic "external" version the "internal" instances are omitted. For a more general description of "internal" connectionism see Figure 2.

When these two assumptions are taken together, a considerable gain in degrees of freedom is obtained as compared with the simple scheme of Figure 1. In the elaborated connectionist scheme a stimulus will no longer trigger a response via a single connection but via a structured pattern of intermediate instances that mediates the final selection of one of several possible responses. In this scheme stimulus representations, intermediate instances, and response programs are connected to each other via a network of pathways as established in previous learning. Models of this type are highly flexible and can be adapted for the description and prediction of response performance in various tasks.

However, with respect to the issue of the linkage between perception and action there is one major shortcoming of all versions of the connectionist view which concerns its unidirectional character. In this approach the organism's activity is in principle reduced to a sequence of responses to a sequence of stimuli, and there is virtually no space to account for intention and voluntary action. In graphic illustrations of connectionist models the arrows will always point from stimuli to responses, and there is usually no way back. Although there may be feedback loops (in cases where the perceptual consequences of a given response act as stimuli for the next response), the basic principle that all activity is triggered by stimuli is never violated. The problem is always to predict the response, given a particular stimulus. There is no way to ask the reverse question and to predict the selection of stimuli (either at the external level of information intake or at the internal level of selective attention), given a particular intention to act. Several additional assumptions are needed to cover these phenomena in the connectionist's framework (cf. Berlyne, 1970; Trabasso & Bower, 1975; Lovejoy, 1968).

2 Matching

These problems can be more easily handled within the conceptual framework underlying the information processing approach to cognition and action. Within this framework it is usually assumed that the basic operation underlying perceptual recognition is a match between the information which forms the transient representation of the stimulus on the one hand (stimulus information) and the information which forms the permanent representations in memory on the other hand (memory representations). By this operation the representation of the actual stimulus event is matched against stored representations of various possible events. If a match occurs between the information provided by the stimulus and the information required for the activation of the pertinent memory representation, the stimulus event is said to be identified as one of the possible alternative events (cf. Neisser, 1967; Sternberg, 1967).

This view regarding the relationship between stimulus and memory representations can be applied to the issue of the linkage between perception and action as well. As illustrated in Figure 2, the assumed match occurs between a stimulus

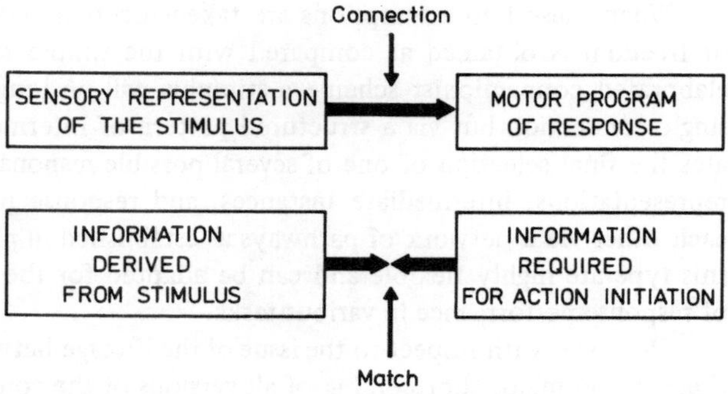

Figure 2. Connection principle and matching principle compared. According to the connection principle, stimulus representations and motor programs can be linked to each other directly. According to the matching principle, the information derived from the stimulus and the information required for action initiation have to be defined at the same level of coding. Note that the boxes in the scheme do not correspond to each other. In the upper panel the boxes are defined by their content. In the lower panel they are defined by their function with respect to the match. See Figure 3 for different mappings of contents to functions.

representation and an action representation, or, more precisely, between a representation of stimulus consequences and a representation of action conditions. The stimulus representation is transient; it contains information derived from actual stimulation. In contrast, the action representations are stored in a permanent way and contain information required for action initiation. It is assumed that the decision to initiate a particular action results from a match between the information derived from the stimulus and the information required for initiating the action. The permanent action conditions specify, for each particular action, the conditions the stimulus (or the information derived thereof) must meet in order to initiate that action. Such condition-action connections have been discussed in more detail by Newell (1973) and Allport (1980).

The match between information given and information required can be described in one of two ways. One possible approach is to think of an extra operation and an extra processing stage for matching these two pieces of information. This line of thinking has proven fruitful in the analysis of speeded reaction tasks with well-defined sets of stimulus and response alternatives (Sternberg 1967/1969; Sanders, 1980). As an alternative to this functional type of interpretation one might also think of a more structural type of description. According to this view the information given and the information required are generated at the same functional locus. They are superimposed within the same representational space, and they match to the degree they overlap in this space. No additional stage or operation for matching is required. This approach has been discussed in more detail elsewhere (Prinz, 1980, 1983).

Within the matching scheme the problem of perceptual selectivity is easier to handle than within the connectionist scheme. This is because the system has two entries instead of only one. To phrase it in oldfashioned language: sensations come in from the left, and intentions come in from the right, and the consequences of stimulation and the conditions of intended actions are matched against each other. The results will always be contingent on both sensational and intentional factors: the occurrence of a match will not only depend on the stimulus-related information – which depends on the actual state of stimulation – but also on the action-related information – which in turn depends on the actual state of the observer's intentions (see Prinz, 1983, for more detailed discussion).

3 Combining the Principles

The connection and the matching principle should not be regarded as mutually exclusive views regarding the nature of the relationship between perception and action. On the contrary, it can be argued that any model which claims to provide a satisfactory account of that relationship must somehow incorporate both principles at the same time because taking either of them alone will lead to serious difficulties.

The difficulties one has to face within the framework of the matching principle concern the complete lack of commensurability between representations of stimulus properties (stimulus representations) on the one hand and motor programs for the execution of actions (action representations) on the other hand. Basically, there is no direct way representations of stimuli and representations of actions can be compared or matched with each other: they are coded in different languages. There may be an exception in the case of highly compatible tasks where some degree of commensurability between sensory representations of stimuli and motor programs of responses may be claimed with respect to their spatial and temporal properties.

If it is correct that stimulus representations and action representations are coded in different languages at different levels, a match between them will only be possible after they have been translated into a single common language. Whether a statement in English is equivalent to a statement in German can only be determined by a person who is capable of translating English into German or vice versa (or of translating both statements into a third language). The translation procedure itself cannot be based on matching the two languages, of course, but must rely instead on existing connections between the two language systems which are defined at various levels of complexity (between words, sentences, grammatical rules etc.). In the same way, the stimulus representations, the motor programs, or both must be translated into a common language by means of connectionistic translation rules before a match can occur. The simple matching scheme will not work unless it is supplemented by connectionistic translations that bridge the gap between different levels of coding.

The difficulties one has to face within the framework of the connection principle are related to the transient nature of the stimulus representation. They were first discussed by Höffding (1889/1890) with respect to the problem of perceptual recognition. Höffding argued that the principle of association by contiguity can only account for the functional bonds between stored cognitive elements. It cannot account, however, for the process of recognition as such, i.e., the process by which the actual stimulus (or its internal representation) contacts the permanent contents of memory. This step between perception and memory, Höffding argued, must be bridged by some other principle. The association between the transient stimulus event and the permanent memory content must be mediated by their similarity. The principle of association by similarity virtually implies the notion of a match between the information derived from the stimulus and the information required for the activation of a particular memory representation or trace.

If Höffding's argument is valid for the relationship between perception and memory, it should also be valid for the linkage between perception and action. The connection principle can only account for connections between stored pieces of information. It cannot be used to bridge the gap between transient stimulus-related information and permanently stored action-related information. For bridging that gap a match (based on similarity) is needed in addition to the connections built in the system (based on contiguity). The simple connectionistic scheme will not work unless it is supplemented by a match where the Höffding step is taken.

If it is accepted that a complete model of the linkage between perception and action must combine both, the connection and the matching principle, there are basically three different ways of doing so in a parsimonious way (i.e., of using one match only): sensory match, motor match and perceptual match (see Figure 3).

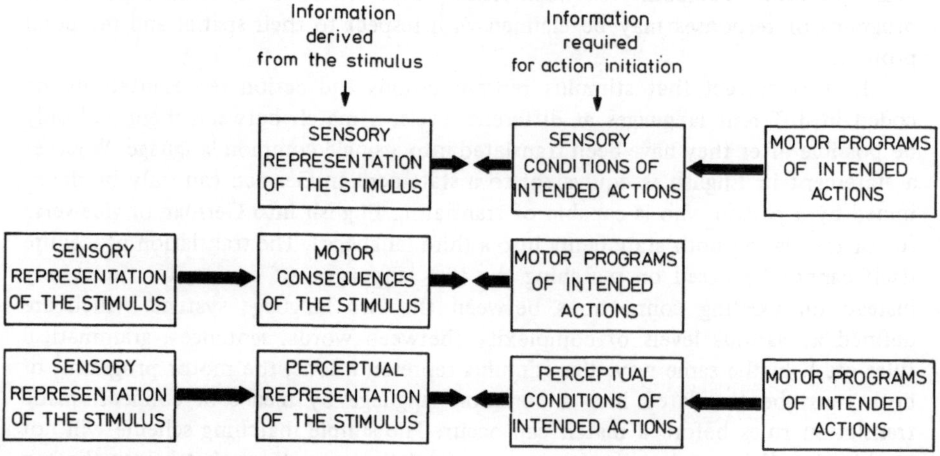

Figure 3. Three combined models of linkage between perception and action (sensory match, motor match, perceptual match from top to bottom).

Sensory Match. According to this model the match between transient stimulus information and stored conditions of action is performed at the level of sensory features. This presupposes that the action conditions are also defined in that language and that each particular action program is connected to a stored set of sensory features which serve as a definition of the conditions the stimulus-related information must meet if the program is to be executed.

A model of this kind has a rather limited range of application. It can only be applied to cases where a valid and sufficient definition of action conditions can be set up in terms of sensory features. This may be valid in some types of simple experimental tasks, for example where a subject presses one of two buttons in response to a red light and the other button in response to a green light. There he can easily learn to establish action conditions in terms of sensory features. It will probably not be valid for the control for more complex actions where it is more reasonable to assume that action conditions are set up in a non-sensory language.

Motor Match. Similar problems arise with respect to a motor match model according to which a match occurs at the level of motor programs. According to this type of model, the information required is defined in terms of the motor programs themselves. This presupposes, of course, that the stimulus-related information which enters the match is also translated into the language of motor programs, so that each particular stimulus representation is connected to a corresponding pattern of motor programs which defines the learned ensemble of motor consequences of the corresponding stimulation. This view is inherent in many of the traditional motor theories of perception. Clearly, the application of this model is also restricted to cases where a rather direct correspondence between the stimulus representation of an event and its motor consequences can be claimed.

This may again be valid in cases of direct sensory-motor correspondence. Notions related to the idea of a motor match have been put forward with respect to topics such as speech perception and ideomotor compatibility where it has been assumed that perception is based on directly encoding the underlying stimulation in terms of its corresponding motor features or programs (see Scheerer, this volume). However, even if one is willing to accept a motor coding interpretation of these phenomena, one may still doubt whether the generation of these codes is really based upon direct connections between the sensory and the motor level as postulated by the motor match model.

Perceptual Match. The obvious shortcomings of these two models are avoided by the perceptual match model. At first glance this model looks like a combination of the other two but it is actually entirely different from either. According to this model, commensurability between perception and action is accomplished by translating both the consequences of stimulation and the conditions of actions into a common third language which is neither sensory nor motor. Instead, this language refers to meaningful objects and events. On the left-hand side of the scheme the sensory stimulus representation is translated into perceived objects and attributes.

These objects and their attributes represent the consequences of stimulation within the postulated intermediate language system. Conversely, on the right-hand side, the motor programs are also connected to representations of meaningful objects and attributes. These objects and attributes represent the conditions of actions within the intermediate language. Both the consequences of stimulation and the conditions of actions are generated and maintained by means of connections that are established between the two peripheral coding systems on the one hand and the intermediate perceptual coding system on the other.

4 Conclusion

The main advantage of the perceptual over the sensory and motor match model seems to be that it provides two possible loci of learning (i.e., establishing connections) instead of only one. This does not only imply a difference in the number of possible connections but rather implies a very important difference in the basic structure of the systems. As long as only one locus of learning is available, all connections that are established must be used to bridge (in one or the other way) the gap between the languages of sensory representations and motor programs. As soon as two loci become available for building connections the system is free to invent a third language and to construct an additional intermediate representational space for matching stimulus consequences against action conditions. It seems reasonable to assume that on both the phylogenetic and the ontogenetic scale the origin of consciousness is linked to the origin of this intermediate space and its neutral language.

References

Adams, J.A. A closed-loop theory of motor learning. *Journal of Motor Behavior,* 1971, *3,* 111–149.

Allport, D.A. Patterns and actions: Cognitive mechanisms are content-specific. In G. Claxton (Ed.), *Cognitive psychology – New directions.* London: Routledge & Kegan Paul, 1980.

Berlyne, D.E. Attention as a problem in behavioral theory. In D.I. Mostofsky (Ed.), *Attention: Contemporary theory and analysis.* New York: Appleton Century Crofts, 1970.

Höffding, H. Ueber Wiedererkennen, Association und psychische Activität. *Vierteljahrsschrift für Wissenschaftliche Philosophie,* 1889, *13,* 420–458; 1890, *14,* 27–54, 167–205, 293–316.

Keele, S.W. Behavioral analysis of movement. In V. Brooks (Ed.), *Handbook of physiology,* (Vol. 3), *Motor control.* Bethesda, MD: American Physiological Society, 1981.

Lovejoy, E. *Attention in discrimination learning.* San Francisco: Holden-Day, 1968.

Neisser, U. *Cognitive psychology.* New York: Appleton Century Crofts, 1967.

Newell, A. Production systems: Models of control structures. In W.G. Chase (Ed.), *Visual information processing.* New York: Academic Press, 1973.

Prinz, W. Selectivity in character classification. In R.S. Nickerson (Ed.), *Attention and performance VIII.* Hillsdale, NJ: Erlbaum, 1980.

Prinz, W. *Wahrnehmung und Tätigkeitssteuerung.* Berlin, Heidelberg, New York: Springer, 1983.

Sanders, A.F. Stage analysis of reaction processes. In G.E. Stelmach & J. Requin (Eds.), *Tutorials in motor behavior.* Amsterdam: North-Holland, 1980.

Schmidt, R.A. A schema theory of discrete motor skill learning. *Psychological Review,* 1975, *82,* 225–260.

Sternberg, S. Two operations in character recognition: Some evidence from reaction time measurements. *Perception and Psychophysics,* 1967, *2,* 45–53.

Sternberg, S. The discovery of processing stages: Extensions of Donders' method. In W.G. Koster (Ed.), *Attention and performance III.* Amsterdam: North-Holland, 1969; *Acta Psychologica,* 1969, *30,* 276–315.

Trabasso, T., & Bower, G.H. *Attention in learning: Theory and research.* Huntington, NY: Krieger, 1975.

Summary. Two basic principles have been proposed to account for the relationship between perception and action with respect to the selection and initiation of responses or actions: connection and matching. According to the connection principle, stimuli and responses (or representations thereof) are linked to each other by stored connections. According to the matching principle, the linkage is due to a match between the information provided by the stimulus and the information required for the control of intended actions. It is argued that the two principles must be combined because none of them provides a satisfactory account of the full problem. The connection principle fails to account for the step between transient stimulus information and permanent information stored in memory. The matching principle fails to account for the lack of commensurability between sensory and motor events. Three models which combine both principles are presented and their implications are discussed.

14 The Contribution of Vision-Based Imagery to the Acquisition and Operation of a Transcription Skill

DANIEL GOPHER

Contents

1 Introduction

This chapter discusses the contribution of an imagery-based code to the acquisition, operation and retention of a complex data entry skill. The power of visual imagery as an aid to memory has been known since the ancient days of the Greek orators. It has been the subject of numerous theoretical debates and extensive experimentation throughout the ages (Yates, 1966) and is still an active topic of research in modern cognitive psychology (Kosslyn, 1980). Modern research can be generally classified into two main categories: Studies interested in the role of imagery in learning processes, memory, thinking, and problem solving (e.g., Paivio, 1971a,b; Hunter, 1977; Kosslyn, 1980), and research into the properties of mental images and their relationship to visual perception (see Haber, 1971; Kosslyn, 1980). A detailed review of this research is beyond the scope of the present chapter. Within the first group of studies, imagery has been shown on numerous occasions to improve memory, comprehension, and problem solving. This improvement is based on improved organization in memory or dual memory codes (Paivio, 1971b; Segal, 1971). The major finding of the second group is that imagery-based performance closely resembles the behavior related to actual processing of visual information (e.g., Podgorny & Shepard, 1978).

The vast majority of current research on visual imagery is primarily dedicated to memorization of verbal material and structuring of abstract knowledge bases. Contribution of imagery to motor performance has been generally neglected as a

Cognition and Motor Processes
Ed. by W. Prinz and A.F. Sanders

topic for systematic research, despite the fact that the links between vision and movement have long been recognized by theorists in both domains (e.g., Hebb, 1949; Werner & Kaplan, 1963; Bernstein, 1967; Turvey, 1977), and the dominance of the visual information channel in guiding movements has been demonstrated in numerous studies (e.g., Lee and Aronson, 1973; Adams, Gopher, & Lintern, 1977). In the last 5 years interest in this issue has begun to emerge. Thus, Chevalier-Girard and Wilberg (Note 1) reported superiority in immediate and delayed recall of ten movement patterns of subjects trained under imagery conditions. Hall (1980) and Hall and Buckolz (1981) showed that the ability of subjects to recall and reproduce movement patterns was correlated with the imagery values of the geometrical forms created by these movements. Carroll and Bandura (1982) report that visual feedback enhances observational learning and later reproduction of novel action patterns that would normally be unobservable.

From the few experiments that have been conducted, it appears that visual imagery may have a role not only in verbal but also in motor processes. We shall examine the nature of this role in the context of the acquisition and operation of a typing skill.

2 Building a Linkage Between Letters and Motor Entries in Typing

Transcription skills such as typing, writing, piano playing or data entry with alphanumeric keyboards, all require the translation of verbal sequences into trains of motor responses. Elements of these skills represent an interesting merger between verbal and motor aspects of behavior. For example, in the acquisition and operation of a data entry skill the physical format in which information is organized and displayed, its content, semantic and grammatical structure, as well as the production control and timing of complex movement patterns, are factors that have a significant influence on the efficiency of data entry performance and affect different components of the skill associated with it. A first step in the acquisition of such a skill is the ability to associate letters and characters with their keyboard locations.

The majority of the existing data entry keyboards follow the design characteristics of the standard English typewriter and concure with the "QWERTY" (or "Sholes", as it is sometimes referred to, after its developer) arrangement of letter-keys (Seibel, 1972). In this design every character is entered by a separate key. Common keyboards can have 30–65 keys, some of which have double functions.

Touch typing with ten fingers is a complex psychomotor skill that requires hundres of training hours to acquire, continuous practice to maintain and rapidly deteriorates if unused (Dvorak, Merrick, Dealey, & Ford, 1936). It entails the formation of association between letters, hands, fingers, and keyboard positions, as well as the learning of movement trajectories of hands and fingers for pairs of characters (e.g., Hirsh, 1970; Norman & Fisher, 1982). Attempts to find solutions to this problem have led, in the last three decades, to several efforts to develop a chord keyboard in which the number of keys is considerably reduced and characters are typed by entering chord combinations of several keys. That is, typing of a single letter requires simultaneous depression of one or more keys together (Conrad & Longman, 1965; Seibel, 1972; Rochester, Bequart, & Sharp, 1978).

By and large, acquisition times and final transcription rates with all chord keyboards did not show a significant advantage over the standard typewriter (see Seibel, 1972, for review). A closer examination may suggest several causes for this finding. Chord keyboards considerably reduced the number of keys and almost eliminated the assignment of more than one key to each finger. However, this was achieved at the expense of sacrificing the one to one mapping of characters to keys and increasing the motor complexity of each letter entry. The performer also has less freedom in parallel or anticipatory preparation movements that signify the performance of experienced typists with regular keyboards (see e.g., Gentner, Note 2). On the cognitive side, chord keyboards have added to the load imposed on the processing system by requiring performers to learn and apply a relatively arbitrary set of rules to link characters with their associated chords, i.e., demanding the acquisition of a new dictionary of motor codes without proposing a simpler or more compatible rationale for translating verbal representations into motor productions.

An innovative principle for pairing letters with motor chords was first explored by Sidorsky (Note 3) who named his single hand chord system "Alpha Dot". The coding principle that the Alpha-Dot exploits is the shape of upper case English letters. Motor entries for each letter are composed of two chords entered in temporal succession. Each chord consists of one, two or three keys pressed simultaneously. Taken together, the three keys and two chords can be conceived to represent an imaginary matrix with six cells in which the first and second chords represent, respectively, the upper and lower parts of each letter. Subjects are required to map letters into their entry codes by integrating in their "mind's eye" the three key locations and the two temporal chords into a single spatial entity which was used by the designer of this code to mimic the distinctive features of the upper and lower sectors forming each letter (see Figures 1 and 2). Sidorsky was undoubtedly the first to capitalize on the shape of letters as a mediator for the translation of letters into motor codes. However, his research was exploratory in nature and driven by economical concerns in system design. In fact, some of the keyboard configurations recommended by him appear to be inconsistent with the imagery match requirements and may have impaired the imagery value of the whole code set. In

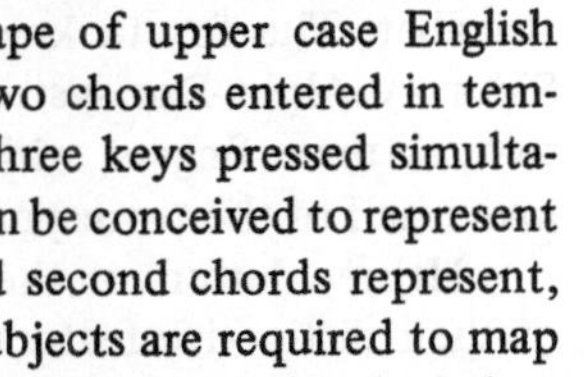

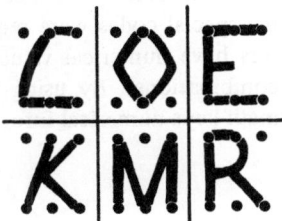

Figure 1. Chord codes of six letters in Sidorsky's Alpha-Dot system (after Sidorsky, Note 3, Figure 3)

addition, a three-by-two array is definitely not the optimal graphical format to illustrate the main features of English letters. For example, consider the problem of describing the six letters illustrated in Figure 1.

Sidorsky's results were brought to my attention by Z. Eilam in 1976, while I was involved in the design of a new data-entry device to be incorporated in airplane cockpits. He also suggested that a 3×2 matrix may be uniquely appropriate to represent the printed format of Hebrew letters. We collaborated on the first experiment, which was highly successful, and instigated a line of studies to investigate the role of imagery in acquisition and operation of this transcription skill.

3 Experiments with the Hebrew Letter-Shape Keyboard

3.1 The System

The Hebrew letter-shape typewriter is a single hand keyboard interfaced with a computer display. It comprises four conventional, light touch, long travel, typing keys arranged as illustrated in Figure 2, and is operated by the thumb and three middle fingers, or little and three middle fingers, in right or left hand operation respectively. Three keys are used to enter all letters and characters of the Hebrew alphabet. The fourth key serves in editing and control functions. Similar to the Sidorsky Alpha-Dot system, each letter is entered by two successive chords of one, two, or three fingers, with the first and second chords representing the upper and lower half of the corresponding letter shape.

Hebrew letters, in their printed format, are based upon square figures and most of their distinctive features are included in the upper and lower horizontal sides of each letter. Consequently, most of these features could be preserved within the six cell imagery matrix created by two successive chords of the three keys. The matching between letter figures and their chord counterparts can be clearly observed for the six letters depicted in Figure 2. The Hebrew alphabet contains 22 letters and an approximately equal level of correspondence could be achieved for two-thirds of them. For the remaining letters, legitimate variations or slightly distorted emphasis of typical features were adopted to encourage visualization of letters with the imaginary six cell grid arrangement.

Control functions were executed by first pressing the thumb key followed by a single chord of the three middle fingers (Figure 2). Here again, effort was made to maintain correspondence between cursor movements and chord representation. No special codes were required to enter digits, because in the Hebrew language letters have numerical values (e.g., dates in a Hebrew calendar are marked by letter combinations). By using the function key to change the entry mode of the three main keys numerical information can be entered.

3.2 Acquisition and Retention of Skill

Since its development the letter-shape typewriter has been used by different groups of subjects under a variety of experimental conditions. A description of some of these experiments can be found in Gopher and Eilam (1979), Brickner and Gopher (Note 4) and Gopher, Brickner, and Navon (1982). All experiments began with a memorization phase in which subjects were presented with a chart of the alphabet, control commands and their chord entries (in the general form depicted in Figure 2). They were required to memorize the codes and be able to produce them upon request before they began actual typing.

Although, based upon our own experience, we were convinced that the acquisition of the chord lexicon with the use of this chart is simple, we were still amazed with the immediacy of the memorization of codes. On average subjects required only 7–10 min to memorize and reproduce all chords in any requested order. At the end of 10 min subjects were essentially beginning to practice touch typing – during the first half hour usually with occasional reference to the chart, thereafter with complete independence of it. At the end of the first training hour subjects reached an average typing rate ranging from 30 to 40 characters/min with 1–7% errors, depending upon the type of material and speed accuracy emphasis (see Gopher & Eilam, 1979). After 6–7 h of training they reached a level of 70–80 characters/min with a 2–4% error (see Figure 3 and upper curve, Figure 4). Entry rates at this level are obtained on a regular typewriter only after several weeks of daily

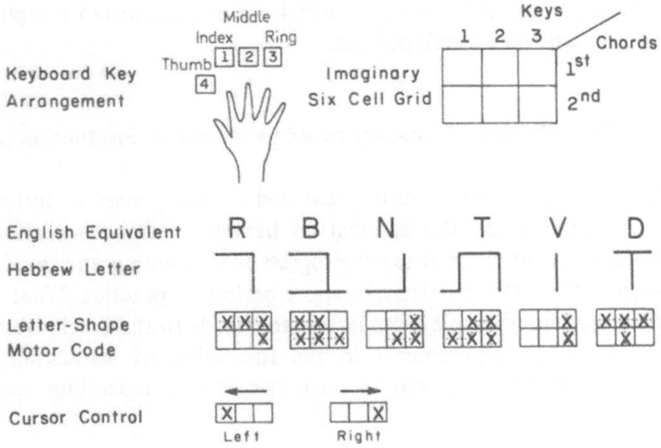

Figure 2. Keyboard arrangement and examples of letter codes in the Hebrew letter-shape typewriter.

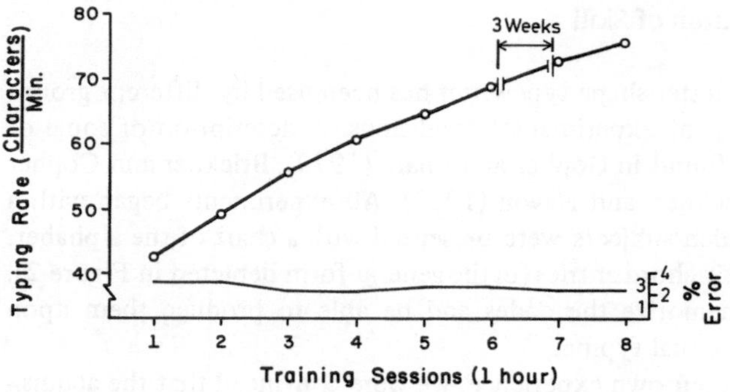

Figure 3. Practice effects on speed and accuracy of typing alpha-numeric messages with the letter-shape keyboard.

training sessions (Dvorak et al., 1936; Seibel, 1972). With 20 h of training, the longest training period that was studied under controlled conditions (Schelach, Note 5), subjects reached with a single hand, an average speed of 123 characters/min and top speed of 140 characters/min, or 28 words/min, and still continue to improve (see also Figure 4).

A typical learning curve is presented in Figure 3. It depicts the average results of six subjects who were trained in typing alphanumeric messages composed of words and digit codes[1]. Messages were presented auditorily from a recorded tape. A steep and smooth progress is clearly shown from one meeting to the other with no indication of an approaching ceiling at the end of the eight meetings. It is important to note that the six and seventh meetings were separated by 3 weeks. Yet, there is no indication of loss of typing speed, and performance immediately continued to improve. Similar results have been reported (Gopher & Eilam, 1979) for 24 h and 6 week break periods.

3.3 Contribution of Imagery to Acquisition and Production of Codes

Taken together the results presented in the previous section show that with the letter-shape code the association between letters and chords is created almost instantly, that there is rapid progress with training and good retention of acquired capabilities after a relatively short period of practice. What is the contribution of the mapping principle from letters to chords to these achievements?

Imagery requirements in the formation of association between letters and chords can be conceived to have two levels, depending upon the mode in which

1 This experiment was conducted at the Israel Air Force in collaboration with Dina Deal and Daniel Zonenshein.

typing material is presented. With visual presentation, subjects can actually see the letters to be typed. The imagery part then consists of superimposing a six cell grid on letters and deciding which cells of the grid should be activated by the first and second chords. With auditory presentations, typing from memory, or composing free texts, letters are retrieved from memory. Would this difference affect performance? A within subject experiment in which the two modes of presentation were alternated failed to reveal differences in the speed of acquisition or rate of entry. Average rates of typing after three hours of training were 51.4 and 50.1 characters/min respectively, for the auditory and visual presentation modes (unpublished experiment, conducted at the Technion as a fourth year student project).

One possible interpretation of this result is to argue that auditory based imagery is as powerful as actual visual information in the control of behavior. This conclusion is in agreement with the long standing claim that whatever representational mechanism underlies visual perception it is the same as that underlying visual imagery (e.g., Neisser, 1967). It is also strengthened by the results of a recent study conducted by Podgorny and Shepard (1978), in which subjects were asked to determine whether probe dots appearing in component squares of a 5 × 5 grid fell on a letter figure. No differences in reaction time were observed when letters were visually presented on the grid as a pattern of darkened squares, only remembered from preceding presentation, or imaginally generated from a verbal code.

An alternative interpretation is that the storage and retrieval mode of letter codes is not related to imagery or to actual shape of letters and is also insensitive to the mode of input presentation. One can further argue that the ease of acquisition and the rapid progress in training with the letter shape keyboard are the sole contribution of the simplification and reduction in the number of keys and the confinement of the total number of chord combinations to 26. Some evidence against this last argument comes from the observation that subjects trained with Sidorsky's Alpha-Dot system, which had inferior imagery quality but the same physical properties and a similar number of chord combinations, needed on average 90 min for initial memorization of codes.

A more direct testing of the contribution of the imagery match was conducted by Schelach (Note 5), who contrasted the performance of two groups of subjects. A visual imagery group (VIM) was trained with the regular letter-shape system. A second group was given an identical inventory of chords, but chord combinations were assigned to letters arbitrarily (ARB). In fact, matching was not completely arbitrary but followed the rule that chord combinations were switched between letters that have similar frequency of occurrence in Hebrew texts. Subjects were then given 20 one-hour training sessions in which they typed paragraphs from a book written by a popular Israeli novelist.

On average, subjects in the VIM group required 7.4 min to memorize all codes, while 49.8 min were needed in the ARB group. Standard variation values were 3.8 and 10.2 respectively. The training curves of the two groups are presented in Figure 4. Both showed a continuous and smooth improvement throughout the 20 training sessions with a clear advantage in the VIM group. Differences between

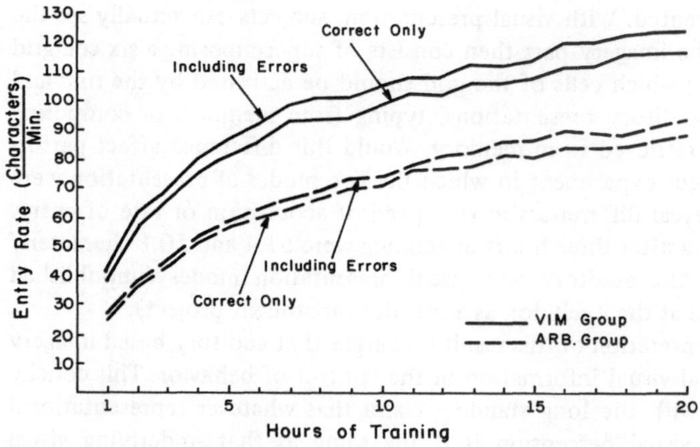

Figure 4. Comparative data entry rates of a group trained with an imagery based code (VIM), and a group trained with arbitrary codes (ARB) (from Schelach, Note 5).

groups increased in the first 7 h of training, thereafter the slopes of the two curves equalized. Similar learning curves were obtained when only the correct responses were plotted for all responses. The final data entry rate reached by the VIM group was 123.8 characters/min (or 247.6 strokes/min) and 93.2 (186.4 strokes/min) for the ARB group. Top speed values were 140.4 and 109.4 for the VIM and ARB groups, respectively, an overall 30% advantage for the visual imagery group. A comparison of the performance levels of the two groups on a neutral tapping task with the same keyboard, which was periodically performed during training, showed that some of the differences between groups should be attributed to a motivational factor favoring the VIM group. However, the main core of the differences is related to the pairing principle. A separate analysis conducted on the training data of the last few sessions failed to reveal a convergence trend between the performance curves of the two groups.

Another finding that may indicate the role of imagery in the acquisition of this skill comes from examining the ability of subjects to switch hands in typing. At the completion of the 20th training session, subjects were asked to switch their typing hand. Because they were initially trained with their dominant hand, a certain decrease in entry speed due to inferior motor capabilities of the nondominant hand was expected. In a control condition, motor speeds in simple tapping with the nondominant hand were found to be the same for the VIM and ARB groups at a level of 82% of the dominant hand capability. However, in text typing, rates of the visual imagery group degraded during the first 15 min to 68.7% of the rate obtained with the trained hand and error percentages were almost doubled. In contrast, the ARB group decreased to only 74.8% of the trained hand rate and suffered from a much smaller increase in the number of errors. That is, the group trained under visual imagery suffered more in the transition from one hand to the other, although both groups revealed a high level of transfer between hands and the VIM group still maintained its advantage. Differences in efficiency between groups were reduced after the first 15 min but small differences, implying lesser transfer between hands in the VIM group, were maintained to the end of this session.

Schelach's results provide a strong support to the contribution of imagery to the acquisition of this transcription skill, while also raising several questions about the nature of this contribution. Imagery clearly improves the ability of subjects to memorize letter codes, as evident from the marked difference between the VIM and ARB groups in their initial memorization time. However, if the provision of a better mnemonic was the sole function of visual imagery, differences between groups should be largest at the beginning of training and then gradually disappear at later stages of training when subjects in the ARB group have mastered the code equally well. With a limited inventory of 26 chords and thousands of trials within each training session, codes should have been mastered and differences between groups disappear long before the end of the 20 h. Yet, differences between the two groups were smallest at the beginning of training, increased during the first 7 h, and then remained constant throughout the next 13 h, while performance of both groups continued to improve. It therefore appears that visual imagery may have an additional role to that of a mnemonic, and a more enduring or permanent effect on performance.

From the general success of subjects to maintain their level of performance in switching hands it can be concluded that the main components of the acquired skill are centrally represented in a form that can be generalized across limbs. At the same time, it appears that some of the motor control loops that are triggered by spatial characteristics of the imagery, acquired a more direct association with specific fingers and these associations were interrupted when the subject changed Hands. In terms of actual keys to be operated, a switch from right to left hand operation does not change the spatial characteristic of the code. The six cell matrix arrangement is not interrupted and the fitting of letter shapes into this format is not changed. What is changed is that the index and ring fingers switch roles. The index that was previously responsible for the left column of the matrix is now responsible for the right column, while the ring finger is now assigned to the left column. Hence, subjects' frequent complaint that the alternation of hands demanded a mirror imaging of letter figures can only be understood under the assumption that fingers have acquired a spatial identity within the coordinates of the imaginary grid. Acquisition of such properties may be an indication of skill automation in which imagery can trigger motor response with no central intervention. The fact that decrements due to the change of finger roles were relatively short-term implies that central mechanisms can regain control and reassign spatial responsibility to local loops in a relatively short interval.

How sensitive is performance to the quality of fit between the letter format and its chord representation? On a rough scale, a comparison between the results obtained by Sidorsky (Note 3) with his Alpha-Dot system with those obtained with the Hebrew letter-shape keyboard shows that quality of representation can be an important and influential variable. In the preliminary experiment of another study (Gopher & Arzi, Note 6) five experienced letter-shape typists were requested to judge the graphic similarity between each letter and its representative chord code on a 7-grade scale. Average scores for codes were correlated with reaction times to single letters obtained for a group of five subjects trained for 3 h. No significant correlations were found between these two measures based upon parametric or non-parametric correlation coefficients (Spearman $r = 0.017$; Pearson $r = 0.14$). This lack of correlation was observed despite the relatively wide range of response times to letters (717 msec to 1011 msec).

One interpretation of these findings is that only the effort to create imagery but not the quality of representation is responsible for all of the differences observed between visually and nonvisually related coding strategies. This interpretation appears rather unlikely in view of the differences between the Hebrew and English keyboards on both of which the creation of imagery was encouraged. It is also in conflict with the consequences of changing hands. Another alternative is that the initial efforts to match chord codes to letter figures in Hebrew were successful to the point that additional differences in fit were too small to be reflected in performance. A related possibility is that subjects were not trained long enough to enable the detection of fine effects on performance. Be what it may, these results are a good reminder for the existence and influence of biomechanical and motor constraints on chord production.

4 Linkage Between Imagery and Movement

As argued in the previous section, the enduring and nonconvergent differences between the performance of the vision-based code group and the group with arbitrarily assigned codes, together with the differential sensitivity of the two groups to the change of typing hands, suggest the existence of an additional factor other than improved mnemonics that influences the efficiency of this manipulation. A natural candidate is the notion of spatial compatibility. With a brief scan of the six letter codes depicted in Figure 2 one is tempted to conclude that the present results are but one more demonstration of the long known fact that a better spatial compatibility leads to a shorter response time (e.g., Fitts & Posner, 1967). Thus, with a vision-based code subjects were simply required to produce a congruent motor response to match a visually presented pattern, while such congruence did not exist for the group with arbitrarily assigned codes. In a general sense the truth of this claim is not debated. However, it would be too hasty to terminate our discussion at this point. First, it should be recognized that the matching between stimulus and response patterns, which was required in the present set of experiments, differed considerably from the traditional tasks that have been employed in spatial compatibility experiments. These typically involved different mapping arrangements of single response keys to single stimulus locations in a stimulus array or reversals in the direction of stimulus and response movements. In our letter-typing the matching task is not as direct and depends to a large extent on imagery. Subjects are first required to analyze the distinctive components of a given visual pattern and then visualize the way in which these components can be reproduced in a sequence of two motor chords. An important element is the imaginary lines that bind together single key presses (the dotted lines connecting key positions in Figure 2). Without these lines different chord combinations are not unique and much less suggestive of the respective letter shape.

Second, compatibility is one of the more vague and ill-defined concepts in human performance theory. To argue that the difference between groups can be attributed to spatial compatibility merely points to a difference in experimental conditions, but contributes very little to our understanding of the underlying processes that govern these differences. What are the origins of the spatial compatibility effect? Why would a better match between a visual form and motor pattern facilitate the creation of a linkage between the two? The ability to answer these questions requires a deeper understanding of the building blocks that link visual percepts to motor acts.

Theoretical postulates that link visual percepts to motor behavior can be dated to the early days of experimental psychology. Leading theorists in developmental psychology have defended the view that imagery consists of internalized imitations paralleling the motor activity involved in perceptual exploration, in which movements "imitate" the contours of the perceived figure (Piaget, 1962; Piaget & Inhelder, 1966). This view is in agreement with propositions developed within the neurophysiological theory of Hebb (1949) who believed that motor components and in particular eye movements play an important role in the establishment of cell assemblies that are the bases of perception and imagery. Margaret Washburn's account of imagery in 1916 was based on the proposition that imagery, like sensation, occurs when a motor response to some external stimulus is inhibited or blocked. She further suggested that impulses diverted from motor centers can spread to other sensory centers and initiate activity in them. In his analysis of the role of operant conditioning in the development of private events and images, Skinner (1953) develops the concepts of classic "conditional seeing" and "operant seeing" that originate in motor experience. Further review of these theories is beyond the scope of this chapter; detailed reviews can be found in Paivio (1971b), Segal (1971), and Kosslyn (1980).

While theoretical accounts of the development and formation of percepts and images emphasize the contribution of motor components, recent theories of action attribute a central role to visual imagery in the intiation and direction of high order action plans that in turn initiate and control trains of motor commands. Turvey (1977) and Kelso and Wallace (1978) have developed a framework to a vision-based theory of action, which is strongly influenced by the work of Russian scientists and in particular Bernstein (1967) who proposed an imagery based theory of motor behavior. Both works have their starting point in the argument that long-term representations of action plans and complex motor sequences should assume a general abstract form rather than be motor specific. Kelso and Wallace (1978) summarize their views saying that: "The role of the image in preselected movement is to generate anticipatory signals that prepare the actor to accept certain kinds of information" (p. 105). The anticipatory signals are argued to be the basis on which motor commands are organized. In a spatial task, for example, we anticipate where our limbs will be. Turvey (1977) reviews a variety of evidence from neurophysiological research with humans and animals and proposes that a body centered visual space is represented by precise topographical mapping in the midbrain in very much the same way in all vertebrates. This map of visual loci also maps a topography of points entry into action systems. As for the representation of high level action plans, he concludes that "it is not so much that the specification of a subset of large coordinative (motor) structures and functions defined on them relates to higher order properties of the optic array, but rather that the description of the (action) plan in perceptual terms are dual statements about the same thing" (1977, p. 258).

Given these views on the relationship between imagery and action, the concept of spatial compatibility and vision-based code can be related to a more general theory of perception and action. By employing imagery as the main coding principle and designing chord combinations that would constitute a spatial match of the letter shape in the motor domain, a common unit base may have been found for the two sets of codes and the amount of transformations required from one set to the other may have been minimized. The basic stimulus response loop was simplified and compatibility maximized. While far from being conclusive, this interpretation can be tested in other instances in which spatial compatibility was found to be a major facilitating factor. Such tests are crucial to the enhancement of our understanding of the relationship between percepts and acts.

Acknowledgments. The helpful comments of Jack Adams, Walter Schneider, Christopher Wickens and two anonymous reviewers are gratefully acknowledged. The paper was written while the author was a visiting scientist at the Cognitive Psychophysiology Research Laboratory at the University of Illinois.

Reference Notes

1. Chevalier-Girard, N, & Wilberg, R.B. *The effects of image and label on the free recall of movement lists.* Paper presented at the Canadian Society for Psychomotor Learning and Sport Psychology, Toronto, November, 1978.
2. Gentner, D.R. *Skilled finger movements in typing.* University of California, San Diego. Center for Human Information Processing, CHIP 104, July, 1981.
3. Sidorsky, A. *Alpha-Dot. A new approach to direct computer entry of battlefield data.* Technical Report 249, U.S. Army Research Institute for the Behavioral and Social Sciences, 1974.
4. Brickner, M., & Gopher, D. *Improving time-sharing performance by enhancing voluntary control on processing resources.* Technion - IIT, Faculty of Industrial Engineering and Management. Research Center for Human Engineering and Work Safety. HEIS, 81-3, 1981.

5. Schelach, A. *The effect of mnemonic aids on performance in Alpha-numeric data entry devices.* Master Thesis, Faculty of Industrial Engineering and Management. Technion, Haifa, Israel, 1982, (Hebrew).
6. Gopher, D., & Arzi, N. *Effects of set size and motor difficulty in the joint performance of typing and classification with manipulation of task emphasis.* In preparation, 1983.

References

Adams, J.A., Gopher, D., & Lintern, G. The effects of visual and proprioceptive feedback on motor learning. *Journal of Motor Behavior,* 1977, *9,* 11–32.

Bernstein, N. *The coordination and regulation of movements.* London: Pergamon Press, 1967.

Carroll, W.R., & Bandura, A. On the role of visual monitoring in observational learning of action patterns: Making the unobservable observable. *Journal of Motor Behavior,* 1982, *14,* 153–167.

Conrad, R., & Longman, D.J.A. Standard typewriter versus chord keyboard: An experimental comparison. *Ergonomics,* 1965, *8,* 77–88.

Dvorak, A., Merrick, N.L., Dealey, W.L., & Ford, G.C. *Typewriting behavior.* New York: American Book Co., 1936.

Fitts, P.M., & Posner, M.I. *Human performance.* Pacific Pallisades: Brooks/Cole, 1967.

Gopher, D., Brickner, M., & Navon, D. Different difficulty manipulation interact differently with task emphasis: Evidence for multiple resources. *Journal of Experimental Psychology: Human Perception and Performance,* 1982, *8,* 146–157.

Gopher, D., & Eilam, Z. Development of the letter-shape keyboard: A new approach to the design of data entry devices. *Proceedings of the Human Factors Society 23rd Meeting.* Santa Monica: The Human Factors Society, 1979.

Haber, R.N. Where are the visions in visual perception. In S.J. Segal (Ed.), *Imagery: Current cognitive approaches.* New York: Academic Press, 1971.

Hall, C.R. Imagery for movement. *Journal of Human Movement Studies,* 1980, *6,* 252–264.

Hall, G.R., & Buckolz, J.B. Recognition memory for movement patterns and their corresponding pictures. *Journal of Mental Imagery,* 1981, *5,* 97–104.

Hebb, D.O. *The organization of behavior.* New York: Wiley, 1949.

Hirsch, R.S. Effect of standard versus alphabetic keyboard formats on typing performance. *Journal of Applied Psychology,* 1970, *54,* 484–490.

Hunter, I.M. Imagery, comprehension and mnemonics. *Journal of Mental Imagery,* 1977, *1,* 65–72.

Kelso, S.J.A., Wallace, S.A. Conscious mechanisms in movement. In G.E. Stelmach (Ed.), *Information processing in motor control and learning.* New York: Academic Press, 1978.

Kosslyn, S.M. *Image and mind.* Cambridge, MA: Harvard University Press, 1980.

Lee, D.N., & Aronson, E. Visual and proprioceptive control of standing in human infants. *Perception and Psychophysics,* 1973, *15,* 529–532.

Norman, D.A., & Fisher, D. Why alphabetic keyboards are not easy to use: Keyboard layout doesn't much matter. *Human Factors,* 1982, *24,* 509–520.

Neisser, V. *Cognitive psychology.* New York: Appleton, 1967.

Paivio, A. Imagery and language. In S.J. Segal (Ed.), *Imagery: Current cognitive approaches.* New York: Academic Press, 1971. (a)

Paivio, A. *Imagery and verbal processes.* New York: Holt, Rinehart, and Winston, 1971. (b)

Piaget, J. *Play, dreams and imitations in childhood.* New York: Norton, 1962.

Piaget, J., & Inhelder, B. *L'image mentale chez l'enfant.* Paris: Presses Universitaires de France, 1966.

Podgorny, P., & Shepard, R.N. Functional representation common to visual perception and imagination. *Journal of Experimental Psychology: Human Perception and Performance,* 1978, *4,* 21–35.

Rochester, N., Bequaert, F.C., & Sharp, E.M. The chord keyboard. *Computer,* December 1978, 57–63.

Segal, S.J. (Ed.) *Imagery: Current cognitive approaches.* New York: Academic Press, 1971.

Seibel, R. Data entry devices and procedures. In H.P. Van-Cott & R.G. Kinkade (Eds.), *Human engineering guide to equipment design,* (pp. 312–344). Washington: U.S. Government Printing Office, 1972.

Skinner, B.F. *Science and human behavior.* New York: MacMillan, 1953.

Turvey, M.T. Preliminaries to a theory of action with reference to vision. In R. Shaw & J. Bransford (Eds.), *Perceiving, acting, and knowing.* Hillsdale, NJ: Erlbaum, 1977.

Washburn, M.F. *Movement and mental imagery.* New York: Houghton Mifflin, 1916.

Werner, H., & Kaplan, B. *Symbol formation: An organismic approach to the psychology of language and expression of thought.* New York: Wiley, 1963.

Yates, F.A. *The art of memory.* Chicago: University of Chicago Press, 1966.

Summary. The linkage between vision and action is discussed in the light of experiments conducted with a single hand chord typewriter for the Hebrew language. This typewriter comprises three main keys, and each letter is entered by typing two successive chords of one to three fingers pressed together. One can thus conceive the code of each letter to be based upon a six cell imaginary matrix consisting of the three keys and the two successive strokes. Codes for letters are designed to mimic the distinctive features of the printed letter in Hebrew, the first and second chords representing the upper and lower halves of each letter shape. Training with the vision-based code shows fast acquisition and enduring superiority in typing speed. These results are interpreted in relation to a theory of action plans based on spatial imagery. It is also argued that they suggest a framework for the interpretation of the long recognized effects of spatial compatability on the efficiency of motor performance.

15 Speech Production and Comprehension: One Lexicon or Two?

D. ALAN ALLPORT

Contents

1 Introduction

Among the modalities of motor control possessed by almost every adult human individual, the control of speech is surely the richest and most exquisitely articulated. This pre-eminence is evident both in terms of the intricacy and precision of the movement dynamics involved, and in terms of the range of cognitive contents to which speech can give expression.

1.1 Phonological Word-Forms

The process of mapping from a prelinguistic, cognitive code (a "message") to the overt movements of speech doubtless involves a complex chain of transformations

Cognition and Motor Processes
Ed. by W. Prinz and A.F. Sanders

(c.f. Garrett, 1982). Within this chain, however, the critical linkage between cognitive codes and motor processes occurs at the level of what I shall call *phonological word-forms.* That is, phonological word-forms are the dominating or highest level *schemata* which can specify and control the overt articulatory gestures of spoken language[1]. "Schema" is used here as illustrated, for example, by Norman (1981).

Word-forms are not to be identified with the linguist's abstract conception of a word, or *lexeme.* Thus two or more homophonous but otherwise unrelated lexemes – such as *berry* (fruit) and *bury* (inter), or *palm* (tree) and *palm* (hand) – will share the same phonological word-form. Further, the term "word-form" is not intended to exclude speech units larger than a grammatical word (e.g., the syntagma discussed by McNeill, 1979). I shall call the structure that embodies the ensemble of phonological word-forms available to an individual the "phonological lexicon" (c.f, Allport & Funnell, 1981).

1.2 One Phonological Lexicon or Two?

The question with which this paper is concerned is simple to state. Are there separate phonological lexicons for speech production and comprehension, or is there a single structure that is used in both operations?

On the face of it, the functions required of a mental lexicon in production and in comprehension are radically different. In production, the input is some specification of the intended meaning (and, perhaps, grammatical category), and the desired output is a pronunciation; whereas in comprehension the input is some representation of a spoken sound sequence, on the basis of which a meaning (and grammatical category?) must be recovered. Since these operations appear to be so obviously different, it seems plausible to suppose that they might be implemented by means of distinct – perhaps structurally distinct – psychological mechanisms. This is, indeed, the classical model of Lichtheim (1885) and of Wernicke (1874/1977), which still forms the accepted framework for much current neuropsychological discussion (e.g., Goodglass & Geschwind, 1976; Hécaen & Albert, 1978).

A theoretical framework for the perception and production of language that has enjoyed considerable recent popularity is Morton's *logogen* concept. Conceptualized abstractly, logogens are rewriting rules (production rules) that specify, for each word-form, a mapping from sound to meaning and from meaning to sound (Morton, 1968). At this abstract level, "input" and "output" logogens manifestly represent different rules. If our theorizing is to remain purely at this abstract, "psychological" level, as Morton (1981) appears to advocate, the question whether these abstract rules are embodied in a separate or in a common structural substrate simply does

1 Phonological word-forms are not the sole controlling schemata. For example, the articulation of word-forms is superimposed on an intonation and stress contour apparently generated in parallel with lexical organization (Butterworth, 1980). However, the discussion in this paper is confined to *lexical* mechanisms only.

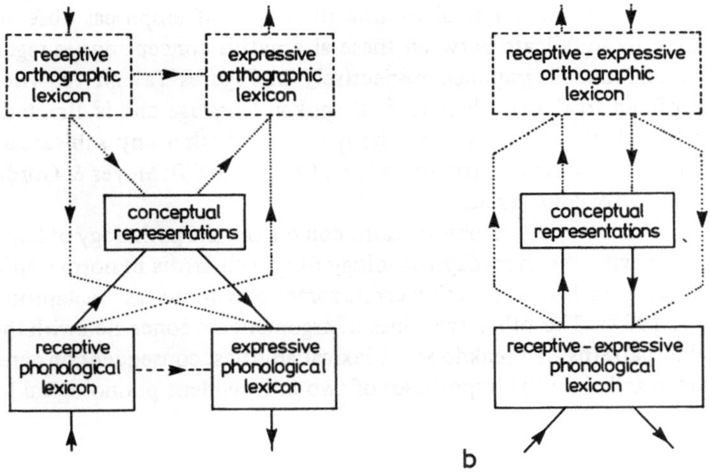

Figure 1. Two theoretical alternatives for the mental lexicon (inventory of word-forms) used in speech perception and production: (a) two separate phonological lexicons; (b) a common lexicon used in both perception and production. [Note: in this paper discussion is confined to the representation of phonological word-forms; alternative arrangements regarding the orthographic lexicon (shown in broken lines: see also Allport & Funnell, 1981) are not considered further in this paper.]

not arise. They just *are* different rules, even if expressions that occur on the left-hand side of one rule may also appear on the right-hand side of another. At some level of analysis, however, there are surely physical structures in the CNS that embody the specification of phonological word-forms. In a succession of recent papers, Morton has discussed, apparently as an empirical issue, whether input and output logogens are functionally, and indeed structurally, independent entities (Morton, 1979, 1980; Morton & Patterson, 1980). His conclusion is that they are (see Figure 1a). *Structural* separation seems to be implied by the notion that, in cerebral injury or disease, input and output logogens can be disconnected, a hypothesis that is central to Morton's account of certain neuropsychological disorders (Morton, 1980; Morton & Patterson, 1980; see also Ellis, 1982).

There are strong grounds for believing that, as between spoken and written language, *phonological* and *orthographic* word-forms are structurally independent, and, moreover, that communication between them can be physically interrupted as a result of cerebral injury (c.f. Coltheart, Patterson, & Marshall, 1980; Allport & Funnell, 1981; Bub & Kertesz, 1982). On the other hand, the arguments for structural independence, within the domain of *spoken* language, of phonological word-forms used in perception and in production seem less compelling (Allport & Funnell, 1981). On grounds of economy alone it might be suspected that the *same* physical structure could serve to represent phonological word-forms used in both speech perception and in production. This, then, is our alternative hypothesis (Figure 1b).

In this paper I shall outline three sets of empirical observations which may serve to adjudicate between these alternative conceptions as regards the phonological lexicon, diagrammed respectively in Figures 1a and 1b. The argument will be confined to the consideration of spoken language and is directed solely to the lexical level of representation. The question whether any sublexical mechanisms may also share common structures (e.g., Cooper, 1970; Meyer & Gordon, 1983) will not be directly considered.

All three sets of observations come from the pathology of language. The first has to do with the everyday pathology of speech errors in normal speakers: specifically the class of lexical selection errors sometimes known as "malapropisms" (Fay & Cutler, 1977). The other two lines of argument are concerned with the different observable patterns of breakdown in lexical abilities, consequent on cerebral injury, which are entailed by the hypotheses of two independent phonological lexicons or one.

2 Evidence from Normal Speakers: Errors of Lexical Selection

The argument outlined in this section is due originally to Fay and Cutler (1977). However, since their argument is insufficiently widely known, it seems worthwhile to recapitulate it here. The interested reader is urged to consult the original paper. First, a preliminary point.

2.1 Organization of a Perceptual Lexicon

There are clear enough reasons why an inventory of word-forms *designed for speech perception* should be organized in terms of phonemic (or acoustic) similarity, since phonemic (or acoustic) cues are the means by which it must be accessed. Moreover, words are often articulated insuffiently distinctly to be uniquely identified in isolation without the preceding or (importantly) the succeeding context (Pollack & Pickett, 1964). It would thus appear advantageous if words having similar or mutu-- ally confusable segments were arranged as functional "neighbors" in the phonological lexicon, so that, pending later contextual information, the momentary range of perceptual uncertainty could be represented as a maximally compact lexical "neighborhood". It is assumed that clear word identification is achieved when the neighborhood can be reduced to just one lexical possibility. Versions of this general idea can be seen in a number of current theories of word recognition, including Marslen-Wilson's word-initial cohort theory (Marslen-Wilson, 1980, 1983), and Klatt's (1980) design for machine recognition of speech, where individual word-forms are represented as unique paths through a network of possible accoustic sequences.

In speech production, on the contrary, the phonological word-forms have to be addressed in terms of syntactic and semantic (conceptual) codes. The optimal arrangement of a store of word-forms designed purely for speech production, Fay and Cutler argue, is therefore surely different from the arrangement that would be optimal for comprehension. In a production or "output" lexicon one might expect to find word-forms organized in terms of *semantic* similarity, but it is not obvious why they should be organized in terms of phonological similarity. This is not to say that an output-lexicon *could* not be organized in terms of phonological neighborhoods. The point is simply that there seems no compelling reason why it should be, *unless,* that is, it were the *same* word-form store as that used in comprehension.

2.2 Word-Substitution Errors: "Malapropisms"

Students of the naturally occurring oral slips made by normal speakers identify three principal types of error involving lexical selection: semantic substitutions, blends, and phonologically similar errors, for which Fay and Cutler adopted the term "malapropisms" (see Fromkin, 1973). Examples of all three types of error are shown in Table 1. (The substitution errors made by Sheridan's redoubtable Mrs. Malaprop were out of ignorance. This is not the case, in any philosophically straightforward sense, of the normal slips under discussion here.)

"Blend" errors, it is commonly assumed, arise when the choice between alternative (usually closely referentially related) word-forms has not been resolved by the time the speaker attempts to utter them. For the other two types of slip, a natural interpretation is that the error reflects some form of mistargeting in retrieval from the lexicon. In the case of semantic errors, the mistargeting is plausibly produced by an incomplete or erroneous specification of the meaning of the sought-for word. In the case of "malapropisms", on the other hand, the erroneous word typically bears no resemblance of meaning to the word intended at that point in the utterance, but a marked resemblance of phonological form, including the number of syllables and stress pattern, as well as its grammatical category (Fay & Cutler,

Table 1. Examples of three, commonly occurring types of lexical selection or substitution errors ("slips") made by normal speakers. (Intended word shown on the left, error on the right of the arrow.)

a)	*Semantic errors*	
	mother → daughter	restaurant → filling station
	oven → fridge	last year → yesterday
b)	*Blends*	
	gripping + grasping → grisping	
	draw + doodle → droodle	
c)	*Malapropisms*	
	equivalent → equivocal	gory → gaudy
	analogy → analysis	collude → collide
	dignitaries → dictionaries	transition → transistor

1977). That is, the error word is characteristically a close, often the closest, phonological neighbor of the intended target word. (For instance there is no other three-syllable word in English more like *equivalent* than *equivocal*.) Moreover, such errors are relatively common occurrences. In Fay and Cutler's sample of just under 400 instances of whole-word substitution errors, almost half could be classified as malapropisms.

Fay and Cutler's argument at this point is simple. If there exist two phonological lexicons, separately accessed in the processes of word comprehension and production, there appears no obvious way to explain why errors made in retrieval from an output lexicon should show such a systematic phonological resemblance to the word being sought. If, on the other hand, the expressive word-finding process takes place in a common inventory of word-forms organized for word perception as well as production, then this is precisely what mistargeting (retrieval of near neighbors of the target word) should lead us to expect.

A large proportion of malapropisms appear to be also semantically or contextually motivated. That is, while malapropisms are (by definition) semantically unrelated to the target word at that position in the utterance, they often bear a clear relation of meaning to some other entity that the speaker had in mind. (For example, pointing with satisfaction to the neat fit of his predictions to the serial position curves he had obtained: "We find some very nice *serial prediction curves.*") Mistargeting would seem a more probable occurrence when the phonological neighbor is already primed semantically. This conjunction of semantic and phonological effects is precisely what we should expect in a single lexicon used for both production and comprehension.

3 Neuropsychological Evidence

Briefly, if the phonological lexicon is divided into two subsystems, specialized respectively for word comprehension and word production (Model 1a) we should expect to find at least some cases in which cerebral accident or disease resulted in deterioration or malfunction of one lexical subsystem leaving the other essentially unimpaired. That is, we should expect to observe, in different individual cases, a "double dissociation" between disorders of the two hypothetical lexicons. Should it turn out that, despite thorough search, we are unable to find any such dissociations, while not strictly refuting model 1a this outcome would at the least greatly weaken its credibility.

On the contrary, if a common structure underlies the inventory of word-forms used in both listening and speaking, dissociations of this kind *at the lexical level* should *never* be observed. Their occurrence would therefore constitute a direct refutational of model 1b.

In principle one might imagine that the structures embodying a subject's phonological lexicon could be completely distroyed through selective injury, leaving other structures essentially intact. In practice the selectivity of cerebral lesion effects is seldom if ever so all-or-nothing, although functional dissociations can be strikingly selective (Shallice, 1979, 1981).

The assumption is made here that, *within* functionally specialized cerebral structures, the representation of memory information is *distributed,* as in matrix or holographic memories (Hinton & Anderson, 1981). In a memory system of this kind each individual engram – for example, each word-form in a phonological lexicon – is represented in a distributed manner *throughout* the structural substrate of storage. Different engrams are thus literally superimposed on one another, and *any* part of the memory structure embodies information about *all* of the engrams (word-forms) stored. There is not space here to expand on this very interesting idea. The most important consequence for the present discussion is that, where memory representation is distributed in this sense, local or partial injury should result not in the loss of individual items (word-forms) but in degradation (loss of specificity or distinctiveness) affecting *all* items (cf. Wood, 1978; Gordon, 1982). Further, if frequency of use affects the quality (distinctiveness? redundancy?) of representation in memory, we might expect a gradient in the severity of lesion effect, varying inversely with frequency of use, as is indeed commonly observed (e.g., Oldfield, 1966).

4 Selective Impairment of a Receptive Phonological Lexicon?

Can evidence be found of impairment of a phonological word-form store that affects only comprehension and not production? There are, undeniably, forms of primarily, or even exclusively, receptive auditory-language deficit that leave speech production relatively intact. However, caution is needed. In evaluating such evidence for its bearing on the hypotheses under consideration, it will be critical to establish whether the underlying impairment occurs at the lexical (word-form) level or not.

4.1 Pure Word-Deafness

The literature contains a number of detailed accounts of a condition known as pure word-deafness (Goldstein, 1974). Without significant hearing loss in the detection of non-speech sounds, the patient is nevertheless profoundly impaired in the auditory comprehension of spoken words; yet his speech production, certainly his naming ability (Howes, 1979), remains relatively intact. Comprehension and production of written language may also be unimpaired. (A patient of Wernicke's was able to continue his career as a journalist.) In some cases, at least, the interpretation of music as well as other non-linguistic sounds is also preserved.

Where appropriate tests are reported, it is evident that such patients have a gross impairment in phonetic (or "prephonemic") discrimination that applies equally to non-lexical syllabic segments and to real words (e.g., Auerbach, Allard, Naeser, Alexander, & Albert, 1982; Okada, Hanada, Hattori, & Shoyama, 1964–64; Saffran, Marin, & Yeni-Komshian, 1976). The observed impairment in speech-sound discrimination, at a sublexical level, thus provides sufficient explanation of these patients' comprehension difficulty, without the need to invoke additional impairment of a receptive lexicon.

4.2 Word-Meaning Deafness

A particularly interesting, although in its pure form infrequently observed, condition is one sometimes referred to as "word-meaning deafness" (e.g., Lichtheim, 1885; Bramwell, 1897; Symonds, 1953), or "transcortical sensory aphasia" (Goldstein, 1915). Luria (1976) has described patients with a similar pattern of disabilities. As in word-deafness, the comprehension of spoken words is radically disturbed while spontaneous speech, together with writing, reading aloud, and silent reading comprehension remain relatively unaffected. However, in contrast with the former condition, in word-meaning deafness auditory-vocal repetition and writing to auditory dictation are also preserved. (In pure word-deafness performance of these tasks is no better than auditory comprehension.) Thus the patient can orally repeat long and complex sentences that he hears, apparently still without comprehension, and he can also write them down. Only then, having written the sentence, is the patient able to understand what was said.

A striking feature of this latter performance is that the patient can correctly transcribe orthographically irregular words whose spelling could only be recovered from a word-specific (i.e., *lexical*) store (e.g., Bramwell, 1897). Thus, to write to dictation words like 'colonel' or 'Edinburgh', or even such familiar words as 'do', 'sew' or 'come', it will not suffice to rely on direct sound-to-letter correspondences (contrast 'few', 'go', 'sum'). Correct transcription must depend on access to word-specific (lexical) specifications of the spelling, and this, in turn, could only be obtained (from auditory input) by access to *functionally intact phonological word-forms.* The performance of such patients is therefore incompatible with the hypothesis of selective loss of a receptive phonological lexicon. On the other hand, the particular combination of preserved abilities and functional deficits seen in word-meaning deafness can be economically accounted for, in terms of the model represented in either Figure 1a or 1b, as an impairment in communication between an intact phonological lexicon and non-linguistic conceptual representations.

4.3 Conclusion

If we exclude the types of dysphasic impairment described in Sections 4.1 and 4.2, which evidently do not require the hypothesis of an impaired receptive phonological lexicon, my contention is that *no* case has yet been described demonstrating a *loss of receptive phonological word-forms without corresponding loss of spoken word production.* At least, I have been unable to find in the literature any case capable of supporting such an interpretation.

On the other hand, it is by no means uncommon to find patients in whom both receptive and expressive lexical impairments are combined. Thus, the available evidence suggests that the neologistic jargon in speech production, characteristic of what might be called the phonological-lexical (rather than the conceptual or semantic) form of Wernicke's aphasia, is invariably accompanied by disordered comprehension of spoken language (e.g., Goodglass & Kaplan, 1972; Buckingham & Kertesz, 1976; Kertesz, 1979). (For an interesting recent discussion of subcategories of Wernicke's aphasia, see Ellis, Miller, & Sin, 1983).

5 Selective Impairment of a "Speech Output Lexicon"?

In contrast with the relatively rare occurrence of dysphasic patients in whom auditory comprehension deficits are observed without accompanying or commensurate expressive disability, reports of expressive, oral naming deficits associated with apparently intact auditory word recognition are commonplace. This, indeed, is the conventional account both of pure *anomia* and of *conduction aphasia* (e.g., Goodglass & Geschwind, 1976; Hécaen & Albert, 1978). *Prima facie* we might therefore look first among these categories of dysphasic patients for evidence favoring the existence of separate phonological lexicons for word comprehension and production.

In Sections 5.1 and 5.2 I consider two issues related to this search in somewhat greater detail. The first considers the type of naming disorder which could be attributed directly to an impaired output lexicon (impaired phonological word-forms), and which will therefore be appropriate to our inquiry. The second considers possible behavioral measures appropriate to the comparison of expressive and receptive lexical impairment.

5.1 Disorders of Naming

Disorders of confrontation-naming appear in practically all cases of aphasia (Howes, 1979). Even if we exclude, at one end, cases with high-level conceptual deficits in the prelinguistic formulation of messages, and, at the other end, relatively low-level disorders in the transmission or execution of articulatory commands, systematic failures of naming can still result from a variety of different functional impairments. For the purpose at hand, however, the most important distinction is between, on the one hand, selective impairments of *access* to otherwise intact phonological word-forms (to which, in the absence of functionally heterogenous multiple deficits, one or more *other* access routes should still be available) and, on the other hand, damage to the phonological word-forms themselves (in which case the deficit will be evident whatever access route is involved). Clearly it is the latter type of impairment which is in principle relevant to the question of one *versus* two phonological lexicons, not the former. That is, a naming deficit in which, *ex hypothesi,* phonological word-forms are intact obviously is not the appropriate condition in which to look for selective injury to only one class of phonological word-form!

Thus, for example, where the naming difficulty results from breakdown in conceptual or semantic access to (intact) phonological word-forms, with no other functional impairment, it should still be possible to access the phonological word-forms from written words, in reading aloud, and in auditory-vocal repetition, where the normal superiority of word over nonword repetition should still be evident.

In contrast, where the naming difficulty results directly from injury to or impoverishment of the phonological (output) lexicon, we should expect the impairment of spoken word production to be evident not only in confrontation naming and spontaneous speech, but also in reading aloud and in auditory-vocal repetition (see Figure 1). Moreover, we might expect the naming disorder in this case to take the form predominantly of phonemic distortions of the target word (phonemic paraphasia, neologisms, malapropisms) and/or, in severe cases, simple failures to provide any overt response ("don't know"). On the contrary, disorder of *access* to the phonological (output) lexicon from a conceptual code might more plausibly result in naming errors that were semantically related to the target but were otherwise phonemically undistorted words (semantic paraphasias) as well as, once again, "don't know"s. In terms of conventional categories of aphasic breakdown, this would suggest that the type of naming disorder suitable for evaluating the one- vs. two-lexicon hypothesis is more nearly that seen in so-called *conduction aphasia* (predominantly phonemic paraphasia; comparable deficit of word-production in oral reading and in auditory-vocal repetition: Green & Howes, 1977) rather than in classical *anomia* (predominantly semantic substitutions, circumlocutions and "don't know" responses; benefit from phonemic cueing; and generally flawless auditory-vocal repetition (Howes, 1979).

Of course, naming disorders may very commonly result from an admixture of both these classes of underlying impairment (Cappa, Cavalloti, & Vignolo, 1981). Nevertheless, we should hope to find individual patients with sufficiently clearcut symptomatology to make testing of the hypothesis possible.

5.2 Comparative Assessment of Receptive and Expressive Lexical Impairment

The apparent asymmetry between expressive and receptive impairment often reported in relation to naming disorders may be a function, at least in part, of the asymmetrical demands on lexical retrieval made by conventional clinical tests of naming and of single-word comprehension (what Howes, 1979, calls "inverse naming") respectively. (Compare a similar suggestion, with respect to reading vs. spelling disabilities, put forward by Frith, 1980.)

In the conventional "confrontation naming" task, the subject is shown an object (or picture) and requested to name it orally. This task involves several elements not generally demanded in inverse naming. To perform correctly, the subject has to find the appropriate word among an unspecified, and indefinitely large vocabulary, and to recover its full phonological form, which in turn must be realized as an articulatory act. Malfunction in any of these aspects, for example partial retrieval of the word-form, as in the tip-of-the-tongue (T.O.T.) state familiar to normal speakers (Brown & McNeill, 1966), or incorrect retrieval of any of its components, will result in failure of correct naming. Parenthetically, the distinction between erroneous naming (as in phonemic paraphasia) and failure to offer any attempt ("don't know") may not be particularly diagnostic, at least without extensive further testing. Normal subjects in the T.O.T. state indicate that they "do not know" the sought-for target. (Indeed, it is a criterion for identifying the T.O.T. state that they do so.) However, when subjects in this state are pressed to make overt attempts at producing the sought-for item, the results are strikingly similar to the successive approximations ("conduite d'approche") – commonly unsuccessful – of pathological phonemic paraphasia.[2]

Word comprehension, on the other hand, is most often clinically assessed by tasks such as picture or object choice where the subject hears a single spoken word and is requested to indicate which among a specified array of alternative pictures (or objects) is best denoted by the word. The task differs from overt naming in that, among other things, it requests a forced choice over a specified (usually small) set of alternatives. Direct comparisons between "recognition" tasks of this kind and associative "recall" tasks, such as naming, are notoriously difficult to interpret. One relevant variable is, of course, the closeness of the distractors. When the names of the alternative, distractor objects are phonemically unrelated to the target word (as in the standard Peabody vocabulary test, for example), the correct choice could, in principle, be made on the basis of access to an extremely degraded or incomplete word-form. (Even when phonemically similar distractors are included, the correct choice may still be possible if features differentiating target and distractor items happen to overlap with even fragmentary information available in the phonological word-form.)

What is needed is a test of auditory word recognition that makes at least approximately comparable demands on the completeness of specification of the phonological word-form to those made by the overt naming task. A candidate for such a test is *auditory lexical decision.* In this task the subject must decide, of a spoken stimulus, whether it is a correctly pronounced word or not. If the mispronunciations are made sufficiently close to the correct form, differing from it, say, by just one articulatory feature, and if they can occur unpredictably on any part of the word, a *receptive* task of this kind could be made at least to approach

2 A similar observation is made by Ellis et al. (1983). See also Goodglass, Kaplan, Weintraub, and Ackerman (1976).

the conventional naming task in sensitivity to pathologically reduced precision or completeness of specification of the word-forms involved. With these considerations in mind, I constructed the following test.

5.3 An Auditory Lexical Decision Test

On each trial the subject heard a spoken word. The word was either correctly pronounced or, on about 50% of trials, one consonant in the word was deliberately altered such that the resulting pronunciation did not form an English word. The subject's primary task was to respond "yes" if the word was correctly pronounced, "no" if not. The words were the names of familiar, picturable objects, of one, two or three syllables in length. Mispronunciations were made in terms of voicing, place and manner of articulation, departing from the standard pronunciation by either 1, 2 or 3 of these articulatory features. Within words of a given syllabic length the point at which a mispronunciation occurred was distributed approximately evenly over possible locations in the word. Only consonantal mispronunciations were used, in view of the greater regional variation in vowel values in British English. Dialect forms were, as far as possible, excluded from the test. The complete test consisted of 153 trials.

On each trial the subject was also confronted with a line drawing of the object named by the experimenter on that trial, together with the drawing of a second object semantically but not phonemically related to the object named. [For example, the word "sock", mispronounced /zak/ ("zock") was accompanied by a picture of a sock and a boot; or "butterfly" pronounced /bʌkəflay/ ("buckerfly"), together with a picture of a butterfly and a wasp.] After saying "yes" or "no" in the pronunciation, or "lexical decision" task, the subject was requested to point to the object named by the experimenter, regardless of whether the word had been correctly pronounced or not. It was emphasized, however, that this aspect of the task was secondary. The subject's primary task was to judge the correctness of pronunciation of each word. The purpose of including the secondary word-picture matching task was to assess the extent to which possibly pathologically-impaired knowledge of phonological word-forms in dysphasic patients might still be sufficient to support correct "word comprehension", in terms of simple picture choice, under identical stimulus conditions.

5.4 Control Subjects: Results

The task was performed by twelve healthy control subjects (7 men, 5 women) recruited from the Oxford University subject pool, whose ages ranged from 22 to 62 years (median 32) and whose years of schooling ranged from 9 years (2 subjects) to college level (3 subjects).

In the primary, lexical decision task, the control subjects missed on average 4.87% of the mispronounced words. The great majority of these were failures to detect mispronunciations of a single distinctive feature. Mispronunciations involving more than one feature were missed overall on no more than 1% of trials. There were very few (0.22%) false alarms. For summary details see Figure 2.

The secondary task, correct picture choice, was evidently extremely easy. The majority of control subjects made no errors in this task: three subjects each made one error; one control subject made a total of four errors. Neither aspect of the test is thus particularly demanding for normal subjects.

5.5 Dysphasic Subjects

Data will be reported here for three dysphasic subjects, selected following the considerations discussed in Section 5.1 above to represent a wide range of severity of naming deficits. All three patients are classified as conduction aphasics. They had fluent spontaneous speech without evident articulatory difficulty, except in the dysfluencies, hesitations and false starts occasioned by their word-finding problems. Auditory comprehension in all three patients was clinically well preserved. Furthermore, all three were of above average intelligence, intellectually well preserved with no clinically obvious conceptual deficits. All three had a relatively severe deficit of auditory-vocal repetition, further details of which are discussed by Allport (1983).

As regards naming, one patient (AL) had essentially total inability to name objects on confrontation or from description; the second (KC) had moderately severe anomic difficulty, while the third (JB) showed only minimal impairment. An important criterion in selecting these patients was that their inability to produce a particular object-name could not simply be overcome on each occasion by presenting the sought-for word in spoken or written form. That is, the type of words that occasioned difficulty in object-naming also showed problems even in immediate oral repetition. In terms of the framework schematized in Figures 1a and 1b, this conjunction of performance deficits and preserved abilities would be most economically accounted for by direct, selective impairment of the phonological lexicon (expressive phonological lexicon, in 1a).

The following three sections are intended to provide a somewhat fuller impression of the individual patients' remaining language capabilities.

5.5.1 Subject AL

AL is a 62-year-old former electrical engineer, who suffered a left-sided cerebrovascular accident (c.v.a.) in December 1977. His most striking resulting symptom is a profound loss of the ability to produce the specific names of objects and actions, in the presence of otherwise fluent, prosodic and well-articulated expressive speech. The quality of his language production is best appreciated by means of an example.

In the following excerpt, AL is describing a picture of a cluttered kitchen, illustrating a variety of incipient domestic accidents. (In the foreground a saucepan is on the point of boiling over, while a second, uncovered gas-ring flares up dangerously. The over door swings open. A cigarette has been left burning on a shelf. Some pills lie scattered around an unstoppered pill-bottle, and so on . . .). AL can generally indicate to the listener, by gesture or mime, the objects and actions to which he is referring. To assist the reader in following this sample I have inserted the inferred target words in the transcript in square brackets.

"You never do that with a place (cooker?) there; you push it and do that (gestures turning down cooker controls). That is the same thing (open?) underneath; there's a little one (knob?) to do (regulate?) that as well . . . that (cigarette?) is – mm – (dropped?) there (on the shelf?) without doing it (stubbing out?) the things (cigarette butts?) that are being done, – you know the thing I mean? . . . What else is there? Oh yes, she is doing (throwing?) some stuff (parafin?) here (onto an already lighted fire) and its really rather . . . bad to do things there. You don't do that, you take it and do it slow. And not that (parafin?) anyway . . .
Oh, of course, I'm sorry, I should have told you that those things (pills?) there were done (spilled?) from – mm – (the bottle?) and put . . . (gestures: scattered?). Can you understand what I'm doing? This is usually made (gestures closing lid) and put somewhere well out of the way . . ."

What is immediately apparent in this short sample of AL's spontaneous speech is the absence of words that should refer to specific categories of objects and events. In their place AL resorts to a restricted vocabulary of the most general, referentially non-specific words in the language; demonstratives, pronouns, nouns like *thing, stuff, place,* verbs such as *do, make, put, tell.* In an extended sample of his speech, no more than 1% of the words were nouns with specific, concrete reference, and these only extremely high-frequency words like *man, girl, car.*

In explicit naming tests AL is almost totally at a loss. For example, in one very simple test of object naming (Coughlan & Warrington, 1978) the experimenter points to each of 15 familiar objects in turn (key, glove, onion, ring, etc.) and asks the patient to name them. With 7 sec allowed for naming each individual object, AL scored zero; even allowing 15 seconds for each item, AL was able to produce the name of only one of the objects. For the rest, his response is a shake of the head. Even when the word is spoken for him by the experimenter, AL still quite commonly fails to repeat it aloud correctly. With low-frequency words (Thorndike-Lorge 1–20 per million) he fails on about 50% of occasions. This score is unaffected by word length: one-syllable words 52%, three-syllable 49% correct. Where *two* common object names were to be repeated in sequence, AL succeeded on only 18% of trials. With words of lower frequency, or less concrete reference, AL's repetition performance declines still further.

In contrast, on conventional tests of *receptive* auditory vocabulary, AL performs well. For example, in the Peabody Picture Vocabulary test, AL scores 130/150, equivalent to a verbal IQ of 121. By this criterion AL thus appears wholly unimpaired in receptive auditory word-recognition, in comparison with his crippling deficit in production.

5.5.2 Subject KC

KC is a retired man of letters, formerly a professional historian, who 4 years previously suffered a left-sided c.v.a. Following his stroke he was initially profoundly aphasic. However, by the time of testing, his spontaneous speech was fluent and clearly articulated, aside from a moderately severe naming difficulty accompanied by numerous phonemic paraphasias. Thus, in Coughlan and Warrington's naming test, KC scored only 6/15. His reading aloud was severely impaired (Masterson, Note 1), as was his auditory-vocal repetition (Allport, 1983). In contrast, his comprehension of spoken language was, for practical purposes, excellent, as was his recognition vocabulary of individual spoken words.

5.5.3 Subject JB

In 1958 JB underwent an operation to remove a large meningioma (5.5 × 5.5 × 3.5 cm) from the region of the left angular gyrus. Since that time she has been the subject of extensive neuropsychological investigation, directed principally towards her deficit in auditory-vocal sequential repetition. She has a repetition span of about three digits, and fewer than three spoken words. Nevertheless, in spontaneous speech, her word fluency, pausing and frequency of paraphasic errors are indistinguishable from that of normal subjects (Shallice & Butterworth, 1977). Her auditory comprehension also appears more or less normal (Warrington, Logue, & Pratt, 1971), and for many years she has been employed in a busy and responsible secretarial post.

JB's naming difficulty is confined to relatively low-frequency words. It is *also* apparent in reading aloud low-frequency (especially abstract) words, and in immediate auditory-vocal repetition. Her errors, in all three tasks, take the form of phonemic paraphasias (mispronunciations in which at least some resemblance to the intended word-form is generally recognizable). Her successive attempts at correct pronunciation, even with repeated auditory presentations of the target word, are frequently unsuccessful; quite often her utterance appears to get captured by a neighboring, phonemically similar word, reminiscent of the malapropisms of normal speakers (Section 2.2) except that, in JB's case, the error could not readily be corrected (examples: paradox – "paradise"; conjecture – "conjunction"; infiltrate – "infantry").

5.5.4 Review

All three patients exhibit the characteristic symptoms of conduction aphasia (Green & Howes, 1977), though in varying degrees of severity. The combination of impaired *expressive* lexical abilities and apparently relatively intact receptive ability, as evaluated by conventional clinical assessment, appears *prima facie* to favor model 1a – independent phonological lexicons for input and output – with impairment (in these cases) of the *output* lexicon alone. The performance of these patients in the auditory lexical decision task, however, suggests otherwise.[3]

5.6 Dysphasic Subjects: Test Results

Scores obtained in the auditory lexical decision task by the three dysphasic patients, together with those for the normal control group, are shown in Figure 2. All three patients reveal marked deficit in their ability to detect distorted word-pronunciations. The patients' error rates, in all three cases, are well outside the range of errors in lexical decision made by the normal controls. Moreover, the severity of this auditory-receptive lexical deficit is ranked in the same order as their difficulty of expressive naming. On the other hand, their word-comprehension, as indexed by

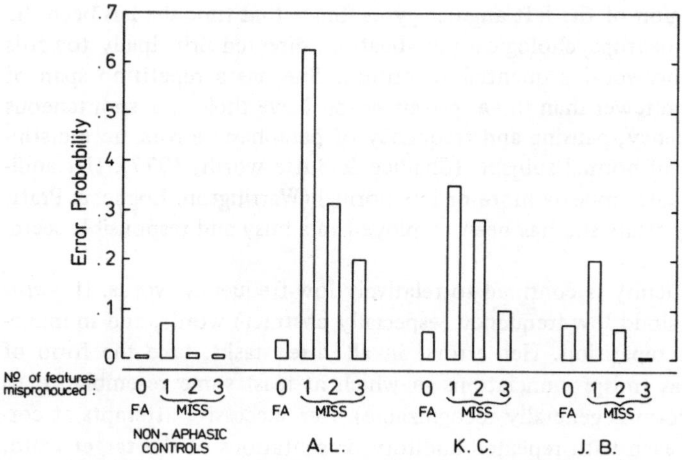

Figure 2. Auditory lexical discrimination: results (errors) for normal controls and for three dysphasic patients with different severity of word production deficit (AL most severe, JB least severe). FA = false alarms.

---

3 Tests of phoneme discrimination in a syllable matching task indicate that, at least in two of these patients (AL and JB), there is also a substantial deficit in *sublexical* speech-sound discrimination, amounting to a mild degree of "word-deafness" (Allport, 1983). This sublexical impairment doubtless contributes to these two patients' poor performance in the auditory lexical-decision task. However, a high error-rate in phoneme discrimination appears to be neither (a) sufficient nor (b) necessary to account for the very low levels of *lexical* discrimination seen. Viz. (a) JB, who makes by far the greatest proportion of errors of the three in immediate phoneme discrimination, is the least impaired in the lexical decision task; (b) KC made only 5% (4/72) errors in immediate phoneme discrimination, nowhere near the rate of omission errors shown by the other two patients (AL, 22.9%; JB, 35.4%); yet KC is very substantially impaired in auditory lexical decision.

correct picture choice in this test, is relatively unimpaired. None of the three patients made more than 5/153 (3%) errors in this aspect of the task.

5.7 Concluding Discussion

It is evident that the expressive naming deficit seen in these three subjects which, I have argued, is attributable to impairment (loss of specificity) at the level of the phonological word-forms *used in speech production* is accompanied in each case by a complementary deficit in auditory lexical discrimination. What appeared, on conventional clinical assessment, to be a dissociation between an impaired phonological lexicon in *production* and an intact phonological lexicon in *reception* turns out, at least in these patients, not to be a dissociation after all.

Clearly, the inferences that we can draw from this result, based on a sample of only three patients, must be at best highly provisional. Other investigators, however (in particular Alajouanine, Lhermitte, Ledoux, Renaud, & Vignolo, 1964), have observed a similar coincidence between phonemic paraphasias in lexical expression and impaired auditory lexical discrimination, in a larger group of conduction aphasics. Nevertheless, failure to find a dissociation, even in a thousand cases, would not show that the dissociation cannot occur. The most that we can say is that the neuropsychological evidence which would unambiguously support the hypothesis of independent phonological lexicons for input and output has thus far proved elusive.

On the other hand, the hypothesis of a *single* inventory of phonological word-forms employed in both reception and production enjoys the advantage not only of theoretical parsimony but also of an attractive empirical vulnerability. Even a single patient, in whom selective, dissociated impairment of either receptive or expressive phonological word-forms could be unambiguously demonstrated, would threaten to disprove it. Two patients, showing between them the double dissociation, would do so conclusively. I shall therefore continue to look for such evidence. For the present, the more economical, as well as more empirically vulnerable hypothesis appears the preferable alternative.

Reference Note

1. Masterson, J. Unpublished doctoral dissertation. Birkbeck College, University of London, Malet St., London, 1983.

References

Alajouanine, T., Lhermitte, F., Ledoux, A., Renaud, D., & Vignolo, L. Les composantes phonémiques et sémantiques de la jargonaphasie. *Revue Neurologique,* 1964, *11,* 5–20.

Allport, D.A. Auditory-verbal short-term memory and conduction aphasia. In H. Bouma & D.G. Bouwhuis (Eds.), *Attention and performance 10: Control of language processes.* Hillsdale, NJ: Erlbaum, 1983.

Allport, D.A., & Funnell, E. Components of the mental lexicon. In D.E. Broadbent, J. Lyons, & C. Longuet-Higgins (Eds.), Psychological mechanisms of language. *Philosophical Transactions of the Royal Society, London, B,* 1981, *295,* 397–410.

Auerbach, S.H., Allard, T., Naeser, M., Alexander, M.P., & Albert, M.L. Pure word deafness: analysis of a case with bilateral lesions and a defect at the prephonemic level. *Brain,* 1982, *105,* 271–300.

Bramwell, B. Illustrative cases of aphasia. *Lancet,* 1897, *1,* 1256–1259.

Brown, R., & McNeill, D. The "tip of the tongue" phenomenon. *Journal of Verbal Learning and Verbal Behavior,* 1966, *5,* 325–337.

Buckingham, H.W., & Kertesz, A. *Neologistic jargon aphasia.* Amsterdam: Swets and Zeitlinger, 1976.

Bub, D., & Kertesz, A. Deep agraphia. *Brain and Language,* 1982, *17,* 146–165.

Butterworth, B. Some constraints on models of language production. In B. Butterworth (Ed.), *Language production,* (Vol. 1). *Speech and talk.* London: Academic Press, 1980.

Cappa, S., Cavalotti, G., & Vignolo, L.A. Phonemic and lexical errors in fluent aphasia: correlation with lesion site. *Neuropsychologia,* 1981, *19,* 171–177.

Coltheart, M., Patterson, K.E., & Marshall, J.C. (Eds.) *Deep dyslexia.* London: Routledge & Kegan Paul, 1980.

Cooper, W.E. *Speech perception and production: studies in selective adaptation.* Norwood, NJ: Ablex, 1980.

Coughlan, A.K., & Warrington, E.K. Ward-comprehension and word-retrieval in patients with localized cerebral lesions. *Brain,* 1978, *101,* 163–185.

Ellis, A.W. Spelling and writing (and reading and speaking). In A.W. Ellis (Ed.), *Normality and pathology in cognitive functions.* London: Academic Press, 1982.

Ellis, A.W., Miller, D., & Sin, G. Wenicke's aphasia and normal language Processing: A case study in cognitive neuropsychology. *Cognition,* 1983.

Fay, D.A., & Cutler, A. Malapropisms and the structure of the mental lexicon. *Linguistic Inquiry,* 1977, *8,* 505–520.

Frith, U. Unexpected spelling problems. In U. Frith (Ed.), *Cognitive processes in spelling.* London: Academic Press, 1980.

Fromkin, V. (Ed.) *Speech errors as linguistic evidence.* The Hague: Mouton, 1973.

Garrett, M.F. Production of speech: observations from normal and pathological language use. In A.W. Ellis (Ed.), *Normality and pathology in cognitive functions.* London: Academic Press, 1982.

Goldstein, K. *Die transkortikale Aphasien.* Jena: Gustav Fischer, 1915.

Goldstein, M.N. Auditory agnosia for speech ('pure word deafness'): a historical review with current implications. *Brain and Language,* 1974, *1,* 195–204.

Goodglass, H., & Geschwind, N. Language disorders (aphasia). In E.C. Carterette & M. Friedman (Eds.), *Language and speech, Handbook of perception,* (Vol. 7). New York: Academic Press, 1976.

Goodglass, H., & Kaplan, E. *Assessment of aphasia and related disorders.* Philadelphia: Lea and Febiger, 1972.

Goodglass, H., Kaplan, E., Weintraub, S., & Ackerman, N. The "tip-of-the-tongue" phenomenon in aphasia. *Cortex,* 1976, *12,* 145–153.

Gordon, B. Confrontation naming: computational model and disconnection simulation. In M.A. Arbib, D. Caplan, & J.C. Marshall (Eds.), *Neural models of language processes.* New York: Academic Press, 1982.

Green, E., & Howes, D.H. The nature of conduction aphasia: a study of anatomic and clinical features and of underlying mechanisms. In H. Whitaker & H.A. Whitaker (Eds.), *Studies in neurolinguistics,* (Vol. 3). New York: Academic Press, 1977.

Hécaen, H., & Albert, M.L. *Human neuropsychology.* New York: Wiley, 1978.

Hinton, G.E., & Anderson, J.A. (Eds.) *Parallel models of associative memory.* Hillsdale, NJ: Erlbaum, 1981.

Howes, D. The naming act and its disruption in aphasia. In D. Aaronson & R.W. Rieber (Eds.), *Psycholinguistic research: implications and applications.* Hillsdale, NJ: Erlbaum, 1979.

Kertesz, A. *Aphasia and associated disorders: taxonomy, localization and recovery.* New York: Grune and Stratton, 1979.

Klatt, D.H. Speech perception: a model of acoustic-phonetic analysis and lexical access. In R.A. Cole (Ed.), *Perception and production of fluent speech.* Hillsdale, NJ: Erlbaum, 1980.

Lichtheim, L. On aphasia. *Brain,* 1885, *1,* 363–484.

Luria, A.R. *Basic problems of neurolingustics.* The Hague: Mouton, 1976.

Marslen-Wilson, W. Speech understanding as a psychological process. In J.C. Simon (Ed.), *Spoken language generation and understanding.* Dordrecht, Holland: Reidel, 1980.

Marslen-Wilson, W. Recognising spoken words: possible words in possible worlds. In H. Bouma & D.G. Bouwhuis (Eds.), *Attention and performance 10: Control of language processes.* Hillsdale, NJ: Erlbaum, 1983.

McNeill, D. *The conceptual basis of language.* Hillsdale, NJ: Erlbaum, 1979.

Meyer, D.E., & Gordon, P.C. Shared mechanisms of perceiving and producing speech. In H. Bouma & D.G. Bowhuis (Eds.), *Attention and performance 10: Control of language processes.* Hillsdale, NJ: Erlbaum, 1983.

Morton, J. Consideration of grammar and computation in language behavior. In J.C. Catford (Ed.), *Studies in language and language behavior.* University of Michigan, Progress Report No. 6, 1968.

Morton, J. Facilitation in word recognition: experiments causing change in the logogen model. In P.A. Kolers, M.E. Wrolstad, & H. Bouma (Eds.), *Processing of visible language,* (Vol. 1). New York: Plenum, 1979.

Morton, J. The logogen model and orthographic structure. In U. Fritz (Ed.), *Cognitive processes in spelling.* London: Academic Press, 1980.

Morton, J. The status of information processing models of language. In D.E. Broadbent, J. Lyons, & C. Longuet-Higgins (Eds.), *Psychological mechanisms of language. Philosophical Transactions of the Royal Society, London, B,* 1981, *295,* 387–396.

Morton, J., & Patterson, K.E. Little words – No! In M. Coltheart, K.E. Patterson, & J.C. Marshall (Eds.), *Deep dyslexia.* London: Routledge and Kegan Paul, 1980.

Norman, D.A. Categorization of action slips. *Psychological Review,* 1981, *88,* 1–15.

Okada, S., Hanada, M., Hattori, H., Shoyama, T. A case of pure word deafness: the relation between auditory perception and recognition of speech sounds. *Studia Phonologica,* 1963–4, *3,* 58–65.

Oldfield, R.C. Things, words and the brain. *Quarterly Journal of Experimental Psychology,* 1966, *18,* 340–353.

Pollack, I., & Pickett, J.M. Intelligibility of excerpts from fluent speech: auditory vs. structural context. *Journal of Verbal Learning and Verbal Behavior,* 1964, *3,* 79–84.

Saffran, E.M., Marin, O.S.M., & Yeni-Komshian, G.H. An analysis of speech perception in word-deafness. *Brain and Language,* 1976, *3,* 209–228.

Shallice, T. Case-study approach in neuropsychological research. *Journal of Clinical Neuropsychology,* 1979, *1,* 183–211.

Shallice, T. Neurological impairment of cognitive processes. *British Medical Bulletin,* 1981, *37,* 187–192.

Shallice, T., & Butterworth, B. Short-term memory impairment and spontaneous speech. *Neuropsychologia,* 1977, *15,* 729–735.

Symonds, C. Aphasia. *Journal of Neurology, Neurosurgery and Psychiatry,* 1953, *16,* 1–6.

Warrington, E.K., Logue, V., & Pratt, R.T.C. The anatomical localization of selective impairment of auditory-verbal short-term memory. *Neuropsychologia,* 1971, *9,* 377–407.

Wernicke, C. Der aphasische Symptomenkomplex. Breslau: Cohn und Weigart, 1874. (English translation in G.H. Eggert, *Wernicke's works on aphasia: A source book and a review.* The Hague: Mouton, 1977.)

Wood, C.C. Variations on a theme by Lashley: lesion experiments on the neural model of Anderson, Silverstein, Ritz and Jones. *Psychological Review,* 1978, *85,* 582–591.

Summary. It is assumed that every language user must possess, in some form, an inventory of phonological word-forms known, or available, to him in his expressive and receptive vocabularies. The structure embodying such knowledge is here referred to as the "phonological lexicon". The question which this paper addresses is: Are there separate phonological lexicons for speech production and comprehension or is a single structure responsible for both?

Two principal types of evidence relevant to these theoretical alternatives are reviewed, concerning (1) speech errors ("slips") by normal speakers and (2) the selective impairment of receptive and expressive language in dysphasic patients. (1) Normal speakers often produce, in place of the intended word, another word resembling it phonemically but not semantically. This type of error, it is argued, is readily accounted for in terms of a single phonological lexicon used in comprehension and production, whereas no obvious explanation is forthcoming in terms of the two-lexicon model. (2) Varieties of dysphasic impairment are reviewed in which either the comprehension or production of spoken words appears to be selectively compromised, and some new data from a test of auditory lexical discrimination, in patients with "expressive" lexical difficulties, are reported. None of the evidence reviewed is consistent with separate impairment of expressive vs. receptive functions at the *lexical* ("word-form") level. The hypothesis of a single phonological lexicon for both the expression and reception of spoken language is preferred.

IV Attention, Cognition, and Skilled Performance

16 S-Oh-R: Oh Stages! Oh Resources!

Daniel Gopher and Andries F. Sanders

Contents

1 Introduction

Current conceptual frameworks for research on human information processing can be roughly divided into two classes, which differ with respect to their view on what is the most productive approach to modelling the human processing system. These classes are characterised, respectively, by the notions of "processing stage" and of "resource". Processing stage models are primarily interested in describing the flow of information through the organism in terms of processing stages between presentation of a signal and completion of a response. Performance variability is considered in terms of number and nature of the required operations in the process of transforming a signal into a response. In contrast, resource models maintain that information processing and performance variability are first of all matters of strategic

---

This paper was written while the second author was on leave from the Institute for Perception, TNO, Soesterberg, The Netherlands, and held the position of Karl T. Compton, Visiting Professor at the Department of Industrial Engineering and Management of the Technion, Israel Institute of Technology, Haifa. Authorship of the paper is equally shared.

The preparation of this paper was supported in part by grant No. AFOSR-82-0069 from the life sciences program of the U.S. Air Force office of scientific research. DM Genevieve Haddad and Major Christofer Lind were the scientific monitors of this grant.

Cognition and Motor Processes
Ed. by W. Prinz and A.F. Sanders

control in allocating scarce processing facilities to tasks in an attempt to deal with the various demands.

In other words, linear stage models emphasize that stimulus and response are connected via a series of transformations. Interpretation and prediction of performance is only possible by way of a proper description of the type and architecture of these transformations. This has been the main logic that guided the early work of Donders (1868/1968) as well as more recent notions on stage analysis (e.g., Sternberg, 1969).

In contrast, resource models are based upon the claim that the human organism acts as a *limited* processor that can be loaded to various degrees. Interpretation and prediction of performance is therefore impossible without studying the limits of the processor and without the development of a measure of task load. It is then immediately apparent, that processing stage studies have been mostly interested in uncovering a multiplicity of processes between stimulus and response (Sternberg, 1969), while the limited capacity approach has traditionally emphasized the search for one general concept of load and a corresponding measure of capacity limits (e.g., Kahneman, 1973). It is important to note that this represents a different approach to the basic question of how to model the internal structure of the human organism in order to explain the relationships between stimuli and responses.

If this basic distinction is accepted, the question arises how the frameworks differ. Do they primarily reflect differences in metatheoretical orientation? Are they merely opposing theories on the same issues, so that one or both can be ultimately rejected? Or do they finally appear to address largely different sets of problems, so that they may both be valid in their own realm? When attempting to address these questions, it is essential to consider the basic concepts and assumptions of either framework and to scrutinize their scope, level of experimental prediction and means of empirical refutation. In addition, one can legitimately raise the question to what extent the frameworks have common elements. For example, is it possible to test their deductions by means of the same experimental methodology?

In the following sections some of these issues will be considered in greater detail. It should be realised that linear stage and resource models are not internally uniform; both contain several variants, some of which are dated while others are viable alternatives. The main emphasis in this paper is on comparing the additive factors version (Sternberg, 1969) of the linear stage models with microeconomic resource models (Navon & Gopher, 1979). The simple reason for this choice is that these versions have the advantage of a clear experimental methodology and an interpretative scheme on the basis of a clear set of theoretical postulates. The next section is devoted to the general stage and resource concepts and to a brief review of their assumptions, methodology and measures. Subsequently there is a discussion of attempts of building theoretical bridges between the frameworks and of developing convergent experimental paradigms. The paper ends with a summary of the outcomes and consequences of the argument.

2 Major Concepts, Basic Assumptions and Measures

2.1 Fixed Structures versus Variable Activation

The major conceptual difference between linear stage and resource notions can be summarized by the difference between fixed structures and variable activation. The fixed structures refer to computational processing mechanisms. Some mechanisms may be more controlled while others may operate more automatically. Again, some may be more and others less cognitive but together they constitute the architecture as well as the "data base" for human mental functioning. In contrast, variable activation refers to activation of the pertinent structures in order to enable efficient processing and to reach optimal adaptation. A mere description of the "structures", as contained in the black box, is insufficient because structures can be selectively occupied or activated at various levels, depending on the momentary task demands.

Implicit in the variable activation notion is the common element to all resource models that "resources are scarce" or that "capacity is limited". Originally, capacity referred to information in a fairly strict sense (Broadbent, 1958). Later, this was transformed to computer capacity (Moray, 1967; Broadbent, 1971), while more recently the capacity concept acquired an energetical content. This is most pronounced in Kahneman's (1973) effort notion which refers to *energetic* sources activating the pertinent structures and controlling its adequate functioning. No performance can occur without allocation of effort or energetic volume to a given task! The microeconomic resource theory (Navon & Gopher, 1979) is more neutral about the nature of resources, although it also clearly adheres to a volume notion and, hence, is not incompatible with an energetic resource interpretation.

Another element that is implicit in most resource models is aspecificity of the "substance" of capacity. This has been assumed in informational as well as in energetic resource theories. What matters is the amount of capacity required rather than the nature of processing, so that the system's architecture plays a minor role. The major reference to the type of demands has been by way of the molar concept of task difficulty, which determines the "average efficiency of resources" (Navon & Gopher, 1979, p. 15). In contrast, the fixed structures of linear stage notions are specific and have strong bearings on the actual direction of behavior but have little to say about energetics, efficiency and attentional strategy.

The distinction between specific processing structures and energetic aspecific resources is a frequently observed dichotomy: Apperception vs. attention (Wundt, 1896), habit vs. drive (Hull, 1943), guiding vs. energising (Hebb, 1955). The dichotomy does not reflect a principal disagreement about these basic dimensions as much as about their relative impact. Stage and resource models have been one-sided in their emphasis. It is surprisingly easy to classify the major research areas in experimental psychology as predominantly belonging to either computational time-based structure or to energetic activation (Sanders, 1977): Capacity limitations, mental load, effort, selective and shared attention and attentional control belong primarily to the resource domain, while linear stages are dominant in problems of encoding, response selection, memory search and motor-programming. In itself this can be considered as a case in favour of the position that the frameworks are engaged with non-overlapping sets of questions. However, the separation should not be exaggerated.

The relations between the frameworks strongly depend on the status of the hypothesis of aspecificity of resources or, in other words, on the assumption that all resources are derived from one common pool. If this assumption is abandoned in favour of a multiple resource notion, the frameworks converge to a considerable extent. The case in point is that the search for multiplicity in linear stage models is hard to reconcile with a unitary capacity notion but can easily be combined with different sets of resources that are specifically related to certain processing operations. It is relevant to note that recent research on arousal tends to favour cognitive-energetic patterns (e.g., Sanders, 1983). Again, microeconomy resource theory explicitly accepts this possibility and discusses ways of testing multiple resource notions (see Section 3.2). Hence the main theme of this paper is the comparison of resource and linear stage notions with emphasis on their common concerns.

2.2 Methodologies and Research Paradigms

For linear stage models the analysis of discrete choice reactions has been dominant since this provides the most natural setting for observing the information flow and its accompanying processing stages. The basic axiom is that information flow is fully reflected by the time taken by the processing stages. Processes that are not reflected in processing duration are beyond the scope of linear stage analysis. There are no time segments where irrelevant processing occurs; the reaction time is the minimal time for relevant processing. It is not surprising, therefore, that reaction time (RT) is the favourite measure in linear stage models, although threshold-type measures are not excluded. They can be used as a tool for studying either the extent of processing as a function of the time a signal is presented or the extent processing relies upon the actual presence of the signal (e.g., Salthouse, 1981). In principle, the task should not be overloading since this is likely to affect the validity of time measures as a reflection of processing; e.g., the issue of speed-accuracy is a tricky problem (Sanders, 1980).

The classic linear stage methodology, as enshrined in the subtraction method (Donders, 1868/1969), simply consisted of subtracting the reaction times obtained at different choice reaction tasks. If two tasks are identical except that one task contains one processing stage more or less than the other, the time taken to complete the processes of that stage can be calculated. This method meets the considerable difficulty of postulating a priori the stages involved in different tasks, without any guarantee or check about the validity of the assumptions. In fact the subtraction method has failed on many occasions (e.g., Posner, 1978), and it is surprising that there have been various recent papers that solely base their deductions upon the subtraction notion (Hunt, 1978). The building blocks may clearly not behave according to our intuitions about their nature and accompanying task effects. This is also the main problem for other intuitive building-block models (e.g., Smith, 1968; Welford, 1967). It is easy to construct a diagram but hard to carry out critical tests.

This is the very reason for the appeal of the additive factor model (Sternberg, 1969), the basic idea of which is that processing stages should be inferred rather than a priori postulated. Thus, additive effects of two variables on reaction time suggest the operation of different stages while interactive effects suggest that the two variables affect at least a common stage (Sternberg, 1969). An additional relevant distinction is between processes and processing stages. Stages comprise integrated structures of processes rather than a single process. The additive factor logic contains also the important shift from drawing conclusions on the basis of differences between tasks to drawing conclusions based upon effects of task variables. Consideration of the effect of task variables instead of tasks is relevant by itself since it allows stronger conclusions (e.g., Sanders, 1979).

The processing stage is merely an operational construction and is certainly in need of converging evidence and further specification (e.g., Salthouse, 1981). It should also be clearly understood that interpretation along the lines of the additive factor method implies accepting extreme strong assumptions that certainly have no general validity (Sanders, 1980). The pertinent issues will be discussed in Section 2.3.

While linear stage models are most suitable for the analysis of discrete choice reactions in single task performance, resource models are best investigated in tests of dual tasks. Examples are the studies on secondary tasks (Bahrick, Noble, & Fitts, 1954; Michon, 1966; Wickens, 1976), on dichotic listening and on shadowing (Cherry, 1953; Broadbent, 1958; Treisman, 1969; Moray, 1967).

The basic rationale of the dual task paradigm is that the volume of resources and the policies of their allocation can only be properly investigated by overloading the organism through multiple tasks and by examining patterns of interference. The problem with the investigation of single tasks is that it is unknown to what extent resources are fully employed or how performance is restricted by non-resource related factors (data limits). The only way single task measures are defendable is when they are accompanied by an independent measure of resources as for instance in the attempts of physiological measurement of resource consumption (Kahneman, 1973). In general, these attempts have been disappointing mainly because of the evidence for strong interrelations between cognitive and energetic physiological processes. Hence the hope of finding one general reliable physiological index for aspecific resource consumption has gradually weakened.

Measurement of secondary task performance as an index of resource expenditure has been quite popular. Theoretically, it was argued that the addition of a second task guaranteed saturation of the capacity, causing performance decrements. The hope has been to find a suitable "standard" secondary task as a yardstick for measuring resources needed by the "primary" task (e.g., Michon, 1966; Jex, 1969). The research has shown two major problems. One is methodological and concerns protection of the primary task against decrement when carried out together with the secondary task. If the primary task is not protected, the value of secondary task performance as a measure is obviously doubtful (Rolfe, 1971; Williges & Weirwille, 1979). The second problem may be even more basic and concerns the finding that a secondary task decrement is not always related to the processing requirements of the primary task. For example Kantowitz and Knight (1974, 1976) combined a self-paced digit naming task with a paced tapping task. The complexity of both tasks was varied and all tasks were carried out singly and in combination. When difficult tapping was paired with simple digit naming, performance in both tasks declined in comparison with their single task levels. However, no further decline of tapping was found when digit naming was made more complex. Similar problems with the "sensitivity" of secondary task performance to variations in primary task complexity have been reported by Israel, Wickens, Chesney, and Donchin (1980) and by Whitaker (1979).

Possible explanations for these problems with secondary task measures include changes in allocation policy from one condition to the other and the operation of non-resource related factors. Non-resource related factors may especially play a role when comparing single and dual task conditions. For example, there may be "structural interference" or the "creation of a new whole" in dual task performance (Kahneman, 1973; Duncan, 1979). The occurrence of such factors renders mere comparison between single and dual task performance dubious, at least as far as conclusions about resource consumption are concerned. Instead, variation in dual task conditions and instructions about priority levels should be compared. Single task measures are only relevant to augment interpretation of dual task trends.

This position is formulated most clearly in the microeconomic theory of resources and has led to the development of the Performance Operating Characteristic (POC) methodology in which both task variables and emphasis are jointly manipulated (Norman & Bobrow, 1975; Navon & Gopher, 1979, 1980).

A POC is a curve depicting all possible dual task performance combinations arising from splitting a common pool of resources among concurrently performed tasks. In other words, it is a performance-performance function which describes the improvement of performance on one task, due to resources released by lowering performance on the other task with which it is time-shared. In this approach the consequence of competition between tasks for common resources is presented as a curve rather than a single point. It is a clear manifestation of the significance and potency attributed in microeconomic resource theory to attention control and allocation policy.

The technique is illustrated in Figure 1, which depicts a derivation of dual task POCs from single task performance-resource functions. Also depicted are the effects of increasing the difficulty of task Y. It should be noted that a POC curve depicts only the consequences of a changing resource allocation policy assuming that all subject-task parameters of both tasks remain constant. When these parameters are changed, so is the efficiency of resources, i.e., the contribution of a unit resource to the level of performance, which in turn requires the construction of a new POC.

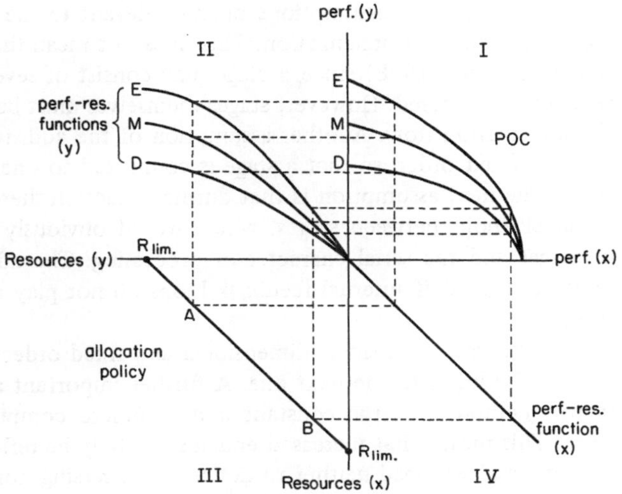

Figure 1. A hypothetical example of a family of performance operating characteristics describing dual task performance of tasks X and Y with manipulation task priorities (allocation policy) and task difficulty. Difficulty of Y is varied at three levels: Easy (*E*), Medium (*M*) and Difficult (*D*), task X is unchanged. Quadrate I depicts the three POCs resulting from the combined performance of task X with the three variants of task Y. Quadrates II, III and IV illustrate how the POCs of joint performance are related to the performance-resource function of each task (Quadrate II for Task Y and Quadrate IV for X). Points A and B in quadrate III show how two different priority levels would affect joint performance. (This illustration was proposed by Arie Melnik of the Department of Economics, University of Haifa.)

This means a family of POCs that together reflect concurrent performance at different levels of a task variable (Figure 1).

Empirical POCs can be obtained by inducing subjects to change the relative emphasis on performing either task by means of payoff or instructions (Norman & Bobrow, 1975, 1976; Navon & Gopher, 1979). Since this is a relatively recent approach, there are only a limited number of experimental studies in the literature. Yet there are some worthwhile examples (Sperling & Melchner, 1978; Gopher & Navon, 1980; Gopher, Brickner, & Navon, 1982; Gopher, 1981; Brickner & Gopher, Note 1).

2.3 Basic Assumptions

The interpretative schemes as embodied in the additive factor and POC techniques are based upon implicit "theories" about human performance. Hence inferences on the basis of either method are only justified as long as the implicit "theory" is valid. In the present section the assumptions will be reviewed and compared.

First, there are assumptions mainly relevant to the additive factor logic. One concerns sequential organization. This does not mean that all processes must occur serially (Posner, 1978) since a stage may consist of several processes that are not sequentially ordered. However, stages themselves must be serial in order to describe the information flow. Another assumption of the additive factor method concerns a single fixed order, so that a stage is connected to one and only one subsequent stage. The next assumption is that during a reaction there are no effects of internal feedback loops between stages, which would obviously abolish the independence of stages and the serial character of processing. The additive factor method can only be applied if internal feedback loops do not play a role in the experimental task.

In the case of a single dimensional and fixed order sequence the output of a stage is the input to the next one. A further important assumption of the additive factor method concerns constant and complete computational analysis in each stage. This means that increased emands will only be reflected in increased processing time and will lead neither to deficient processing nor to premature transfer of information to the next stage. Only in the case of constant stage output, it follows that the differences in demands on stages are fully manifest in processing duration. If this assumption is weakened, relations between effects of experimental variables on reaction time become immediately multi-interpretable (e.g., McClelland, 1979).

In contrast, the microeconomic resource theory as developed by Navon and Gopher (1979, 1980) is relatively indifferent to the organizational structure and output quality of processing mechanisms. The important factor is the required level of activation to achieve a certain performance level, irrespective of the arrangement of the processing mechanisms or the state of their output. The critical question concerns the amount of resources made available to task performance and the efficiency of these resources (Figure 1). Therefore, response quality and response time are equally valued measures.

Two additional very important assumptions of linear stage models are that subjects are assumed to dedicate their resources fully to performance of the experimental task and that the available resource volume is fixed unless intentionally manipulated. In other words, allocation is invariant and maximal unless intentionally varied and investigated. Only through this assumption is it possible to interpret effects of task variables on CRT – means as well as higher moments – in terms of the structure and efficiency of processing stages.

Full resource dedication and a fixed resource volume are also mandatory when applying the POC technique. Interpretation of performance operating characteristics in terms of resource scarcity is only possible under the assumption that all relevant resources are consumed and that their volume is fixed (Navon & Gopher, 1979). It is interesting to note that the microeconomical theory and Kahneman's (1973) effort theory deviate in this regard. It is well-known that Kahneman assumed elasticity of capacity. It grows or shrinks as a function of task demands. The problem is that this notion raises serious doubts about behavioural measures of resource consumption. On the one hand resource allocation determines the quality of performance, but on the other hand performance demands affect the availability of resources. Hence, behavioural measures become basically unreliable.

A final assumption that is also clearly shared by additive factor and POC methodology concerns process invariance. In the additive factor logic this means that, if the effects of the two variables have a certain mutual relation, i.e., add or interact, and a third variable is added that affects similar stages as the first two variables, the mutual relation between the first two variables should not change. In other words, addition of a third variable should not lead to a different picture of stages, as inferred from the relation between the first two variables. If this were usual, the applicability of the additive factor logic would be seriously undermined, since it would suggest that the structure of stages depends on the *total* of variables involved. The requirement of stage robustness is in fact a major criterion for deciding whether the logic is properly applied. In the POC methodology process invariance means that two tasks do not lose their independence when carried out together and that subjects do not change their basic strategies in performing each individual task when task variables are manipulated, i.e., different dual task priority combinations *only* reflect changes in the amounts of allocated resources. If there is either task integration or a major change in task strategy, effects of priority changes are hard to interpret. From the assumption of process invariance it follows that resource demands of joint dual task performance are the additive sum of their separate task demands with no extra costs or benefits for concurrent performance which are unrelated to resources.

There are two more assumptions that are only relevant to the POC approach, since they both relate to dual task performance. They concern the sensitivity of performance to resource investment and the ability of the performer to control the allocation of his resources.

Performance is assumed to be a monotonic non-decreasing function of resource investment until it reaches the limits of a performance scale or a data limitation, so that further resource investment does not yield improvement of performance (Norman & Bobrow, 1975). The slope of the performance-resource function at any given point represents the sensitivity of performance to the amount of invested resources (Figure 1). If the marginal efficiency of resources is constant on both tasks, then resources removed from one task will yield a fixed rate of improvement in the other and we may expect a linear POC. If the marginal efficiency of resources decreases, the performance-resource function and the resultant POC will negatively accelerate, show increased curvature until it reaches an asymptote at the data limited region, where the marginal efficiency of resources drops to zero. Performance is assumed to be a continuous function of resources through the performance sensitive region, such that every unit of resources released by decreasing performance on one task may be used to increase performance on the other task. Furthermore, subjects are assumed to be able to manage their resources and allocate them in different shares among the concurrently performed tasks.

If these assumptions do not hold, i.e., if performance cannot benefit from each unit of allocated resources or if subjects cannot carry out a fine adjustment of resource allocation, POCs will show step changes or discontinuities. Some resources may remain idle at various levels so that the full capacity assumption is violated.

The "theory" implied in the POC technique concerns resource allocation in terms of distribution of resource volume over different simultaneously performed tasks. A shift in resource allocation causes a *quantitative* change in the amount of work carried out on each task. This very clearly excludes the notion of resource allocation in the sense of bringing about *qualitative* changes in the way of performing a task or a combination of tasks. The emphasis on quantitative changes in allocated resource volume versus qualitative strategical shifts is the basic difference between the POC model and Rabbitt's (1979) views about qualitative strategical changes in performing tasks when resources are redistributed between tasks or when resources are otherwise affected as in the case of environmental stress. Rabbitt's definition of resources encompasses all the acquired knowledge of an individual about a specific environment. He further assumes frequent shifts in the ways in which this knowledge is used to meet task demands.

At first sight, Rabbitt's view seems very attractive since it operates on a minimum of assumptions. In fact it does not need any of the assumptions required for either the linear stage or POC techniques. This means that it is not limited to situations where these assumptions apply and that it can be used as a starting point for any experiment. However, it is a framework without constraints and, hence, without predictive power. Once one allows for *qualitative* changes when executing a task, any result fits the model as another "qualitative change". It is quite hard to formulate meaningful constraints in this way.

3 Evaluation and Relative Merits

3.1 The Scope of the Models

It is evident that the additive factor logic as well as the POC methodology are based upon a considerable number of strong assumptions. Comparatively, the assumptions of the additive factor logic are more restrictive. This is the price paid for allowing strong and relatively straightforward interpretation of simple time measures. It should be noted that the stronger the assumptions, the easier the interpretation but also the more limited the realm where a conceptual framework can be properly applied. In this sense the additive factor logic is more restricted than the POC methodology which, in turn, is less widely applicable than a resource strategy notion. Thus, at one extreme there are conceptual models with strong assumptions and, therefore, with limited applicability. At the other extreme there are models which are so loose that they are almost always correct, whatever the experimental outcome. They are widely applicable, but hard to submit to a conclusive scientific test.

It is easy to illustrate the limited validity of the additive factor logic, even within the analysis of CRT, it cherished experimental paradigm. For example,

Rabbitt's (1967) findings concerning very rapid error detection times strongly suggests the presence of internal feedback loops in choice reactions which are not allowed by the additive factor logic. The results on processing intense signals (Sanders, 1977, 1980; v.d. Molen & Keuss, 1981) suggest parallel processing loops which violate the assumption of unidimensionality of the choice reaction process. The constant stage output assumption is seriously challenged by alternative linear stage models, suggesting overlapping stages (Taylor, 1976) or processes in cascade (McClelland, 1979).

Despite these problems, it is striking that the additive factor logic still does a fair job in summarizing the results of a considerable number of choice reaction experiments. The observed relations between most experimental variables are fairly robust. As briefly mentioned (p. 239) robustness is the main criterion for deciding whether the additive factor method is properly applied. Robustness means that observed relations remain the same, irrespective of the state of confounded variables. Thus, if the effects of two variables are additive, they should remain additive when studied in combination with the effect of a third variable. To claim process invariance a fair degree of robustness is required, and in fact observed (Sanders, 1980). It provides the additive factor method with some post hoc justification and serves as an aid in establishing the conditions to which it applies.

The POC methodology also has clear limits. Thus, Neisser (1976) has argued that tasks are integrated under dual task conditions and that such an integration may lead to a reduced demand for resources. Hirst, Spelke, Reaves, Caharack, and Neisser (1980) have conducted several experiments with simultaneous reading and writing to show how such integration may develop even without automaticity. Concurrence costs in dual task performance have been demonstrated in many experiments (e.g., Gopher & North, 1974; Wickens & Gopher, 1977; Kantowitz & Knight, 1974, 1976; Wickens, Derrick, Gill, & Donchin, Note 2; Noble, Sanders, & Trumbo, 1981). Concurrence benefits have been found in studies on word superiority (Johnson & McClelland, 1974; Reicher, 1969), or on configural superiority in part-whole studies (Pomerantz, Sager, & Stoever, 1977). In addition, problems of attention control have been demonstrated by Brickner and Gopher (Note 1) and by Gopher (1981) and strategic changes that violate process invariance are discussed by Rabbitt (1979). Nevertheless, the fact that violations of assumptions can occur and may even be common does not reduce the importance of spelling out the assumptions and identifying the nature of the violations and the instances at which they occur. The robustness of results obtained in several experimental situations using the same variables may enable one to assign proper weights to the consequences of different assumptions. Robustness of results may also allow some relaxation of certain constraints, while becoming more apprehensive about others.

The question is obviously not which model is "true" or "not true" but rather how to study the generality of results across models. Answering such questions requires communication lines between frameworks. Two possible approaches will be briefly considered in the remainder of this paper. First, we shall examine convergence of theoretical notions in the sense that suggestions derived from the one framework serve indeed as hypotheses for the other. Second, we will discuss attempts towards developing a common experimental methodology.

3.2 Convergence of Stage and Resource Thinking

Stage and resource models have developed in relative isolation. Hence attempts to make connecting hypotheses have been rare and mutual incorporation of their main constructs has been slow. Recent attempts to remedy this weakness have led to the development of the multiple resource approach and of a cognitive-energetic stage model.

Multiple Resources. The concept of multiple resources has emerged from failure of the main prediction of the single resource model that, irrespective of their characteristics, all tasks compete with one another when demands for resources increase. This prediction has been repeatedly refuted in dual task experiments. The results show that some tasks interfere with one type of task but not with another, while others are equally affected by both types of tasks. For example, mental arithmetic is rarely impaired when performed together with pursuit tracking but is severely impaired when paired with choice reactions to visually presented digits. In contrast, the visual choice reaction task was equally impaired when performed concurrently with either mental arithmetic or with pursuit tracking (see Williges & Weirwille, 1979; Ogden, Levine, & Eisner, 1979; Rolfe, 1971). Even more relevant are the findings that some manipulations of task variables within the same pairs of tasks affect joint performance while others affect performance on one task only (Gopher & Navon, 1980; Gopher et al., 1982; Wickens & Kessel, 1980). These results are difficult to interpret for a single resource model but can easily be accounted for by the hypothesis that the human processing system is composed of several relatively independent resources and that tasks compete with one another only to the extent that they overlap in their demands for common resources (Navon & Gopher, 1979, 1980; Wickens, 1980).

It is clear that attribution of separate energetic resources to specific structure provides a natural bridge between resource and stage models. As suggested by Wickens (1980) a group of processes served by a common resource might be identical to or at least meaningfully related to a processing stage. Experimental work to identify the nature and number of separate resources is still in its infancy. Findings from recent studies seem to converge on the existence of at least two kinds of resource: one associated with perceptual and another with motor demands (Gopher & Navon, 1980; Gopher et al., 1982; Israel, Chesney, Wickens, & Donchin, 1980; Wickens & Kessel, 1980).

A Cognitive-Energetic Stage Model. An attempt to incorporate energetic concepts in stage thinking has emerged from the study of the effects of stresses on choice reaction tasks and interpreted by way of the additive factor logic (Frowein, 1981;

Sanders, Wijnen, & v. Arkel, 1982). There, the conclusion is reached that effects of stress such as sleep-loss, time-on-task and psychoactive drugs on CRT are selective rather than general. For instance, the effect of amphetamine was only found to interact with those of motor variables, while the effect of a barbiturate only interacted with that of perceptual feature extraction. Together the stress results suggest the existence of at least two energetic mechanisms, the operations of which are coordinated by a third. This suggestion is in line with Pribram and McGuinness' (1975) neurophysiological theory on the control of attention, which assumes two "basal" energetic mechanisms, one related to input (arousal), one to output (activation), and in addition a coordinating "effort" mechanism. The effort mechanism has the task to secure the optimal state and cooperation of the basal mechanisms and, in addition, has a direct relation to the operation of response choice in CRT tasks. Whether the effort mechanism itself is in an optimal state depends on evaluation of the state of adaptation of the organism to environmental demands. In other words the investment of effort depends to a large extent on motivation. The cognitive-energetic stage model (Sanders, 1983) is illustrated in Figure 2. There is a striking resemblance between suggestions from the multiple resource approach and those of the cognitive-energetic stage model. Thus, the association of arousal and activation with, respectively, early and late processing stages, agrees with the multiple resource distinction between perceptual and motor related resources. In addition, the introduction of a voluntarily controlled effort source together with an evaluation mechanism are not much different from the "resource allocation" mechanism denoted by the resource approach proponents. Hence, at least on the conceptual level, the two frameworks show a promising convergence in their views on the human processing system.

An important feature of the cognitive-energetic model is that it provides a clear hypothesis about the relation between primarily basal energetic mechanisms (arousal, activation) and coordinating control (effort). In many resource models this relation is not considered. In the original conception of Kahneman (1973)

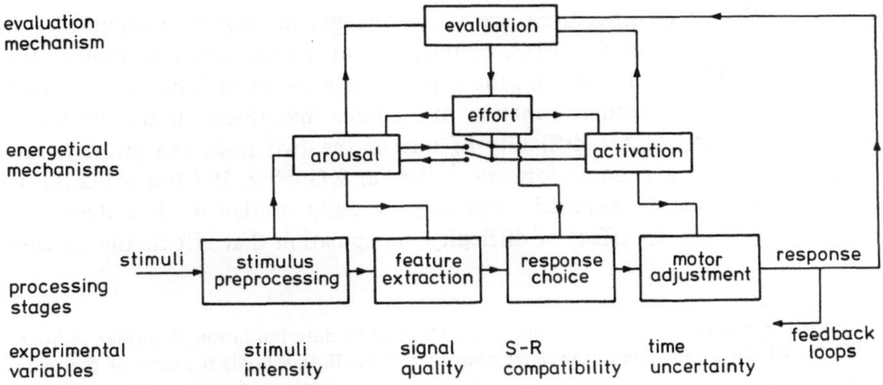

Figure 2. Block diagram of the cognitive-energetic stage model (from Sanders, 1983).

arousal and effort were not too different, except that arousal was affected by "what happens to the subject", i.e., his basal state, while effort was not. Yet at a low level of arousal, the effort volume decreased. In later resource models (Norman & Bobrow, 1975; Navon & Gopher, 1979; Wickens, 1980), the relation between voluntary and basal energetic components was hardly mentioned.

The similarity between the results obtained through the multiple resource and cognitive-energetic stage approach should not mislead one to think that the differences between frameworks have been eliminated. Since the approaches are largely complementary this would also not be the aim. Yet the emergence of conceptual agreement is highly relevant: it raises the hope that differences between frameworks do not result from basic theoretical disagreement but rather from different constraints imposed by their methodologies. As argued in Section 3.1 the question is whether similar types of processing stages or resources emerge if the basic assumptions of a framework are relaxed or if different methodologies are employed.

The ability to carry the conclusions of this analysis beyond the conceptual level at which it has so far been conducted greatly depends on the ability of the two frameworks to co-ordinate their methodologies. This is our next topic of discussion.

3.3 Addivitity and Interactions

Central to the derivation of processing stages from the results of additive factor experiments and of demand compositions of resources from dual-task performance is the argument that interaction between experimental variables indicate their association with the same processing mechanism, while additive effects imply that variables are related to separate mechanisms (Sternberg, 1969; Navon & Gopher, 1980). To consider cross-experimentation of the two approaches it is important to examine the basic rationale for this argument.

In resource theory the additivity-interaction issue is related to the assumed relationship between manipulation of task variables and resource efficiency. Task difficulty is defined as equivalent to average resource efficiency (Navon & Gopher, 1979). Difficulty manipulations are those changes in task parameters that also affect the efficiency of resources, or the slope of the performance-resource function (Fig. 1). The easier the task, the larger the average slope. Since performance is assumed to be a monotonic function of resource investment, it follows that the joint manipulation of the difficulty of one of the two tasks and priority should have interactive effects on performance[1] (Navon & Gopher, 1980) (see Fig. 1). This implies that, when time-shared tasks only partially overlap in their demand for common resources, an effect of difficulty manipulation that affects the efficiency

1 This is if performance does not approach its ceiling or data limitations (Norman & Bobrow, 1975). Within a multiple resource framework this prediction holds for each of the separate capacities.

of a resource, shared by both tasks, should interact with the effect of priority level (or amount of allocated resources). If, in contrast, manipulation of difficulty deploys a "new" resource, then the joint effects of difficulty and priority change should be additive.

In the additive factor methodology, interpretation of interactions and additivities in the effects of task variables is exclusively limited to processing duration in single task conditions[2]. However, once it is accepted that linear stage models do not refute the notion of activation levels, but assume them to be constant and optimal, changes in processing duration can be linked to the resource efficiency notion. For instance, whatever the units of the feature extraction stage may be, degrading stimulus quality decreases the efficiency of these units and increases processing time in order to obtain constant stage output. Similar decrements occur in response choice efficiency when S-R compatibility is degraded. The joint effects of these variables on performance is additive because they affect the efficiency of different units. When, however, two variables affect units of the same stage, their effects on unit efficiency interact in much the same way as in performance-resource functions. If in addition prolongation of processing duration is assumed to be a monotonic function of stage efficiency, the similarity between the two frameworks in the logic of interpretation is apparent.

We can therefore conclude that the basic rationale for deriving theoretical constructs from empirical results is very similar in linear stage models and in the multiple resource approach. At the same time, one should bear in mind that such a derivation is only valid within the constraints imposed by the respective methods. Hence, interpretation of interactions and additivity from additive factor experiments is valid only as long as the assumption of process invariance and constant stage output is not violated. Interpretation of interference patterns obtained in dual-task performance in terms of resource demands is only meaningful to the extent that structural or concurrence cost effects and qualitative changes in response strategy have been ruled out. As we shall demonstrate, these constraints are the major obstacle in combining additive factor logic and dual-task methodology in actual experiments.

---

2 Furthermore, additive factor theorists use real physical time units, so that additivity and interaction are directly interpretable in terms of these units (Sternberg, 1969). Interpretation of experimental effects under resource theory is complicated by the fact that it allows multiplicity of performance measures, and scale transformation of raw performance. Resource theorists, therefore, often face the additional initial requirement to justify their unit and scale of measurement before additivities and interactions can be interpreted in the framework of the theory predictions. The requirement to justify the selection of performance measures is not unique to resource models and is, naturally, separate from predictions based upon the relationship between theoretical constructs.

3.4 Conducting Convergent Experiments

Three recent sets of studies have attempted to investigate resource notions in additive factor studies. They all have in common that dual task demand – or sometimes single versus dual task performance – was considered as a task variable in an additive factor experiment. First, Logan (1978, 1979, 1980) carried out a series of experiments in which subjects were given a list of items for later serial recall. Between presentation and recall a rehearsal period was introduced during which a choice reaction trial was carried out. The idea underlying this research was to use memory load (4 vs. 8 items) as an additional additive factor variable, determine its pattern of additivity and interaction with other CRT variables and, from there, decide which stages are affected by rehearsal. More important, rehearsal was considered as a standard controlled resource requiring process, so that processes in stages that show additive effects with the memory load variable were supposed to run off automatically.

Second Whitaker (1979) combined tracking in two difficulty versions and choice reactions in which S-R compatibility and number of alternatives were varied to determine whether effects of dual task performance are specific or general. Third, Wickens et al. (Note 2) conducted a series of experiments in which two variants of a compensatory tracking task were combined with four variants of a letter classification task. Tracking difficulty was manipulated by changing the control dynamic (velocity vs. acceleration) or by increasing the frequency bandwidth of the disturbance function that had to be corrected. Among the variables in the letter task, there were display quality (perceptual load) and response complexity (response load). The effects of each manipulation on letter classification time was studied under single task conditions and in joint performance with tracking. The aim was to reveal the locus of the load in the different variants of tracking by applying the additive factor method.

Wickens et al. (Note 2) explained the rationale of their combined design arguing that: "When the CRT task is imposed as a secondary task, a manipulation that produces a greater prolongation of CRT in the presence of the difficult as opposed to the easy version of the primary (tracking) task – i.e., a positive interaction between the primary and secondary task difficulty variables – can be interpreted as overlapping in resource demands with the difficult primary task. Similarly, if the secondary task manipulation is amplified in the presence of the primary task, relative to its absence (single task control), then the primary task is assumed to demand the resources consumed by the secondary task manipulation." This rationale can be generalized to fit the work of Logan and Whitaker since both also employed a single to dual task comparison as a two level factor in their additive factor analysis. The first question is whether this comparison constitutes an acceptable variable.

According to the basic axioms of the additive factor method only those variables qualify which prolong or reduce the time taken by the stages. This is a reasonable assumption for variables such as signal quality, S-R compatibility, etc., but at least dubious for variables such as concurrent tracking or simultaneous memory load. There are many reasons to believe that constant stage output as well as process invariance can be violated in the transition from single to dual task conditions, i.e., the number and combination of stages or the type and amount of resources may change as well as the output quality in letter classification. This may be damaging to additive factors as well as to resource interpretations. Even if such changes do not occur, interpretation of results greatly depends on the model of interference between the tasks, which is not explicity discussed by either Logan, Whitaker or Wickens.

One possible model is that primary and secondary tasks cannot be simultaneously performed. Instead, the primary task is interrupted at some convenient moment to enable a choice reaction as if it were a single task. This would mean that complexity of the primary task will generally have a main effect on CRT, since it is reasonable to assume that an easy primary task can be more easily interrupted than a difficult one. Yet, once processing of the choice reaction has started, it runs off unhampered so that no interactions are found between CRT variables and primary task difficulty. This could explain Logan's (1978) finding of dominant additive effects of CRT variables and memory task difficulty. His conclusion, then, is unwarranted that variables with additive effects to that of task difficulty are related to automatic component stages.

A second possible model is equal to the previous one with the exception that the length of the period that the primary task can be interrupted is limited and more so as the primary task load is heavier. In that case CRT processes may run off undisturbed if they take either only a short time or a longer time but the concurrent task load is low. Thus there arises only a problem for the choice reactions in the case of a high concurrent task load *and* a reaction time exceeding a certain duration limit. As a consequence, the effects of choice reaction and primary task variables will interact but in a fairly aspecific way: It does not matter *which* CRT variable causes the total RT to exceed the limit. As long as the limit is exceeded, interactions will show up. This model can explain Whitaker's (1979) major finding that a compatible reaction time task was not affected when paired with a tracking task while an incompatible reaction task showed a step increment in response time when tracking was added. It may also account for Logan's (1979) observation that the effects of rehearsal load were sometimes additive to those of variables affecting a particular stage, while interactive effects were observed on other occasions. It can also accommodate his evidence for interactions with practice and, in particular, his finding that interactions tend to disappear after practice in the double task. Apparently, subjects learn to time the interval of interrupting the primary task. None of these results is decisive with regard to his inferences about automatic or controlled processing in component stages.

Such an inference is only permitted when a third model is assumed, which suggests distribution of resources over both tasks *at any specific moment* in time. In that case a dual task or a more difficult dual task can selectively affect the energy supply to the active processes involved in the CRT task so that the time taken to complete these processes is prolonged. This is probably the model that resource theorists such as Logan and Wickens would promote. Yet, the two previous interpretations cannot be ignored because structural interference and concurrence costs are most likely to occur in the transition from single to dual task conditions (Duncan, 1979). Moreover, none of these experiments has changed task priorities or manipulated resource allocation in any other way, so that interpretation of their results cannot exclude the possibility of concomitant shifts between conditions in allocation policy.

The results of Wickens et al. (Note 2) are particularly bothersome in this regard. Their experiments showed that, under dual task conditions, a change in the control order of tracking from a velocity to an acceleration controller interacted positively with the effects of degrading the quality of letter presentations (perceptual load) on the latency of letter classification. In contrast, incrementing the response complexity (response load) of the letter task had only an additive effect on those of the difficulty manipulation of tracking, irrespective of the nature of this manipulation (control order or frequency or target movements). Based upon their argument that a positive interaction with primary task difficulty implies competition for a common resource, the authors conclude that the locus of load in the manipulation of control order appears to reside in the perceptual or central processing stages and not in the response stage. They also argue that the tracking task target band-width is not associated with either of these stages.

The validity of the above interpretation becomes questionable in the light of other aspects of their data. One finding was that the addition of the easy tracking version to the performance of letter classification tempered the effects of the letter display quality manipulation. That is, display quality had stronger effects on the latency of letter classification when this task was performed alone than under dual task conditions with easy tracking. If tracking performance is assumed to compete with letter classification for the same processing resources, such under additivity is hard to comprehend. Furthermore, how can underadditivity occur with the easy version of tracking and be changed to a positive interaction with the difficult version of this task? This reversal raises the strong possibility of a change in strategy or violation of the processing invariance requirement. Another puzzling outcome was a small but significant improvement in tracking performance when the difficulty of the jointly performed letter task was increased. Tracking also improved in the transition from single to dual task conditions. These findings suggest a change of task emphasis, or resource allocation policy effects, as viable alternatives to the locus of load interpretation proposed by Wickens and his colleagues.

3.5 Implications for Research

From the analysis of the three examples it becomes fairly clear that applications of the additive factor logic and POC methodology are restricted to their parent research paradigms, not so much because of their theoretical basis but because of the ability to interpret data within the constraints imposed by the methodology itself. This conclusion is hardly surprising when one considers their different starting points. It is nevertheless discouraging, because it raises doubts regarding the generalization of the theoretical constructs proposed by either approach. Viability of such constructs would have been much strengthened if they could have been derived from two different methodologies. This last statement may lead to a way of conducting combined experiments, which may enable us to end this discussion on a more positive tone.

Instead of efforts to develop one combined paradigm, it might be better to conduct a full additive factor experiment together with a separate POC based dual task experiment, back-to-back on the same pair of tasks. Stages derived from one experiment will be tested by the other and vice versa. For example, a back-to-back experiment on the same pair of tasks employed by Wickens may be conducted according to the following procedure: First, an additive factor experiment is conducted on the letter classification task with manipulation of the same task variables as used by Wickens. The aim of such an experiment would be to establish an additive factor stage structure of the letter classification task. Subsequently, a POC based dual task experiment is run in which letter classification and tracking performance are combined in order to investigate the demand composition of the letter classification task, viewed from a resource framework. In each session of this experiment a fixed-difficulty tracking task (second order control or high frequency band-width) should be paired with different versions of the letter classification task under several dual-task priority conditions. Agreeement on corresponding stage and resource structures developed in such a study will provide much stronger support and generality to the building boxes of the human processing system. Studies conducted according to this logic may be laborious and more complex to design, but we believe that their merit may outweight these difficulties.

4 Conclusions

Various conclusions have been drawn in earlier sections of this chapter. They can be summarized in four statements. The first concerns the question whether stage and resource models are merely opposing or whether they address different sets of problems. The answer strongly tends to the latter. Throughout this paper, stage and resource theory have proven to have different interests, different methodologies and to be equipped for different types of problems. In a way the frameworks can be considered as modern versions of structuralism and functionalism with the important addition that they have now fairly well defined methodological and empirical tools, as well as criteria for their application.

The second, which directly follows from the first, is that in order to be fruitful the frameworks must recognize their limited area of validity. Neither of them is "the" universal psychological model. Yet the fact that this is not the case does not mean that they should be abandoned. One of the problems of present-day psychology is that frameworks are too easily abandoned without sufficient knowledge about their area of application or without a sufficiently convincing higher order alternative. At various places in this paper the warning has been issued not to consider loose and unconstrained strategical notions as such an alternative. This is despite the attractiveness of seemingly avoiding the trodden path of S-R paradigms and of gaining ecological validity.

The third conclusion is that one should be extremely careful in attempting a combination of experimental methodologies of both frameworks for the simple reason that one is liable to violate assumptions which are essential for interpretation. It was argued that it seems more promising to look for converging evidence in areas where the frameworks overlap. One of these areas concerns the types and nature of energety resources. It would be relevant to use the outcome of linear stage models as hypotheses in POC studies and vice versa.

The fourth major conclusion is that models with strict constraints have the advantage of strong interpretative power but run the risk of being quite limited in their realm of application. We argue that it is preferable to establish the limits of the validity of more constrained models and subsequently study the generality of their conclusions beyond their realm of applicability. To put it more concretely in the context of stage and resource theory, the question is whether in situations where strategical effects dominate – and where consequently the additive factor assumptions do not apply – the basic stage structure still constitutes an essential constraint for strategical manipulations. Again, in situations where strategical shifts within tasks change the nature of the operations – thus violating the process invariance assumption of the POC methodology – the question remains to what extent conclusions drawn on the basis of the POC technique are relevant. If the conclusions are irrelevant, the POC technique loses much of its attraction. Similarly, if the stage structures cannot be shown to make sense beyond the realm of additive factor application and if they do not in some way refer to more complex processing structures, they do not advance performance theory. Various recent authors would strongly endorse these last statements (e.g., Shaffer, 1980), while we tend to suspend judgment until the limits of additive factor and POC applicability have been better explored and until it has been determined to what extent additive factors and POC results provide useful hypotheses for each other as well as anchors for broader theories.

Reference Notes

1. Brickner, M., & Gopher, D. *Improving time-sharing performance by enhancing voluntary control on processing resources.* Technion, Israel Institute of Technology, Research Center for Work Safety and Human Engineering, AFOSR-77-3131C, HEIS-81-3, 1981.
2. Wickens, C.D., Derrick, W.A., Gill, R., & Donchin, E. *The processing demands of higher order manual control: Convergent evidence for event related brain potentials and additive factors methodology.* Dept. of Psychology, University of Illinois, Urbana Champaign, IL 61820, 1981.

References

Bahrick, H.P., Noble, M.E., & Fitts, P.M. Extra-task performance as a measure of learning a primary task. *Journal of Experimental Psychology,* 1954, *48,* 298–302.

Broadbent, D.E. *Perception and communication.* London: Pergamon Press, 1958.

Broadbent, D.E. *Decision and stress.* London: Academic Press, 1971.

Cherry, E.C. Some experiments on recognition of speech with one and two ears. *Journal of Acoustical Society of America,* 1953, *25,* 975–979.

Donders, F.C. On the speed of mental processes. Translation by W.G. Koster. *Acta Psychologica,* 1868/1969, *30,* 412–431.

Duncan, J. Divided attention: The whole is more than the sum of its parts. *Journal of Experimental Psychology: Human Perception and Performance,* 1979, *5,* 216–228.

Frowein, H.W. Selective effects of barbiturate and amphetamine on information processing and response execution. *Acta Psychologica,* 1981, *47,* 105–115.

Gopher, D. Performance trade-offs under time-sharing conditions: The ability of human operators to release resources by lowering their standards of performance. *Proceedings of the International Conference on Cybernetics and Society,* 1981, 609–614. IEEE 81 DH 1683-2. Library of Congress: 74-170870, Iss. No. 360-8913.

Gopher, D., Brickner, M., & Navon, D. Different difficulty manipulations interact differently with task emphasis: Evidence for multiple resources. *Journal of Experimental Psychology: Human Perception and Performance,* 1982, *8,* 146–157.

Gopher, D., & Navon, D. How is performance limited: Testing the notion of central capacity. *Acta Psychologica,* 1980, *46,* 161–180.

Gopher, D., & North, R.A. The measurement of capacity limitation through single and dual-task performance. *Proceedings of the 18th Annual Meeting of the Human Factors Society,* pp. 480–485. Santa Monica, CA: The Human Factors Society, 1974.

Hebb, D.O. Drives and the conceptual nervous system. *Psychological Review,* 1955, *62,* 243–254.

Hirst, W., Spelke, E.S., Reaves, C.C., Caharack, G., & Neisser, U. Dividing attention without alternation or automaticity. *Journal of Experimental Psychology: General,* 1980, *109,* 98–117.

Hull, C.L. *Principles of behavior.* New York: Appleton Century, 1943.

Hunt, E. Mechanics of verbal ability. *Psychological Review,* 1978, *85,* 109–130.

Israel, J.B., Chesney, G.L., Wickens, C.D., & Donchin, E. P_{300} and tracking difficulty: Evidence for multiple resources in dual-task performance. *Psychophysiology,* 1980, *17,* 259–273.

Israel, J.B., Wickens, C.D., Chesney, G.L., & Donchin, E. The event related brain potential as an index of display monitoring workload. *Human Factors,* 1980, *22,* 212–214.

Jex, H. Two applications of a critical instability task to secondary workoad research. *IEEE Transactions on Human Factors in Electronics,* 1969, *HFE-8,* 279–282.

Johnson, J.C., & McClelland, J.L. Perception of letters in words: Seek not and ye shall find. *Science,* 1974, *184,* 1192–1194.

Kahneman, D. *Attention and effort.* Englewood Cliffs, NJ: Prentice Hall, 1973.

Kantowitz, B.H., & Knight, J.L. Testing tapping time-sharing. *Journal of Experimental Psychology,* 1974, *103,* 331–336.

Kantowitz, B.H., & Knight, J.L. Testing tapping time sharing. II: Auditory secondary tasks. *Acta Psychologica,* 1976, *40,* 343–362.

Logan, G.D. Attention in character classification: Evidence for the automaticity of component stages. *Journal of Experimental Psychology: General,* 1978, *107,* 32–63.

Logan, G.D. On the use of a concurrent memory load to measure attention and automaticity. *Journal of Experimental Psychology: Human Perception and Performance,* 1979, *5,* 189–207.

Logan, G.D. Attention and automaticity in Stroop and priming tasks: Theory and data. *Cognitive Psychology*, 1980, *12*, 523–533.

McClelland, J.L. On the time relations of mental processes: An examination of systems of processes in cascade. *Psychological Review*, 1979, *86*, 287–330.

Michon, J.A. Tapping regularity as a measure of perceptual motor load. *Ergonomics*, 1966, *9*, 401–412.

Molen, M.W. v.d., & Keuss, P.J.G. Response selection and the processing of auditory intensity. *Quarterly Journal of Experimental Psychology*, 1981, *33*, 177–184.

Moray, N. Where is capacity limited? A survey and a model. *Acta Psychologica*, 1967, *27*, 84–92.

Navon, D., & Gopher, D. On the economy of the human-processing system. *Psychological Review*, 1979, *86*, 214–255.

Navon, D., & Gopher, D. Task difficulty, resources and dual-task performance. In R.S. Nickerson (Ed.), *Attention and Performance 8*. Hillsdale, NJ: Erlbaum, 1980.

Neisser, U. *Cognition and reality. Principles and implications of cognitive psychology.* San Francisco: Freeman, 1976.

Noble, M.E., Sanders, A.F., & Trumbo, D.A. Concurrence costs in double stimulation tasks. *Acta Psychologica*, 1981, *49*, 141–158.

Norman, D.A., Bobrow, D.J. On data-limited and resource-limited processes. *Cognitive Psychology*, 1975, *7*, 44–64.

Norman, D.A., & Bobrow, D.J. On the analysis of performance operating characterisitcs. *Psychological Review*, 1976, *83*, 508–510.

Ogden, G.D., Levine, J.M., & Eisner, E.J. Measurement of workload by secondary tasks. *Human Factors*, 1979, *5*, 529–548.

Pomerantz, J.R., Sager, L.C., & Stoever, R. Perception of words and their component parts: Some configural superiority effects. *Journal of Experimental Psychology: Human Perception and Performance*, 1977, *3*, 422–435.

Posner, M.I. *Chronometric explorations of mind.* Hillsdale, NJ: Erlbaum, 1978.

Pribram, K.H., McGuinness, D. Arousal, activation and effort in the control of attention. *Psychological Review*, 1975, *82*, 116–149.

Rabbitt, P.M.A. Time to detect errors as a function of factors affecting choice response time. *Acta Psychologica*, 1967, *27*, 131–142.

Rabbitt, P.M.A. Current paradigms and models in human information processing. In V. Hamilton & D.M. Warburton (Eds.), *Human stress and cognition.* New York: Wiley, 1979.

Reicher, G.M. Perceptual recognition as a function of meaningfulness of stimulus material. *Journal of Experimental Psychology*, 1969, *81*, 274–280.

Rolfe, J.M. The secondary task as a measure of mental load. In W.T. Singleton, R.S. Easterby, & D.E. Whitfield (Eds.), *Measurement of man at work.* London: Taylor and Francis, 1971.

Salthouse, T.A. Converging evidence for information processing stages: A comparative influence stage-analysis method. *Acta Psychologica*, 1981, *47*, 39–61.

Sanders, A.F. Structural and functional aspects of the reaction process. In S. Dornic (Ed.), *Attention and performance 6.* Hillsdale, NJ: Erlbaum, 1977.

Sanders, A.F. Some remarks on mental load. In N. Moray (Ed.), *Mental workload.* New York: Plenum, 1979.

Sanders, A.F. Stage analysis of reaction processes. In G.E. Stelmach & J. Requin (Eds.), *Tutorials in motor behavior.* Amsterdam: North-Holland, 1980.

Sanders, A.F. Ten symposia on attention and performance: Some issues and trends. In H. Bouma & D. Bouwhuis (Eds.), *Attention and performance 10.* Hillsdale, NJ: Erlbaum, 1983.

Sanders, A.F. Towards a model of stress and human performance. *Acta Psychologica*, 1983, *53*, 61–97.

Sanders, A.F., Wijnen, J.L.C., & v. Arkel, A.E. An additive factor analysis of the effects of sleep-loss on reaction processes. *Acta Psychologica*, 1982, *51*, 41–59.

Shaffer, L.H. Analysing piano performance: A study of concert pianists. In G.E. Stelmach & J. Requin (Eds.), *Tutorials in motor behavior.* Amsterdam: North-Holland, 1980.

Smith, E.E. Choice reaction time: An analysis of the major theoretical positions. *Psychological Bulletin,* 1968, *69,* 77–110.

Sperling, G., & Melchner, M.J. The attention operating characteristic: Examples from visual search. *Science,* 1978, *202,* 315–318.

Sternberg, S. On the discovery of processing stages: Some extensions of Donders' method. *Acta Psychologica,* 1969, *30,* 276–315.

Taylor, D.A. Stage analysis of reaction time. *Psychological Bulletin,* 1976, *83,* 161–191.

Treisman, A.M. Strategies and models of selective attention. *Psychological Review,* 1969, *76,* 282–299.

Welford, A.T. Single channel operation in the brain. *Acta Psychologica,* 1967, *27,* 5–22.

Whitaker, L. Dual task interference as a function of cognitive processing load. *Acta Psychologica,* 1979, *43,* 71–84.

Wickens, C.D. The effects of divided attention on information processing in tracking. *Journal Experimental Psychology: Human Perception and Performance,* 1976, *1,* 1–13.

Wickens, C.D. The structure of attentional resources. In R. Nickerson (Ed.), *Attention and performance 8.* Hillsdale, NJ: Erlbaum, 1980.

Wickens, C.D., & Gopher, D. Control theory measures of tracking as indices of attention allocation strategies. *Human Factors,* 1977, *19,* 349–356.

Wickens, C.D., & Kessel, C. Processing resource demands of failure detection in dynamic systems. *Journal of Experimental Psychology: Human Perception and Performance,* 1980, *6,* 564–577.

Williges, R.C., & Weirwille, W.W. Behavioral measures of air crew mental workload. *Human Factors,* 1979, *5,* 549–575.

Wundt, W. *Grundriss der Psychologie.* Leipzig, Engelmann, 1896.

Summary. In the literature on human information processing and performance two major conceptual frameworks can be distinguished based upon linear stage and capacity allocation notions. Both frameworks have a long history and have usually been considered as mutually exclusive. This paper analyses nature, assumptions, interests and predictive potential of current forms of both frameworks. It is concluded that to a considerable extent they are concerned with different questions and, therefore, should be regarded as largely complementary. Yet, they can render useful services to each other by way of generating hypotheses for research and tests of their limits. Finally, the paper concentrates on conceptual and empirical points of contact between the two approaches. It is argued that efforts to develop a single experimental paradigm for testing hypotheses derived from either framework should be avoided. Instead, back-to-back experiments are recommended to obtain convergent evidence.

17 Automatic Processing: A Review of Recent Findings and a Plea for an Old Theory

ODMAR NEUMANN

Contents

1 Introduction

The rediscovery of the old distinction between automatic and consciously controlled mental processes has been one of the major developments in attentional theory during the last decade. According to the "two process" approach (e.g., LaBerge & Samuels, 1974; Neumann, Note 1; Posner & Snyder, 1975a; Shiffrin & Schneider, 1977), mental operations can function in two different modes. Processes in the first mode occur as a passive consequence of stimulation and take place in a parallel, capacity-free manner, whereas processes in the second mode are controlled by the person's conscious intentions and are subject to capacity limitations. This distinction has stimulated a wealth of research, some of which has been summarized by LaBerge (1981) and Posner (1978, 1982). The present paper reviews part of these recent findings on automaticity in an attempt to answer two questions: First, what are the functional properties of automatic as opposed to non-

Cognition and Motor Processes
Ed. By W. Prinz and A.F. Sanders

automatic processes? Second, what kind of theory is best suited to explain these properties?

The main part of the paper will be devoted to the first question. I will argue that the automatic-nonautomatic distinction is valid in the sense that it captures qualitative differences and not just a difference in degree (e.g., different efficiency of resources, as suggested by Navon & Gopher, 1979). However, the functional properties of automatic processing that emerge from the data are not quite those that were originally formulated by two-process theories. After a brief review of the three main criteria of automaticity put forward by these theories (Section 2) I will argue that at least two of them need major modification:

(a) Most processes considered to be automatic are not generally free from suffering or producing *interference.* However, interference-free processing is found with well-practiced skills if certain task conditions are fulfilled. Quite another matter is the finding that mere states of readiness do not produce interference as long as they are not involved in actual processing (Section 3).
(b) Automatic processes are not independent of a person's current *intentions* and the direction of attention. However, there are processes that depend on an intention without being explicitly intended, and there are cases where a process, though depending upon an intention, does not exactly conform to it (Section 4).

Adopting these modifications means, of course, destroying the beautifully simple notion of automaticity that has been one reason for the attractiveness of the two-process approach. However, a new integration of the findings seems possible. In Sections 5 and 6 I will outline some ideas on what it could look like. The basic suggestion will be to reconsider a different notion of automaticity, which (like the current two-process approach) dates back to nineteenth century psychology, particularly to Wilhelm Wundt. It conceives automatic processing not as lacking control, but as being controlled at levels below the level of conscious awareness. This line of thinking on automaticity has been developed mainly in the context of psychomotor research, and it fits into current ideas about hierarchical control of action. In my view, it also offers a promising possibility to integrate much of the cognitive data.

2 The Concept of Automaticity in Current Two-Process Theories

The information-processing approach which began to dominate cognitive psychology in the late 1950s and the 1960s (e.g., Broadbent, 1958, 1963; Haber, 1967; Sperling, 1967; Waugh & Norman, 1965) centered around two ideas: (1) There are subsystems (channels, stores, or stages) whose built-in arrangement determines the possible pathways of information flow; (2) One of the essential features of the processing system is its limited capacity. The groundwork for the two-process approach was laid when it became clear that these assumptions provide a less complete description of human information processing than had been assumed by these early theories.

As to the first idea, it left open how the processing system is being used to accomplish a given task. There was no explicit account of the control processes that determine which of the possible routes of information flow is actually taken. Of course, some kind of control had to be tacitly assumed: The filter concept only made sense if it was taken for granted that the subject could set the filter according to instructions; the rehearsal loop had to be used by the subject in order to become effective, etc. But the corresponding control processes were not included in explicit theorizing. This shortcoming began to be recognized towards the end of the 1960s. Considerations about control appeared in Broadbent's 1971 revision of filter theory (e.g., Broadbent, 1971, pp. 11ff. and 476ff.). In memory research, the distinction between structural features of memory systems and control processes was introduced by Atkinson and Shiffrin (1968). An overview of the subsequent development of the control process idea can be found in Underwood (1978).

The assumptions about the ubiquity of capacity requirements were likewise found to need modification. In the early models, information was thought to be represented in a capacity-free manner only at the system's periphery in the form of sensory storage (i.e., before processing proper has been initiated). This view was challenged by several findings, e.g., that choice RT is independent of the number of alternatives in highly compatible or extensively practiced S-R mappings (Leonard, 1959; Mowbray & Rhoades, 1959), that target set size has practically no effect on speed of visual search in trained subjects (Neisser, 1963), or that probe RT is apparently unaffected by simultaneous stimulus encoding (Posner & Boies, 1971). Data of this kind led a number of theorists to suggest that processes such a stimulus encoding (Posner & Boies, 1971) and memory retrieval (Keele, 1973) do not demand capacity. Evidence to support the distinction between capacity-demanding and capacity-free processing was summarized by Kerr (1973).

Thus, at the beginning of the 1970s, two distinctions had emerged within the information-processing framework: Processing could be either controlled by the subject, or determined by the structural features of the system; and it could either require capacity or be free from capacity requirements. What the two-process theories did when they appeared some years later, in the midseventies (e.g., LaBerge & Samuels, 1974; Neumann, Note 1; Posner & Snyder, 1975a; Shiffrin & Schneider, 1977), was essentially to combine these two modifications and to link them by means of the phenomenological distinction between conscious processes and processes that lack conscious awareness.

The concept of automaticity, as put forward by these theories, thus refers to a construct that is defined by converging operations. It states a functional linkage between three aspects of processing:

1. A mode of operation: Automatic processes operate without capacity and they thus neither suffer nor cause interference;

2. A mode of control: Automatic processes are under the control of stimulation rather than under the control of the intentions (strategies, expectancies, plans) of the person;
3. A mode of representation: Automatic processes do not necessarily give rise to conscious awareness.

These are what may be called the *primary criteria* of automaticity upon which most versions of the two-process approach more or less agree, although there are, of course, some variants[1]. In addition, there are attributes of automatic processes which are stated or implied by most theorists, but which are not generally regarded as defining the concept. These *secondary criteria* may be collected under the general idea that the course of an automatic process is determined by relatively permanent structural connections, either wired-in or acquired through practice. Thus, automatic processing is believed to be relatively simple (e.g., unable to integrate semantic information to extract the deep structure of a sentence, e.g., MacKay, 1973; Neisser, Hirst, & Spelke, 1981; Underwood, 1977), but rapid (e.g., Neely, 1977; Posner & Snyder, 1975b) and relatively inflexible in that it can be modified only through extended specific training (e.g., LaBerge & Samuels, 1974; Shiffrin & Schneider, 1977).

The validity of this notion of automaticity obviously depends on (a) whether these criteria of automaticity grasp functionally relevant properties of mental operations, (b) whether they are connected in the way predicted, i.e., whether the converging operations really do converge. The following discussion of these issues will focus on the three primary criteria of automaticity and their interrelationship.

3 Mode of Operation: The Interference Criterion

The general way to operationalize the notion that a process does not need capacity is to demonstrate lack of interference with other, simultaneous processes. This has

1 These are mainly of three kinds. First, some authors prefer less rigorous formulations of the three criteria. For example, according to Posner (1978, p. 91), an automatic process "... *may* occur without intention, without giving rise to conscious awareness and without producing interference". In Jonides' (1981, p. 189) formulation of the capacity criterion, automatization means "... ever *lessening* demands on attentive resources" (both italics mine). Formulations like these hint at weaker versions of the concept of automaticity (e.g., the possibility of partial automaticity, or a continuum of more or less automaticity) but these are rarely discussed explicitly (for one brief discussion see Jonides & Irwin, 1981). Second, authors do not put equal emphasis on all three criteria, and some do not consider all of them to be *defining* attributes of automaticity (e.g., LaBerge, 1981). Third, there have recently been some critical comments which, in line with the argument put forward in the present paper, call in question the existence of automaticity as defined by the above three criteria (e.g., Neisser et al., 1981; Paap & Ogden, 1981; Regan, 1981).

been done in three ways, which differ considerably in what is meant by "process": (1) In dual-task experiments the subject has to carry out two actions simultaneously; e.g., respond vocally to acoustic stimuli while performing a tracking task. (2) In monitoring and search experiments there is usually only a single task. The processes assumed to occur in parallel are operations performed on simultaneously presented stimuli within that task (e.g., words presented concurrently to the left and right ear). (3) In the third paradigm, there are neither simultaneous tasks nor simultaneous stimuli. What is shown to be interference-free in these experiments is the preparation for one stimulus and the response to a different, unprepared stimulus. For example, priming a certain group of semantically related words does not interfere with a lexical decision response to a semantically unrelated word.

In this section, I will briefly review these three paradigms. I will argue that (a) there are indeed differences between what are called automtic and nonautomatic processes in all three cases, (b) that these do not, however, justify the conclusion that an interference-free mode of operation is an intrinsic property of automatic processing.

3.1 Interference-Free Dual-Task Performance

There are two experimental traditions that link the concept of automaticity to dual-task performance. One is concerned with changes in performance during skill acquisition (e.g., Bahrick & Shelly, 1958; Mohnkopf, 1933; Salthouse & Somberg, 1982; Solomons & Stein, 1896). The second line of research has grown out of attempts to measure attention demands of mental operations by way of dual tasks (for reviews see Kerr, 1973; Posner, 1978). Capacity-free operations at a processing stage are assumed either if a concurrent secondary task (e.g., probe RT) does not suffer interference (e.g., Posner & Boies, 1971), or if task parameters that are believed to be operative at the stage in question do not interact with the difficulty of the concurrent task (e.g., Logan, 1978).

The latter approach, which has been dominant in present-day research, suffers from several weaknesses. A fundamental fallacy in Logan's (1978, 1979) reasoning has been pointed out by Gopher and Sanders (this volume; see also Hawkins, Church, & deLemos, Note 2): If two factors manipulated within a single task have additive effects on reaction time, this indicates, according to the additive factors' logic (e.g., Sternberg, 1969), that they affect different stages of processing. This reasoning cannot, however, be transferred to a dual-task situation, since here additivity is also to be expected if the subject interrupts task B in order to perform task A. In this case, just as in the case of independent stages, the two factors will have additive effects because they affect processing time one after the other. But this temporal sequence reflects the order in which the two tasks are performed, and not an order of processing stages *within* each task.

Besides this problem with the application of the Sternberg logic there is a shortcoming in Logan's paradigm that it shares with all dual-task experiments: What can be demonstrated at best is that *two particular tasks* do not interfere at a certain processing stage. However, this does not prove that this stage is "interference-free" in the sense of an intrinsic functional property which it will exhibit in any other task combination. Such an extrapolation can only be tentative, and there is always the possibility that it will be refuted by a counterexample.

From a theoretical point of view, the extrapolation is, of course, reasonable if one believes in limited undifferentiated capacity as the only source of interference. Under this assumption, a lack of interference is likely to be due to a lack of capacity requirements. There has, however, been ample evidence during the last decade that this notion is insuffient and that at least part of the dual-task interference is specific in nature (Allport, 1980b; Allport, Antonis, & Reynolds, 1972; Friedman & Polson, 1981; McLeod, 1977; Navon & Gopher, 1979, 1980; Sanders, 1979; Wickens, 1980). If one admits this possibility, there is no justification for the above conclusion. All that can be said is that if two tasks fail to show interference there is probably little or no functional overlap between them.

If this is correct, processes whose classification as automatic has been based on evidence from a particular task combination should at least in some cases exhibit interference if the task combination is changed. This is indeed what has been found. For example, a shift of attention to a sudden peripheral signal suffers little interference from a concurrent memory task and has hence been classified as automatic (Jonides, 1981). However, such a shift interferes heavily with a secondary task that demands the processing of visual information at a different location (Neumann, 1979). Likewise, Briggs, Peters, and Fisher (1972) found no interaction between memory set size in a Sternberg task and the absence or presence of concurrent tracking, and concluded from this that there is no divided attention effect at the comparison stage. In Logan's (1978) study, by contrast, memory set size was actually the only factor that interacted with secondary task difficulty, leading to the conclusion that this stage is *not* automatic. (This type of conclusion is not affected by Gopher & Sanders' criticism mentioned above.) In the present view, this apparent disagreement is due to the fact that Logan used concurrent short-term memory rather than concurrent tracking as the secondary task.

To my knowledge, no type of process or stage of processing has so far been reliably identified as interference-free across task combinations. For some time, it was a widely held belief that this was true for the stimulus encoding stage (e.g., Keele, 1972, 1973; Kerr, 1973; Posner, 1978; Posner & Boies, 1971). In the meantime it has been shown, however, that stimulus encoding *will* interfere with probe reaction time under appropriate conditions [e.g., if the stimulus is followed by a mask (Comstock, 1973) or if the probe occurs unexpectedly in the auditory modality (Proctor & Proctor, 1979)]. Even under the conditions originally used by Posner and Boies (1971), interference is found if an appropriate control condition is used (Paap & Ogden, 1981).

One might resort to the position that automatic processes, while being prone to specific interference, do not demand general, undifferentiated processing capacity. (Two recent papers by Shiffrin, Dumais, & Schneider, 1981, and Posner, 1982, may be understood to imply this.) This is hardly a way out, however. First, the necessity to assume this kind of capacity in addition to specific resources has yet to be demonstrated (see Allport, 1980a,b, for an excellent discussion of this point); and second, granted the existence of unspecific capacity, how should one determine that it is *not* involved in producing a given case of interference?

Thus, the dual-task literature which we have so far been considering does not support the notion that automatic processing is interference-free in the sense of an *intrinsic* property of operations at a certain stage of processing. However, there is still the possibility that it is an *emergent* property which depends on task factors as well as on the characteristics of the processing system. Likewise, if no processing stages have been found at which processing is interference-free *throughout,* this does not preclude that *some* processes (perhaps not confined to particular stages) exhibit this property.

There is support for these alternatives in the data from the second line of research mentioned above. Here, the dual task method has been used to investigate changes in performance during skill acquisition. The general trend in the data is that, as training proceeds, dual-task interference is reduced or may even disappear (e.g., Bahrick & Shelly, 1958; Hirst, Spelke, Reaves, Caharack, & Neisser, 1980; Mohnkopf, 1933; Spelke, Hirst, & Neisser, 1976). Likewise, there is often little or no interference if at least one of the concurrent activities has been extensively practiced prior to the experiment. Examples include piano playing (Allport, Antonis, & Reynolds, 1972), typing (Shaffer, 1975), and calculating with the aid of an abacus (Hatano, Miyake, & Binks, 1977).

Quite obviously, these examples do not fit into the concept of automaticity as defined by the two-process theories: The processes that function in an apparently interference-free mode are not operations at a certain stage of processing, but skills that include sensory as well as motor components. Moreover, there is evidence that even such well-practices skills are not interference-free under all conditions and in every respect. There seem, however, to be some interesting regularities in the patterns of interference and non-interference:

1. In the majority of the cases where interference disappeared or was reduced very substantially during training, at least one of the tasks was of the tracking or "skilled transcription" type (I take the latter term from Allport, 1980b); i.e., there was a stream of information which the subject had to follow by producing a continuous stream of responses. Examples of pre-experimentally acquired skills include sightreading piano music and typing, as already mentioned. Among the laboratory tasks that belong to this category are tracking (e.g., McLeod, 1977) and writing down dictated words (Spelke et al., 1976; Hirst et al., 1980). In all these cases there is a rule relating input to output which the subject acquires during practice, and which remains unchanged. This is equally true of tasks that do not exactly fit into the present category, such as naming individually presented words (Lison, Note 3), and producing motor reactions to cutaneous stimuli (Mohnkopf, 1933).

2. Both the stimuli and the reactions were usually very different for the two tasks. In most cases, stimuli were presented in different modalities, and responses

were either vocal and manual, or one task did not demand continuous overt responding. For example, Mohnkopf (1933) used visual/cutaneous stimuli and vocal/motór responding. In the experiments of the Hirst, Spelke, and Neisser group, stimuli were presented visually/acoustically, and only the acoustic stimuli had to be responded to continuously. By contrast, even well-practiced tasks apparently cannot be carried out simultaneously with a second task if the stimuli are similar, as in dichotic listening (Underwood, 1974) or in visual search (Shiffrin & Schneider, 1977, exp. 4). Likewise, McLeod (1977) found little interference between tracking and vocal responses to tones, but substantial interference if responding was manual in both tasks.

3. There seems to be a difference between *initiating* and *maintaining* an activity. The former is apparently much more prone to interference than the latter. For example, Salthouse and Somberg (1982) trained subjects in a task comprised of several parts such as detection, memory scanning and visual discrimination over 51 sessions. In some sessions, a probe reaction time task was added. While probe RT decreased somewhat in the course of training, it remained substantially above the single-task control condition throughout the experiment, the difference being 160 msec in session 5, and still 102 msec in session 45. Posner and Cohen (1980) conducted a probe stimulus experiment with a saccadic eye movement towards a visual target as the primary task. Though eye movements are certainly highly practiced, there was considerable interference with the manual probe RT task. In an unpublished exeriment by McLeod and Posner, reported by Posner (1982), interference was found between discrete manual and vocal responses. As Posner (1982) points out, this contrasts with McLeod's (1977) finding of only slight interference if the manual task is continuous.

Taken together, these findings suggest that lack of interference is not an intrinsic, invariant property of processes. Rather, an action can be performed without interference in a dual-task situation if certain conditions are fulfilled: It must have been practiced; there must be information available suited to guide its course; this information must be unique in the sense that no alternative, equally suitable information is present from which the relevant information must be selected (as in the case of dichotic listening). Interference will usually occur if a new action has to be initiated, if information selection in the sense just mentioned is necessary, and/or if the course of the action cannot be determined by applying a previously acquired skill to the information present. The latter is the case in all task situations which demand the usage of new or changing rules, or which have to be carried out without continuous support from incoming information.

These are tentative generalizations, on the basis of which it cannot yet be decided whether the case of interference-free processing should be termed "automatic". I will return to this question in Section 6.

3.2 Parallel Monitoring and Search

Unlike the experiments discussed in the previous section, these paradigms comprise only a single task. The subject is asked to monitor or search for target stimuli in the presence of more or less numerous nontarget elements (e.g., visual search, auditory monitoring of dichotic stimuli). Lack of interference and hence capacity-free operation is assumed if performance is essentially independent of the number of simultaneously presented elements which the subject has to examine (e.g., Egeth, Jonides, & Wall, 1972; Ostry, Moray, & Marks, 1976; Poltrock, Lansman, & Hunt, 1982; Schneider & Shiffrin, 1977; for reviews see Allport, 1980b; Duncan, 1980; Egeth, 1977; Fisher, 1982; Neumann, Note 4). This performance pattern is usually reached after more or less extended practice, and it is found with content-specified target categories such as digits embedded in letters (e.g., Duncan, 1980) or animal names (Ostry et al., 1976) as well as with targets that are defined by simple physical properties. Target sets composed of stimuli which do not form a preexperimentally established category (e.g., nine randomly selected letters as targets, with nine other letters as background items; Shiffrin & Schneider, 1977, part II) can acquire the same mode of processing through extended practice. What is critical is not practice per se, but its consistency: targets have to remain targets throughout training, and not be targets on some and background items on other trials (e.g., Schneider & Shiffrin, 1977; Schneider & Fisk, 1982b).

Although this pattern of results has often been replicated, its generality is still difficult to assess. It may or may not be obtainable for targets defined by feature conjunctions (for example, an item that is both red and vertical in a context of items some of which possess one of these features; e.g., Treisman, 1982; Treisman & Gelade, 1980; Shiffrin et al., 1981). There seem also to be difficulties in producing it with the partial-report technique (Francolini & Egeth, 1979). Let us assume, however, that the effect is genuine and fairly general. In what sense does it prove the existence of automatic, capacity-free processing? There are at least two (not mutually exclusive) possibilities, both of which have been put forward by proponents of the two-process approach.

First, that which is automatic may be the *parallel identification* of *all stimuli* at least up to the point where the attributes that specify the target become available (e.g., Duncan, 1980, 1981; Shiffrin, 1975). Second, the automatic process may be what has been called an "automatic-attention response" (Schneider & Shiffrin, 1977) which leads to the *attentional selection* of only the *target*. Both kinds of automaticity have been assumed in many "late selection" models (e.g., Deutsch & Deutsch, 1963; Norman, 1968; Shiffrin & Schneider, 1977), but they are logically independent. Parallel identification, of course, does not necessarily produce an automatic attention response; and, perhaps less obviously, the assumption of an automatic attention response does not imply parallel identification of nontargets as well as of the target.

To elaborate on the last point, let us briefly consider a line of reasoning that, since Moray's (1959) early paper on one famous case of the "breakthrough of the unattended" [to borrow a phrase from Broadbent's (1982) recent review], has been put forward over and over again: Stimuli can be selected by virtue of complex properties, such as being a digit rather than a letter, or being one's own name. Since this property is the criterion for selection, it must have been represented before selection occurred. This implies that all stimuli impinging on the receptor surface are being processes to a level where this kind of property is represented. (For an early and a recent version of this argument see Deutsch & Deutsch, 1963, and Duncan, 1980). This line of argumentation is, however, far from conclusive. There is no compelling reason why stimuli must be identified before they can be selected for attention, even if selection is based on complex properties. From Treisman's (1960) "dictionary units" and Morton's (e.g., 1970) "logogens" to Neisser's (1976) "schemata", processing units have been described which serve to perform identification *and* selection at the same time. If access to such a unit is not limited to a particular position in the stimulus array, performance may be essentially independent of the number of stimuli, although there is no parallel identification. All that has to be assumed is that features like orientation, spatial frequency, brightness and the like are registered and grouped in parallel (cf. Neisser, 1967; Treisman, 1982). Identification, i.e., the activation of a recognition unit by a feature combination, need not be parallel at all in parallel search. Moreover, it cannot even be concluded from the data that the processing and grouping of features occurs independently of the direction of attention. As Fisher (1982) has pointed out, most of the relevant experiments have used only a small number of background items (usually not more than six, including the target). Moreover, the subjects *were* attending to the whole stimulus field, although attention was not directed at any particular item. Hence, it would be at least premature to conclude even for the feature level that all information in all sense modalities can be processed in parallel.

What cannot be rejected is the second assumption cited above, i.e., that stimuli can, in this mode of search and monitoring, elicit an "automatic attention response". However, this response is not interference-free. It is by now abundantly clear that, in this kind of experiment, any focussing of attention on one target, whether automatic or not, will interfere with the detection of any other target (see the reviews by Allport, 1980b; Duncan, 1980; Neumann, Note 4). Interference also occurs, though to a lesser degree, if only one of the targets is reported (e.g., Dennis, 1977; Ostry et al., 1976). This has been even demonstrated in the original research by Schneider and Shiffrin (see Shiffrin & Schneider, 1977, exp. 4) and has since led these authors to propose as one definition of automaticity: "Any process that demands resources in response to external stimulus inputs, regardless of the subject's attempts to ignore the distraction, is automatic" (Shiffrin et al., 1981, p. 228). This is, of course, tantamount to the elimination of the capacity criterion from the general definition of automaticity.

To sum up, what the monitoring and search experiments indicate is that targets can be identified although attention has not already been focussed on them, and that this results in attention being attracted to them. This may be called automatic detection, and it is certainly functionally different from controlled, sequential search. But there is no evidence that this automatic process is free from interference and hence, presumably, from capacity requirements: As to identification, we need not assume that it takes place in parallel for all stimuli; and as to the attraction of attention, it is definitely interfering.

While they are thus in disagreement with the two-process approach or at least fail to support it, the monitoring and search data fit well into the tentative generalizations which have emerged from the preceding section. First, what is automatized as a result of practice is not processing at one stage, but a skill which includes a sensory and, at least during practice, a motor component. After practice, the response may remain covert, but it is still – as Schneider and Shiffrin's term correctly suggests – an attentional *response* connected to the particular target stimuli. Second, our conclusion from the dual-task data that even highly practiced skills display interference if the stimuli are similar is supported by the finding that attentional (and, more so, overt) responses to simultaneous targets cannot occur without interference.

3.3 Preparation Without Interference

The experiments to be reviewed in this section are quite different from those discussed so far. There are neither dual tasks nor multiple stimuli. All the subject has to do is to perform a choice reaction – usually pressing one of two buttons – to the appearance of a stimulus, usually a word. Yet, somewhat surprisingly, this kind of experimental setup has been extensively used in the study of what is believed to be interference-free processing. Since there is only one task and only one stimulus at a time, there can, of course, be no interference between simultaneous processing operations. However, *preparation* might disturb processing if the stimulus actually presented is different from that which the subject is prepared to process. The absence of this kind of disruption is taken to indicate interference-free "processing".

The most often used type of experiment is a priming paradigm in which the subject has to perform some task on a target stimulus (e.g., make a lexical decision), and there is a prime or cue stimulus which usually precedes the target. If the prime and the target are related, semantic priming is often found, i.e., reaction time is shorter with related primes than in a control condition with a neutral prime (e.g., a row of Xs). This benefit is sometimes associated with a cost in those trials where prime and target are unrelated (i.e., reaction time is longer than in control trials), and sometimes there is only benefit, but no cost (for recent reviews see e.g., Becker, 1980; de Groot, Thomassen, & Hudson, 1982; Fischler, 1981; Posner, 1982; Stanovich & West, 1981). In two influential papers, Posner and Snyder (1975a,b) have proposed that these two patterns of results are indicative of, respectively, consciously controlled processing and automatic activation. Later research has rendered it doubtful that the second type of effect meets all criteria of automaticity originally put forward by Poser and Snyder; for example, it may (Neely,

1977; Stanovich & West, 1979) or may not (e.g., Antos, 1979; Fischler & Bloom, 1980; Myers & Lorch, 1980) be faster than the second type of processing (as to the mode of control criterion, see Section 4.2). The existence of the two different patterns of results, however, seems by now well established.

So this operationalization of the interference criterion, unlike those discussed previously, seems to capture a basic functional difference. There is apparently a kind of preparation which helps if it can be used, but which does not hinder if it cannot. However, as already mentioned, it is misleading to call this automatic *processing*. What is free from producing or suffering from interference is rather a state of readiness for particular processing operations, and possibly also the corresponding change in readiness (the "spreading" of activation). These should be clearly distinguished from processing operations which involve the uptake and evaluation of information and the planning and execution of actions (Neumann & Kautz, Note 5; see also Broadbent, 1982, for a similar argument). The lack of interference in the priming paradigm is strictly confined to the former category. As soon as the prime's effects go beyond influencing the state of readiness, they can produce interference as well as facilitation. For example, an invalid prime causes interference if it primes a stimulus alternative which (because of common visual features) receives activation from the stimulus presented, and can thus pass from mere readiness to being actually involved in processing (Neumann & Kautz, Note 5). Likewise, conscious strategies of setting up expectations, making predictions etc. involve processing operations and not simply a change in readiness and may thus cause interference.

Distinguishing changes in state of readiness from processing proper may also help to clarify a result recently reported by Schneider and Fisk (1982a; see also Shiffrin et al., 1981). These authors used a visual search task similar to that of Schneider and Shiffrin (1977) described in Section 3.2, with the modification that the subjects had to search for a variable target, which was specified before each trial, and in addition for a consistent-mapping target which had consistently served as a target in previous practice and could be considered to be detectable in an automatic fashion. The two targets could not both occur on a given trial; so the two kinds of search had to be prepared together, but did not have to be executed simultaneously. There was no concurrence cost, in contrast to findings from both a control experiment by Schneider and Fisk with two variable-mapping targets, and similar results from other areas which indicate that simultaneous preparation for two tasks produces interference even if both tasks do not have to be executed together (e.g., Adler, Note 6; Noble, Sanders, & Trumbo, 1981).

What is interesting about this experiment in the present context is that one necessary condition for the disappearance of interference was that the subjects did *not* actively prepare their search for the consistent-mapping target. They were "strongly encouraged to place all their emphasis on the VM (variable mapping) task" (Schneider & Fisk, 1982a, p. 264), and even then, it took some thousands of trials before subjects had learned not to "waste resources" by actively preparing for both types of targets. Only then did performance become interference-free. Thus, as in the priming paradigm, any kind of mental activity which transcends the mere establishment of a state of readiness is apt to produce interference.

To sum up the conclusions which have emerged from this and the two preceding sections, the interference criterion as defined by the two-process approach does not adequately account for the majority of the findings. Lack of interference as an intrinsic property, independent of circumstances, seems to be restricted to mere states of readiness (this section) and, possibly, early feature extraction (Section 3.2). Within the realm of processing proper, (i.e., taking up and evaluating information, planning and executing action), no processes are interference-free in this absolute sense (Sections 3.1 and 3.2). However, interference may be absent given certain conditions which probably include extensive practice, no concurrent tasks which use similar input and/or output, the presence of a continuous stream of information suitable for guiding the action, and no need to initiate a new action.

4 Mode of Control: The Intentionality Criterion

This criterion of automaticity is, like the previous one, usually believed to be essentially dichotomous: Processes are either under intentional control, or else they are "invariant" (Posner, 1978), "unavoidable" (LaBerge, 1981), or "controlled by the stimuli in the task environment" (Logan, 1981, p. 205). I will argue in this section that this distinction does not cover all the relevant functional differences which actually exist. Specifically, it fails to account for the very frequent case in which a process, though not conforming exactly to a person's intentions, nevertheless depends on an intention in the sense that it would not have taken place without it. Furthermore, the distinction neglects the difference between what is explicitly intended and what is implied by an intention. In the following sections three types of evidence will be reviewed which are thought to demonstrate processing without intention: The unavoidability of interference in Stroop-type tasks; priming based on a semantic relationship between target and prime; and indirect indications that stimuli are processed semantically without the person's intention.

4.1 Unavoidability of Interference

The statement that a process is unavoidable can mean various things. One sense of unavoidability is that the process will occur whenever the proper stimulus conditions exist ("invariance" as defined, e.g., by Posner & Rogers, 1978). Alternatively, processes can be unavoidable in the sense of being conditional upon stimulation and, in addition, appropriate internal conditions, which, however, cannot be fully controlled by the person. For example, schizophrenics may exhibit an abnormally high susceptibility to distraction from external stimuli (see Broen, 1977; Venables, 1978). Third, processes may be caused by an intention, but the person cannot avoid that its execution departs more or less from the intended goal. Speech errors, which usually bear phonological and often semantic similarity to what the person intended to say (see, e.g., Allport, this volume; Fromkin, 1973, 1980), are a case in point. Fourth, the execution of a process may be unavoidable in the sense that it cannot be stopped once it has been started. Here unavoidability does not mean departure from an intention, but rather the inability to depart from the initial plan of execution. These different meanings of the notion of unavoidability are usually not clearly distinguished, with the consequence that experimental results have often been overinterpreted.

As to the first notion of unavoidability, it can be easily dismissed. There are even many reflexes which do not exhibit invariance from internal conditions. Indeed, the strict definition of invariance, according to which "isolable codes (are) activated uniformly by a given stimulus" (Posner & Rogers, 1978, p. 161) has disappeared from Posner's 1978 book, to be replaced by a more liberal concept of invariance which includes the possibility that patterns of automatic activation may in some aspects be varied by context or intentions. The notion that automatic processes are unavoidable in that they cannot be stopped after initiation is intuitively more attractive, but it has scarcely been investigated and is even less supported by empirical evidence (Logan, 1981, 1982). Its possible significance remains to be seen. So there remain the second and third possibilities.

The conclusion that a process is unavoidable is only valid if it can be shown to take place under conditions where it is certain that avoidance was actually attempted. The easiest way to meet this requirement is to look for interference as an indicator of unavoidability. The classic example is the Stroop experiment (Stroop, 1935) in which color naming is interfered with by the presence of incongruent color words (for reviews see, e.g., Dyer, 1973; Jensen & Rohwer, 1966; Neumann, Note 7; Schulz, 1978). Other examples of the same basic design include naming pictures with words as interference stimuli (e.g., Lupker & Katz, 1981; Neumann & Kautz, Note 5), naming letters which are part of a global structure forming a different letter (e.g., Hoffman, 1980; Navon, 1977), identifying letters flanked by different letters (e.g., Eriksen & Schultz, 1979); classifying words flanked by different words (e.g., Shaffer & LaBerge, 1979), and counting the number of digits which name a different number (Morton, 1969). The usual result is that the task-irrelevant aspects or parts of the stimulus display cause interference despite the subject's attempts to ignore them. Is this unavoidability in the sense that the process is independent of intentions, or in the sense that it depends on intentions, but does not exactly conform to them? What the subject intends to do is to attend to a certain stimulus, and to carry out a certain prescribed task. The question, then, is whether or not the interference depends on attentional and task factors. The evidence clearly shows that it does.

Consider first, attention. There is growing evidence that the focussing of visual attention involves the selection of a position in visual space (e.g., Broadbent, 1977; Hoffman & Nelson, 1981; Posner, 1980; Posner, Snyder, & Davidson, 1980; Wolff, Note 8), where a "position" may vary in size, depending, in a structured visual field, on perceptual grouping (Kahneman & Henik, 1981; Neisser & Becklen, 1975; Treisman, 1982; Wolff, Note 8). If Stroop-type interference depends on the focussing of attention, it should consequently be strongly influenced by spatial and grouping factors. This is actually the case. In the color-naming experiment, interference is strongest if the word is printed in colored ink, i.e., if target and distractor are two dimensions of the same visual object. There is a considerable reduction in interference if the word is instead written on a colored patch, and even less interference if the word and the patch are spatially adjacent (Neumann, Notes 7, 9; Tecce & Happ, 1964). If the target and the distractor are moved away from one another, interference goes further down, and at some angular separation, which depends on the kind of task and on the experimental conditions, it may drop to near zero (Egeth, 1977; Gatti & Egeth, 1978; Kahneman & Henik, 1981; Iwasaki, 1978). With an angular separation of 7°, no interference has been found even if the distractor is viewed foveally and the target appears either to its left or right in an unpredictable sequence (Goolkasian, 1981). The importance of grouping factors has also been shown in an experiment by Flowers and Stoup (1977), in which interference vanished in the course of training when target and distractor were spatially separate, but not when they belonged to the same perceptual object. Likewise, Francolini and Egeth (1980) found that subjects can escape interference if targets and distractors are perceptually separated by different coloring.

Direct evidence for the involvement of attentional factors in Stroop-type interference has recently been presented by Ward (1982) and by Lowe and Mitterer (1982). Ward used a simplified variant of a paradigm introduced by Navon (1977). The stimuli were large forms (either an X or a plus) made up of small forms (either Xs or pluses). Depending on instructions, subjects had, on a given trial, to identify either the large (global) or the small (local) form by pressing a button. Interference from the task-irrelevant form(s) was found only on those trials where the immediately preceding trial had required a different hierarchical level of attention (i.e., if, from one trial to the next, attention had to be shifted from global to local or vice versa). In the study by Lowe and Mitterer (1982), target and distractor were two words to the right and left of fixation. Here the degree of interference as well as overall latency depended on the proportion of trials in which the distractor was congruent, i.e., in which it was potentially helpful to attend to both stimuli rather than focussing attention on the target.

These data indicate quite conclusively that processing the distractor is not unavoidable in the sense that a subject was unable to avoid it by attentional focussing; rather, it becomes unavoidable *because of* attentional focussing.

The evidence is equally clear-cut with respect to task requirements. To a large degree, a distractor causes interference not because of its intrinsic properties but because it is related to the intended action. For example, the interference which color words cause in a color classification task depends on whether or not the task requires implicit naming of the colors (Flowers & Blair, 1976; Flowers & Dutch, 1976). Interference also declines if implicit naming is impeded by a concurrent articulation task (Martin, 1978). If the task includes number naming as well as color naming, number words interfere with color naming at least equally strong as color words (Neumann, Note 7). Likewise, what types of letters are efficient distractors in a letter matching task depends on whether physical matching or name matching is required (Keren, O'Hara, & Skelton, 1977).

Further evidence for the importance of task requirements and the relative unimportance of task-independent stimulus properties comes from a study by Regan (1981), which employed a modification of the Navon (1977) paradigm mentioned above. In one experiment, subjects had to name both English and Armenian letters which combined together to form larger letters, both English or Armenian, giving four possible combinations. These were presented in a mixed sequence. Thus, English as well as Armenian letters were relevant within the overall task. Armenian distractor letters interfered almost to the same degree as English letters both with naming English and with naming Armenian letters. This was true although it had in a previous experiment been established that, measured by a capacity criterion, the English letters showed automaticity of processing, and the Armenian letters did not. Moreover, the interference pattern was essentially the same across two different practice levels in the naming task.

It seems safe to conclude that Stroop-type interference is not unavoidable in the sense of occurring as an invariant consequence of stimulation, independent of the person's intentions. Quite to the contrary, it depends heavily on where attention is directed, which task the subject is required to perform, and what strategies are used. However, interference often *is* unavoidable in the sense that the subject does not succeed in executing the intended action as planned. This is the third of the four meanings of unavoidability discussed at the beginning of the present section. It is obviously quite different from what is proposed by most current versions of the two-process approach.

A further aspect of interference deserves a brief discussion. As in the case of the automatic-attention response discussed in Section 3.2, processing the distracting stimuli in a Stroop-type experiment is considered to be automatic, and yet it is obviously not interference-free. To circumvent this difficulty, it has been argued by Posner and Snyder (1975a) that there are two different processing stages involved: An initial stage (reading the word and naming the color), which is automatic and interference-free, and a subsequent stage (response competition) which produces the interference. Apart from the fact that there is certainly more in Stroop interference than mere response competition (see e.g., Glaser & Glaser, 1982; Neumann, Note 7; Neumann & Kautz, Note 5; Stirling, 1979; Seymour, 1977), this notion fails to explain how the unwanted information gets from the first stage to the second. Either the false response tendency is triggered automatically, then we have once again a process which is automatic and yet interfering; or else it is not, then Stroop interference is not indicative of automatic processing.

The discussion in the present section has been based on evidence from Stroop-type experiments, because since the Posner and Snyder (1975a) paper this has been the paradigm most widely believed to demonstrate the unavoidability of automatic processing. For the sake of completeness it should be noted that there probably *are* situations where processing a stimulus is unavoidable in the sense that it is independent of current intentions. This occurs obviously in case of a sudden, unexpected and sufficiently intense change in the environment which elicits an orienting reflex (e.g., Lynn, 1966; Sokolov, 1963). It has also been demonstrated that verbal stimuli which are unrelated to the current intentions of the person can sometimes attract attention by virtue of their meaning (e.g., Flowers, Polansky, & Kerl, 1981; Moray, 1959; Wolford & Morrison, 1980). However, these stimuli have usually been presented under conditions where they were not easy to separate from the attended stimuli. Whether this kind of effect can be obtained under less favourable conditions is still an open question.

4.2 Semantic Priming

The data to be reviewed in this section come from the priming paradigm described in Section 3.3. We now have to consider these findings with respect to the intentionality issue. In discussing these data, two points should be kept in mind. First, as pointed out earlier, the distinction between stimulus-controlled ("invariant") and consciously controlled processes is too crude to provide an adequate framework for categorizing the findings on automatic processing. To understand the interference data, it was considered necessary to distinguish between processes that depend on an intention and those that, in addition, conform to the intended course of execution. A further subdivision will help to clarify the priming data: Only some of the processes occurring in the execution of an intention are explicitly intended. For example, I have just intended to type this sentence. Alternatively, a process may be implied by an intention without being explicitly intended. When intending to type the sentence beginning with 'For example' I did not explicitly intend to type 'e' as the 10th letter. Yet this was of course implied in my intention. Clearly, low-level details of skilled action are not part of conscious intentions. As Bryan and Harter put it more than 80 years ago, a habit "... tends to lose itself in habits of higher order when it appears as an element therein" (Bryan & Harter, 1899, p. 361).

The second issue to bear in mind is that the logic of the priming paradigm is different from that of the interference paradigm. If there is interference, one may conclude that some processing is taking place which does not conform to the person's intentions, since presumably no subject intends to suffer interference. Conversely, priming is helpful, and hence the possibility that it results from a conscious strategy can never be ruled out.

Taken together, these two considerations suggest that one should carefully consider the possibility that priming is due to processes that are either consciously intended or implied by an intention, before deciding that it is a mere passive consequence of presenting the prime. Recent data suggest that there may be at least three kinds of such processes.

First, subjects will prepare for related words if they have some reason to expect them. One way to induce expectation is to increase the proportion of related prime-target pairs relative to unrelated or neutral pairs. As Tweedy, Lapinski, and Schvaneveldt (1977) have shown, this increases the amount of facilitation without significantly affecting RT to unrelated words. A further possibility is to employ a type of semantic relationship that allows a high chance of success when predicting the target from the prime. Using antonym pairs such as HOT-COLD or DRY-WET, Becker (1980) found an increase in facilitation and a reduction in inhibition as compared with a condition with less predictable targets. Third, there is of course the procedure of simply telling the subject what to expect, which Neely (1977) has shown to be effective in producing facilitation even in the absence of semantic relatedness. In this condition, inhibition was found when the expected type of target did not occur.

In these cases subjects presumably made intentional use of the prime-target relationship. There is, however, clear evidence that not all instances of semantic facilitation are of this kind. Neely (1977) found facilitation at short prime-target intervals even when the subject did not expect a semantically related pair. Fischler (1977a) investigated facilitation for a critical prime-target pair which one group of subjects received at the end of a list containing only unrelated pairs. Facilitation did not differ from that found in a second group of subjects who had seen several related pairs prior to the critical pair. Similarly, Tweedy and Lapinski (1981) found that the amount of semantic facilitation was the same regardless of whether or not a biasing list preceding the test items contained related prime-target pairs.

These findings suggest that there may be a *second* type of facilitation which occurs even when the subject does not expect semantically related stimulus pairs. However, this does not imply that facilitation is independent of intentions. As discussed above, processes may be implied by intentions, even though they are not explicitly intended. The subject's intentions in this tape of experiment are essentially determined by a task which involves linguistic processing of the target, e.g., naming it or deciding whether it is a meaningful word. To accomplish this, a person has to make use of previously acquired linguistic skills, which normally serve to relate words and concepts, and not merely to identify words. This kind of facilitation could thus be considered to be "the purest form of predictive reading" (Levy, 1981, p. 7), rather than the result of a quasi-physiological spread of excitation along associative pathways.

If this interpretation is correct, the degree of facilitation should depend more on linguistic factors than on the 'semantic distance' between prime and target as measured, for example, by normative associative strength or by category dominance. Indeed, the latter factor has been found to be of surprisingly little importance. In the majority of studies, it had no consistent effect at all (Becker, 1980; Fischler, 1977b; Fischler & Bloom, 1980; Koriat, 1981; Neely, 1977; Warren, 1977). Of two studies that did find an effect of associative strength, one (de Groot et al., 1982) used extremely infrequent associates (less than 3%) in the "weakly associated" condition, while the other (Fischler & Goodman, 1978) only found a clearcut difference between weak and strong associates in a condition where the prime was masked by the target. By way of contrast, experiments in which a sentence context rather than a single word has been used as a prime, have revealed strong effects of the degree to which the word is predicted by the context (Fischler & Bloom, 1980; Forster, 1981; Kleiman, 1980; Morton, 1964; Schuberth, Spoehr, & Lane, 1981; Underwood, 1977). Moreover, it has been demonstrated that even purely syntactic relationships (e.g., the prime "we" followed by the target "agreed") produce facilitation (Goodman, McClelland, & Gibbs, 1981; Schmidt, Note 10). Further strong support for the importance of linguistic factors has been provided by Foss (1982) who found that priming occurred over a distance of more than 10 words in a sentence context, while even one or two intervening words abolished priming with lists of isolated words. This is clear evidence against a passive "spread of excitation" interpretation.

The application of reading skills to the experimental task and the formation of expectancies can explain much, but not all of the priming data. Two lines of evidence suggest the existence of a *third* mechanism. First, it has recently been shown that pictures as well as words may produce priming, for both word and picture targets (Carr, McCauley, Sperber, & Parmelee, 1982; Lupker & Katz, 1982; McCauley, Parmelee, Sperber, & Carr, 1980; Sperber, McCauley, Ragain, & Weil, 1979). Second, some of the effects found in lexical decision tasks disappear when a naming task is employed (Forster, 1981; West & Stanovich, 1982). Since naming a word and making a lexical decision both demand lexical access (the possibility of a direct grapheme-phoneme-translation can be discarded, see Neumann, Note 7), there must be postlexical factors that are specific to the lexical decision task. They can neither be due to the application of reading skills nor due to expectancies, which should affect both tasks similarly. Likewise, the effectiveness of picture primes can obviously not be due to reading. Conscious expectancies are also excluded as an explanation, since priming is preserved when the pictures are presented below the threshold for conscious identification (e.g., Carr et al., 1982; also see below, Section 4.3).

There is a common explanation for both types of finding which is suggested by the interference data discussed in the preceding section. As is apparent from Stroop-like interference, the subject does not always succeed in preventing unwanted information from being processed and influencing the course of the intended action. This failure in selectivity is likely to occur if relevant and irrelevant stimuli are presented in close spatial proximity (see Section 3.3). Likewise, interference extends over a range of *temporal* proximity in the order of some hundreds of milliseconds (e.g., Eriksen & Schultz, 1979; Glaser & Glaser, 1982; Neumann, Notes 7, 11). It is known from the Stroop literature that, while this inability to filter out irrelevant information causes interference in the case of incongruity, it facilitates processing if the irrelevant information is congruent to the relevant information.

Since congruent Stroop stimuli are often indistinguishable from the prime-target pairs that are used in the priming paradigm (see Logan, 1980; Neumann, Notes 7, 12), it is safe to conclude that the lack of selectivity characteristic of Stroop experiments will likewise be found to some extent in the priming studies.

Note that, according to this explanation, there is *no* task-independent "spread of activation" from prime to target. Rather, information from both sources is being *used* in the course of executing an intention. If the prime precedes the target, at least part of the information from it must be assumed to be stored temporarily. However, there are no facilitative processes of any kind prior to the appearance of the target. An identical conclusion has been reached by Koriat (1981) based on the finding that backward associates are at least as effective as forward associates in producing priming. The present interpretation also explains why practically no "associative latency" is needed for the prime to become effective (Fischler & Goodman, 1978; Neumann, Note 7; Warren, 1977).

Semantic congruity and incongruity effects can also explain the difference between data from naming and from lexical decision experiments. A naming response, although dependent on lexical access, can be derived from a phonological code without necessarily requiring semantic analysis. Conversely, lexical decisions are more likely to involve semantic processing, especially if a sentence context is used. As West and Stanovich (1982) have pointed out, it can be assumed that this will lead to the detection of incongruity if the target does not fit the context, interfering with the semantic decision. A similar mechanism has been suggested by de Groot et al. (1982) for interference effects found with single word stimuli.

By way of summary, we may conclude that the priming data reviewed in this section do not point to an invariant process of "spreading activation". Priming effects are partially due to conscious intentions. Second, subjects may make use of linguistic skills, which implies taking into account semantic as well as syntactic relationships between successive words. Finally, part of the results can be explained along the same lines as the interference data discussed in the preceding section. In all three cases, the processes leading to facilitation depend on the subject's intentions. However, in the second case, they follow implicitly from that which has been intended, and in the third case the processes that actually take place may not exactly conform to what has been intended.

4.3 Indirect Indicators of Semantic Processing

The last decade has produced ample evidence that stimuli can be processed up to at least some representation of their meaning without awareness. One line of evidence stems from dichotic listening experiments in which conditioned responses can apparently be elicited by unattended words and even generalize to semantically similar words (Corteen & Wood, 1972). Despite one unsuccessful replication (Wardlaw & Kroll, 1976) this effect seems to be established (Corteen & Dunn, 1974; Dawson & Schell, 1982; Forster & Govier, 1978; Govier & Pitts, 1982; v. Wright, Anderson, & Stenman, 1975). It has further been shown that shadowing latency is affected by the semantic content of the nonattended message (Lewis, 1970; Underwood, 1977), and that the semantic interpretation of ambiguous words in the attended channel may depend on which of their meanings is primed by a word in the rejected message (MacKay, 1973). Corresponding examples of what was once called "subliminal perception" (see Dixon 1971, for a summary of the early research) have been found in the visual domain. Thus, semantic priming of a naming or a lexical decision response has been reported to occur even if the prime is masked (Carr et al., 1982; Fischler & Goodman, 1978; Fowler, Wolford, Slade, & Tassinary, 1981; Marcel, 1978, 1980; Marcel & Patterson, 1978; McCauley et al., 1980).

At first sight, these phenomena seem to be rather different from the effects discussed in the preceding sections. Indeed, they seem to provide much more compelling evidence for a type of processing which is independent of attentional and task factors. However, closer scrutiny suggests that they may in fact be functionally similar to the effects found when the subject is aware of all stimuli.

To begin with, attentional factors may well be of critical importance in these experiments. In the visual experiments the subjects, while unaware of the prime, still directed their attention to it since it appeared in a position where it was followed (or even preceded and followed) by a target stimulus. Processing did not lead to conscious awareness of the prime because of masking, but the prime did not occur outside of attention. It is known from experiments with unmasked primes that the degree of facilitation depends on the spatial separation between prime and target, though perhaps to a lesser degree than interference does (Gatti & Egeth, 1978; Goolkasian, 1981, exp. 2; Iwasaki, 1978). Masked and unmasked primes are about equally effective (but their effects are not identical in every respect; see Marcel, 1980) and thus probably share common mechanisms. So the effectiveness of the masked primes may well depend on attention being directed to them. This consideration gains support from experiments by Inhoff (1982), Inhoff and Rayner (1980) and Klintman (1971). Here, the primes surrounded or flanked the target rather than appearing in the same spatial position. Thus, attention was focussed away from them. No facilitation was obtained from primes of which the subjects remained unaware, i.e., which they were unable to report.

The evidence is even more clear-cut for the dichotic listening experiments. Treisman, Squire, and Green (1974) replicated the Lewis (1970) experiment and found that shadowing latency was affected by the semantic content of the rejected message only during the early part of the experiment, where attentional focussing on the to-be-shadowed channel had presumably not yet been fully achieved. The dependence of the Lewis effect on attentional selectivity has equally been shown by Johnston and Heinz (1979). Newstead and Dennis (1979) obtained a similar result when they tried to replicate MacKay's (1973) finding that unattended words may prime the interpretation of the attended message. They report that this effect depends decisively on the specific conditions used by MacKay, *viz.* presenting the prime without preceding and succeeding words in the unattended channel, and separating the to-be-shadowed sentences by pauses. These are conditions where the subjects may have had a chance to attend to memory traces of the primes during pauses, although there is no conclusive evidence that they were actually aware of them. In MacKay's original study there was probably little awareness of the prime (MacKay, 1973, study II).

Of special interest are two recent experiments by Johnston and Wilson (1980) and by Dawson and Schell (1982). The latter authors repeated the Corteen and Wood experiment, but controlled for shifts of attention to the nonshadowed channel by using indicators such as shadowing errors. A separate analysis of the trials where no such shift could be detected revealed that galvanic skin responses to conditioned stimuli in the unattended message occurred only when that message was presented to the left ear and the subject was shadowing the right ear message, but not vice versa. Similarly, Johnston and Wilson (1980) failed to find semantic priming in an experiment similar to that of MacKay (1973) where the subjects attended to the left message and the primes occurred in the right channel. Since speech processing takes place in the left hemisphere for most right-handed persons, this may mean that semantic processing of items in the unattended message is possible if they arrive at the right hemisphere (which is known to have some linguistic capabilities; see Searleman, 1977), but not if they arrive at the left hemisphere which is already busy with the shadowing task. (For a more detailed discussion see Dawson & Schell, 1982.) This is again evidence that the semantic processing of unattended stimuli is selective and does not occur in a stimulus-driven fashion for all inputs.

It seems safe to conclude that, like the interference data discussed in Section 4.1, the results reviewed in the present section are in all probability not independent of attentional factors. Moreover, they probably also depend on what activity the subject is engaged in, though positive evidence to this effect is still scarce. It has, however, been suggested by Neumann (Note 7) and more recently by Johnston and Dark (1982) that the Lewis effect may occur because words in the attended message prime semantically similar words in the unattended message, which then in turn may interfere with the shadowing task by attracting attention. The same may apply to the MacKay effect. As to the Corteen and Wood effect, Johnston and Dark (1982) have suggested that this effect could also be task-dependent in that it may be due to persistence of priming from the conditioning phase of the experiment.

Further research is needed to clarify these possibilities. For the time being, it looks as if the effects reviewed in the present section are rather similar to those discussed previously. Not all processes are explicitly intended, and not all of them conform exactly to intentions. It may sometimes be impossible to focus attention as sharply as intended. This is one of the reasons for Stroop interference, and it explains the dichotic listening findings discussed in this section. Further, processes during the execution of an intention may remain implicit. They need not be explicitly intended, as discussed in the preceding section. The present section has added the possibility that they may even escape awareness. All this only demonstrates that processes may depend on intentions in different ways. It does not prove that they occur independent of intentions.

5 Mode of Representation: The Awareness Criterion

We are obviously consciously aware of some, but not all of the processes that go on while perceiving and acting. Posner and Snyder (1975a) were among the first to reintroduce the notion into psychology that introspective accessibility is associated with functional differences. Despite the popularity of the two-process approach, there has been relatively little systematic research into the awareness issue. The present section will therefore be mainly concerned with data that have already been mentioned in one of the earlier sections. The question to be asked is whether differences in awareness are associated with the different kinds of processes which have been discussed in the sections on interference and on intentionality.

5.1 Three Kinds of Unawareness

To begin with, it will be helpful to remember that the term "process" can mean various things (see introduction to Section 3). We may ask (1) whether brain processes that are not directly related to ongoing actions lack awareness, (2) whether there are processes within the execution of an action which escape awareness, and (3) whether an action as a whole (e.g., performing one of the tasks in a dual-task situation) can be carried out without awareness.

The first question can obviously be answered in the affirmative. For example, the contents of long-term memory and the changes taking place during forgetting lack conscious awareness. The same is true for the central control of most bodily functions.

The evidence regarding the second question is equally positive. Section 4.3 contains numerous examples of processes that occur when a subject performs a task, but that do not reach conscious awareness. While these examples stem mainly from the cognitive areas of research, there is no doubt that motor control likewise involves many processes of this kind. Indeed, it is possible that motor processes as such are not conscious at all, and that their representation in awareness is only in terms of the intentions that guide them and the sensory feedback that they produce (e.g., Bernstein, 1967; Wolff, this volume).

Central processing may likewise lack conscious awareness. One famous example is the "tip-of-the-tongue" phenomenon: A long-term memory item such as a person's name which cannot be recalled at a given moment will often come to mind spontaneously some time later, indicating that memory search has been going on in the meantime. To give another example, almost a century ago Münsterberg (1889) conducted a comprehensive series of vocal reaction time experiments in which the subjects answered complex questions such as "Who is the more outstanding philosopher – Kant or Hume?" Münsterberg found that under a speed instruction, subjects responded without any conscious awareness of the underlying processes, implying that "even the more complicated decision processes must thus sometimes be capable of taking place without their component parts entering . . . consciousness" (Münsterberg, 1889, p. 114; translated). Thus, complex operations during the execution of an intention may escape awareness. Likewise, an intended action may be started without awareness of the stimulus. This is demonstrated by the Fehrer-Raab effect where a completely masked stimulus triggers an intended motor reaction (e.g., Fehrer & Raab, 1962; Neumann, Note 13). A further example is the "blindsight" phenomenon (Weiskrantz, Warrington, Sanders, & Marshall, 1974).

Undoubtedly, "unconscious" processes of this kind are involved in all kinds of human activity. Indeed, there is every reason to believe that conscious representation is the exception rather than the rule in human information processing (see Allport, 1980a; N.F. Dixon, 1981, and Neumann, Note 14, for a more detailed discussion). However, in the examples discussed so far, lack of awareness is confined to *component* processes of an action which, as a whole, is represented in conscious awareness. This should be distinguished from the question of whether an action as a whole may remain without representation in awareness. (Third question mentioned above.) Here the evidence is less unequivocal. While common experience (Neisser, 1967) and anecdotic evidence (Reason, 1979) suggest that acting without awareness is probably a normal part of routine behavior, this phenomenon is not easy to reproduce in a laboratory situation where the subject must be given an explicit task, and where some measurement of performance has to be taken. Perhaps the most convincing example of this kind of unawareness to be found in recent experimental research is an experiment by Hirst et al. (1980), where it was demonstrated that, after extended practice, a person who is engaged in reading a text for comprehension can write down and even to a degree understand simultaneously dictated sentences unrelated to the text with little or no awareness.

5.2 Awareness, Interference, and Intentionality

Having distinguished between these three kinds of unawareness, we may ask how they relate to the distinctions that have emerged from the discussion of the interference and intentionality issues. Indeed, there seem to be some connections:

First, processes seem to exist that function essentially independently of the person's intentions and of whatever actions are currently taking place. These are not represented in awareness, and lack of interference is one of their intrinsic properties. Examples include early feature extraction (Section 3.2) and the control of bodily functions (preceding section). Thus, we have here processes which fulfill all three criteria of automaticity put forward by the two-process approach. However, they make up only a small portion of what is usually considered to be automatic processing.

Second, there are processes that occur within the functional framework of an ongoing action and depend upon intentions, but that are not explicitly intended (see Section 4.2). Many of these processes take place with little or no awareness. Examples include finger movements in skilled typing (Section 4.2), predictive reading (Section 4.2), the semantic processing of words in an unattended channel

(Section 4.3) and the Münsterberg experiment mentioned in Section 5.1. However, awareness seems to arise where this type of processing involves selecting information from a particular position in space by means of attentional focussing. Among the examples mentioned previously are search and monitoring experiments (Section 3.2) and the Stroop paradigm (Section 4.1).

Third, an action as a whole may take place outside of awareness. The conditions where this happens seem to be very similar to those which are a prerequisite for interference-free dual-task performance (Section 3.1): There must be a well-practiced skill, information must be available for continuous guidance of the action, and competing stimuli among which that information has to be selected must be absent. An example is provided by the experiment by Hirst et al. (1980) mentioned in Section 5.1. Everyday examples such as walking, dressing, or preparing tea without awareness (see Reason, 1979) belong to the same category.

These regularities suggest that the notion of automaticity may still be of value in arranging the data, despite the failure of the two-process approach in its present form. An alternative concept of automaticity will be outlined in the next two sections.

6 Some Theoretical Conclusions

6.1 Direction of Processing vs. Level of Control

The two-process view of mental functioning as reviewed in Section 2 is a modern version of an old doctrine. Its basic idea seems to date back at least to Protagoras (ca. 485–415 B.C.) who, according to Plato (see Siebeck, 1879/1961) taught that cognition arises when two movements meet, one originating in the environment and the other originating in the mind. This distinction between what is nowadays called bottom-up and top-down processing has taken many forms in the history of prescientific psychology, both in the area of perception (Neumann, 1972) and attention (Neumann, 1971). When psychology entered its scientific stage, it found its most elaborate formulation in Wundt's doctrine of apperception, which bears a striking resemblance to modern two-process theories[2], though Wundt did not use the term "automatic" for the bottom-up type of processing.

2 In Wundt's system cognitive processes fell into two categories. The first comprised sensations (*Empfindungen*) and representations (*Vorstellungen*), among which he rated perceptions (*Wahrnehmungen*). The sensations were determined exclusively by stimulation (plus, of course, the wired-in properties of the sensory systems), while perceptions depended in addition on perceptual learning. This was the bottom-up portion of Wundt's system. The top-down portion was provided by a different category of cognitive processes which went under the heading of apperception. Apperceptive processes served to bring perceptions into the focus of conscious awareness, to analyze them and to create new connections between them. Active apperception (there was also passive apperception) was intentional; in fact Wundt conceived it as an inner voluntary action. And it was sharply limited in capacity. Wundt believed that active apperception could carry out only one action at a time (e.g., Wundt, 1903; especially Chapters 18 and 19; see also Neumann, 1971; Rappard, 1980; and Rappard, Sanders, & Swart, 1980, for a more detailed discussion of the notion of apperception).

As we have seen, only a small portion of the data can be accounted for by this type of theory. There is, however, a second line of thought about automaticity which may be termed the "level of control" approach. Again it was Wilhelm Wundt who gave one of the clearest formulations of this notion of automaticity. What may be automatic, according to him, is a *coordination between a sensory event and a motor action.* Without automaticity, the two have to be linked by a conscious act which includes the apperception of the sensory event and the choice of the action. Automatization occurs during practice if there is an invariant relationship between the sensory event and the subsequent motor action. This results in the establishment of direct neuronal connections between them, so that the sensory event can lead to the proper action without any conscious mediation (Wundt, 1903, Chap. 18.2). What distinguishes automatic from reflex movements, according to Wundt, is that the latter are elicited by a stimulus, whereas automatic movements depend upon inner conditions and are usually components of voluntary actions; i.e., they are conditional upon a goal represented in consciousness although they are executed without conscious guidance (Wundt, 1903, Chap. 17.2). Thus, they are not purely bottom-up processes.

Some decades later, Ach, following Wundt's lead, suggested using time-sharing as an index of automatization (as it was called later by Bahrick & Shelly, 1958; for a summary of the work carried out in his Göttingen laboratory see Ach, 1935). Perhaps the most outstanding subsequent contribution, both in method and in theory, was that of Bernstein (1967, originally published 1947), who carefully studied the modifications which movement sequences undergo during skill acquisition. One of the main conclusions was that this involves a change in how the degrees of freedom of the movements are being controlled. Early in practice, the subject attempts to keep fixed as many degrees of freedom as possible. This involves much effort and is inefficient, because the movement only roughly follows the intended path. As training proceeds, this strenuous tension can gradually be released, because the subject learns to use feedback, and finally to incorporate reactive forces into the control of the movement. At this final stage, the movement is experienced as taking place automatically, without effort or need for attention.

Thus, according to the "level of control" approach, the difference between automatic and consciously controlled processes is not that the former are controlled by stimuli and the latter by intentions. Instead, there is a difference in the way intentions and stimuli work together in shaping action. This notion of automaticity provides, in my view, the right starting point for an integration of the findings that have emerged from the present review. In conclusion, let me briefly outline a theoretical framework that can account for the major aspects of the data.

6.2 Automaticity as a Mode of Parameter Specification

An action can only be performed, if its parameters are specified, either in advance or during execution. This is necessary for all processes which are part of the action, both peripheral and central. For example, naming the color of an incongruent Stroop stimulus (see Section 4.1) requires (a) selecting a particular aspect (the color rather than the word) of a particular perceptual object (the Stroop stimulus rather than, say, the screen on which it is displayed); (b) retrieving specific information (the color's name in English rather than, e.g., the German word naming the category 'color'); (c) and carrying out a certain movement sequence (e.g., pronouncing 'green' as quickly as possible). All this could, of course, be done differently; but the action could not be carried out at all, if any of its parameters remained unspecified.

Where do the necessary specifications come from? Apart from the constraints imposed by biomechanics (see Pew, this volume) there are essentially three sources: First, procedures for carrying out actions or parts of actions are stored in long-term memory. I will use the term *skill* for such permanently stored procedures. Second, *input information* is usually available, which can be used to specify parameters of the action. Third, there must be additional mechanisms whose function is to provide the specifications that cannot be obtained by linking input information to skills. These mechanisms may be termed *attentional.*

The way in which these three sources of constraint work together can be conceived as follows:

Skills have essentially two functions. First, a skill usually specifies part of the action's parameters. For example, a skilled typist will type a given text by carrying out a highly predictable spatio-temporal sequence of finger movements (e.g., Rumelhart & Norman, 1982; Shaffer, 1978). Second, a skill can be used to pick up and integrate the information needed to specify the action. Skills of information pick-up have been investigated e.g., in chess players (Chase & Simon, 1973), Go players (Reitman, 1976), and electronics technicians (Egan & Schwartz, 1979). Complex skills (e.g., reading, car driving, skilled tracking) usually incorporate both types of functions. They specify parameters directly, and they provide the schemata (Neisser, 1976) to pick up information for further parameter specification.

Additional mechanisms are needed, if these two sources of constraint do not specify all parameters of an action completely and unequivocally. Two cases can be distinguished. First, there can obviously be underspecification because appropriate input information is lacking and/or because the available skills are not specific enough to provide the necessary constraints or to pick up the information present. Second, there can be what may be called overspecification. In this case several mutually exclusive specifications are provided by the input information available. To illustrate this, take the action of speaking. Free speech is an example of underspecification since speech and language skills only specify how to speak, but do not specify what to say. Neither are the missing constraints provided by input information. By contrast, the latter is the case in oral reading. However, presenting a literate person with an array of written words still does not specify *what* particular word to utter at a given moment in time. So this is an example of overspecification.

The obvious solution in the case of overspecification is input selection. One powerful mechanism probably available to all organisms that possess movable sense organs (see Trevarthen, 1978) is spatial focussing which can take the form of peripheral adjustments or of a purely central shift of attention.

On the other hand, if there is underspecification, the problem is not to prevent information from being linked to action, but rather to provide additional constraints so that the action can be determined unequivocally. As the example taken from Bernstein (1967) in Section 6.1 suggests, one possibility is to fix motor parameters. If this type of constraint is put into effect prior to execution, we have a case of motor preparation (see e.g., Gaillard, Note 15; Gottsdanker, 1975; Requin 1980; and Sanders, 1983, for some psychological and physiological data).

A further possibility – at least in humans – is parameter specification by concrete anticipatory planning, i.e., by specifying in advance low-level details of the action, or by setting up algorithms which determine how input information has to be transformed in order to specify the desired parameters. Three interesting empirical approaches to this topic have recently been provided by P. Dixon (1981), Logan and Zbrodoff (1982) and Rosenbaum (1980).

A discussion of these proposed attentional modes of parameter specification is beyond the scope of the present paper and can be found elsewhere (Neumann, Note 14). What is important in the present context is the definition of automaticity which is suggested by the above considerations: *A process is automatic, if its parameters are specified by a skill in conjunction with input information.* If this is not possible, one or several of the attentional mechanisms for parameter specification must come into play. They are responsible for interference and give rise to conscious awareness.

According to this view, the process of "automatization" is the acquisition of a *specific* skill. This is similar to the idea of displacement and structural constriction discussed by Heuer (this volume). Related ideas have been expressed by Allport (1980a), Anderson (1982), MacKay (1982) and Schaeffer (1975). The availability of a specific skill, however, does not yet constitute a sufficient condition for automaticity. A further prerequisite is the presence of input information which unequivocally specifies those parameters not already specified by the skill. Thus, automaticity is not an intrinsic property of processes, but an emergent property depending both on the processing system and the situational context.

This simple notion of automaticity can cover most of the findings reported in the preceding sections:

1. Interference-free processing is not universally found at certain processing stages. Rather, it depends on whether or not there has been extended, specific practice (Section 3.1). This follows immediately from the present concept of automaticity.

2. The finding that even extensively practiced actions will exhibit interference under unfavorable task conditions (Section 3.1) is also obviously in agreement with that concept. More specifically, as reported in Section 3.1, there must be sufficient, usually continuous stimulation to guide the action, and further, the information must be unique. If the first condition is not met, guidance must be provided by

concrete anticipatory planning. If the second condition is not fulfilled, there is a need for spatial focussing. Both are attentional mechanisms, and hence automaticity is lost.

3. While spatial focussing can be made more efficient by acquiring a specific skill of information pick-up, it cannot be completely dispensed with, if the relevant stimulus has to be selected from among similar stimuli. This explains why sensory interference is found after prolonged constant-mapping training in a search task even though there is no effect of the number of alternatives, indicating that the task does not demand advance planning (Section 3.2).

4. A skill as defined above is a procedure which can be used to perform an action or part of an action; it is not an input-output-connection which becomes activated by the mere presence of appropriate stimulation. This is why unavoidability is *not* a criterion of automaticity. As the findings reviewed in Sections 4.1 and 4.3 indicate, processes can be unavoidable because spatial focussing is not sufficiently effective, whether they are automatic or not. Furthermore, since skills cannot be put into effect by the mere presence of stimulation, initiating a new action can never be completely automatic (Section 3.1).

5. Most laboratory tasks are performed with the help of advance planning, i.e., they are intentional. However, depending on the availability of specific skills, the contribution of advance planning to specifying the action will vary in magnitude. What is controlled by specific skills need not be intended explicitly. This helps to clarify the nature of the processes "implied" by an intention, as discussed in Section 4.2. Since skills are procedures stored in long-term memory, they will remain without effect as long as they are not being used. By contrast, advance planning is a *process* which may produce interference, even if the plan is not carried out. Thus, the difference between active preparation and mere changes in readiness as discussed in Sections 3.3 and 4.2 can be incorporated in the present concept of automaticity.

6. The awareness criterion seems to be closely related to the functional differences discussed so far (Section 5.2). It looks as if some kind of awareness arises whenever an attentional mechanism is needed to specify parameters which cannot be determined by linking input information to existing skills. What is behind this apparent connection between mode of representation and mode of parameter specification? One might speculate that parameter specification by means of attentional mechanisms must have at least two functional properties which correspond to phenomenological properties of conscious awareness. First, in order to determine parameters of an action by means of information provided by mechanisms which reside in different parts of the system (e.g., focussing on that part of a visual scene one is talking about; or performing a movement sequence which has been described verbally), there must be free traffic between all subsystems involved in this mode of control. This corresponds to what is phenomenologically the unity of consciousness. Second, if this is not to lead to chaos, there must be extremely powerful inhibitory mechanisms which prevent all but the currently needed functional connections from gaining access to action (see Shallice, 1978). The cor-

responding phenomenological aspect may be said to be the narrowness of consciousness.

In conclusion, a "level of control" notion of automaticity seems to be more in line with the majority of findings than the "direction of processing" approach favored by current two-process thinking. That does not mean that automaticity in the sense of genuine bottom-up processing does not exist at all. Two (very different) examples mentioned earlier are feature extraction (Section 3.2) and passive attention shifts of the "orienting response" type (Section 4.1). However, one would, in my view, be mistaken to consider these cases to be the prototypes of automatic processes.

Acknowledgments. I thank Herbert Heuer, Adam Reeves, Andries Sanders and an anonymous reviewer for helpful comments on an earlier draft.

Reference Notes

1. Neumann, O. *Über den Unterschied zwischen Lesen und Benennen.* Unpublished Manuscript, Psychologisches Institut der Ruhr-Universität Bochum, 1973. Reprint: Bericht Nr. 9/1979, Psychologisches Institut der Ruhr-Universität Bochum, Arbeitseinheit Kognitionspsychologie.
2. Hawkins, H.L., Church, M., & deLemos, J. *Time-sharing is not a unitary ability.* (Techn. Rep. No. 2, N0014-77-C 0643 NR 150-407). Oregon: Center for Cognitive and Perceptual Research, University of Oregon, 1978.
3. Lison, E. *Phonetische Kodierung und zentrale Verarbeitungskapazität.* Unpublished thesis. Bochum, FRG: Psychologisches Institut der Ruhr-Universität Bochum, 1976.
4. Neumann, O. *Zum Mechanismus der Interferenz beim dichotischen Hören.* Bericht Nr. 5/1978, Psychologisches Institut der Ruhr-Universität Bochum, Arbeitseinheit Kognitionspsychologie.
5. Neumann, O., & Kautz, L. *Semantische Förderung und semantische Interferenz im Benennungsexperiment.* Bericht Nr. 23/1982, Psychologisches Institut der Ruhr-Universität Bochum, Arbeitseinheit Kognitionspsychologie.
6. Adler, A. *Die Wirkung eines zusätzlichen Reizes im Gesichtsfeld auf den Verlauf der Metakontrast-Funktion: III. Ein Kontrollexperiment mit objektiven Leistungsmaßen.* Unpublished thesis. Bochum, FRG: Psychologisches Institut der Ruhr-Universität Bochum, 1979.
7. Neumann, O. *Informationsselektion und Handlungssteuerung. Untersuchungen zur Funktionsgrundlage des Stroop-Interferenzphänomens.* Dissertation. Bochum, 1980.
8. Wolff, P. *Entnahme der Identitäts- und Positionsinformation bei der Identifikation tachistoskopischer Buchstabenreize. Ein theoretischer und experimenteller Beitrag zur Grundlagenforschung des Lesens.* Dissertation, Bochum, 1977.
9. Neumann, O. *Steuerung der Informationsselektion durch visuelle und 'semantische' Reizmerkmale.* Bericht Nr. 2/1977, Psychologisches Institut der Ruhr-Universität Bochum, Arbeitseinheit Kognitionspsychologie.
10. Schmidt, R. *Die Bedeutung des Genus für den Zugriff zu Wortrepräsentationen im Gedächtnis.* Paper presented at the Conference on Memory and Information Processing, Center for Interdisciplinary Research, Bielefeld, September 30 – October 2, 1977.

11. Neumann, O. *Zeitliche und funktionale Asymmetrien beim Stroop-Effekt.* Bericht Nr. 7/1979, Psychologisches Institut der Ruhr-Universität Bochum, Arbeitseinheit Kognitionspsychologie.
12. Neumann, O. *Eine Umkehrung des 'semantischen Gradienten' beim Benennen von Stroop-Reizen.* Bericht Nr. 1/1977, Psychologisches Institut der Ruhr-Universität Bochum, Arbeitseinheit Kognitionspsychologie.
13. Neumann, O. *Experimente zum Fehrer-Raab-Effekt und das „Wetterwart"-Modell der visuellen Maskierung.* Bericht Nr. 24/1982, Psychologisches Institut der Ruhr-Universität Bochum, Arbeitseinheit Kognitionspsychologie.
14. Neumann, O. *Über den Zusammenhang zwischen Enge und Selektivität der Aufmerksamkeit.* Bericht Nr. 19/1983, Psychologisches Institut der Ruhr-Universität Bochum, Arbeitseinheit Kognitionspsychologie.
15. Gaillard, A. *Slow brain potentials preceding task performance.* Institute for Perception TNO, Soesterberg, 1978.

References

Ach, N. Analyse des Willens. In E. Abderhalden (Ed.), *Handbuch der biologischen Arbeitsmethoden,* Section 6, Part E. Berlin, Vienna: Urban & Schwarzenberg, 1935.

Allport, D.A. Patterns and actions: Cognitive mechanisms are content-specific. In G. Claxton (Ed.), *Cognitive psychology – New directions.* London: Routledge & Kegan Paul, 1980. (a)

Allport, D.A. Attention and performance. In G. Claxton (Ed.), *Cognitive psychology – New directions.* London: Routledge & Kegan Paul, 1980. (b)

Allport, D.A., Antonis, B., & Reynolds, P. On the division of attention: A disproof of the single channel hypothesis. *Quarterly Journal of Experimental Psychology,* 1972, *24,* 225–235.

Anderson, J.R. Acquisition of cognitive skill. *Psychological Review,* 1982, *89,* 369–406.

Antos, S.J. Processing facilitation in a lexical decision task. *Journal of Experimental Psychology: Human Perception and Performance,* 1979, *5,* 527–545.

Atkinson, R.C., & Shiffrin, R.M. Human memory: A proposed system and its control processes. In K.W. Spence & J.T. Spence (Eds.), *The psychology of learning and motivation: Advances in research and theory,* (Vol. 2). New York: Academic Press, 1968.

Bahrick, H.P., & Shelly, C. Time-sharing as an index of automatization. *Journal of Experimental Psychology,* 1958, *56,* 288–293.

Becker, C.A. Semantic context effects in visual word recognition: An analysis of semantic strategies. *Memory & Cognition,* 1980, *8,* 493–512.

Bernstein, N.A. *The coordination and regulation of movements.* New York: Pergamon Press, 1967.

Briggs, G.E., Peters, G.L., & Fisher, R.P. On the locus of the divided-attention effects. *Perception & Psychophysics,* 1972, *11,* 315–320.

Broadbent, D.E. *Perception and communication.* New York: Pergamon Press, 1958.

Broadbent, D.E. Flow of information within the organism. *Journal of Verbal Learning and Verbal Behavior,* 1963, *2,* 34–39.

Broadbent, D.E. *Decision and stress.* New York: Academic Press, 1971.

Broadbent, D.E. Colour, localization and perceptual selection, In *Psychologie éxperimentale et comparée: Hommage á Paul Fraisse.* Paris: Presses Universitaires de France, 1977.

Broadbent, D.E. Task combination and selective intake of information. *Acta Psychologica,* 1982, *50,* 253–290.

Broen, W.E., jr. Attention in schizophrenia. In B.B. Wolman (Ed.), *International encyclopedia of psychiatry, psychology, psychoanalysis, and neurology,* (Vol. 2). New York: Aesculapius, 1977.

Bryan, W.L., & Harter, N. Studies on the telegraphic language: The acquisition of a hierarchy of habits. *Psychological Review,* 1899, *6,* 345–375.

Carr, T.H., McCauley, Ch., Sperber, R.D., & Parmelee, C.M. Words, pictures, and priming: On semantic activation, conscious identification, and the automaticity of information processing. *Journal of Experimental Psychology: Human Perception and Performance,* 1982, *8,* 757–777.

Chase, W.G., & Simon, H.A. Perception in chess. *Cognitive Psychology,* 1973, *4,* 55–81.

Comstock, E. Processing capacity in a letter matching task. *Journal of Experimental Psychology,* 1973, *100,* 63–72.

Corteen, R.S., & Dunn, D. Shock-associated words in a nonattended message: A test of momentary awareness. *Journal of Experimental Psychology,* 1974, *102,* 1143–1144.

Corteen, R.S., & Wood, B. Autonomous responses to shock associated words in an unattended channel. *Journal of Experimental Psychology,* 1972, *94,* 308–313.

Dawson, M.E., & Schell, A.M. Electrodermal responses to attended and unattended significant stimuli during dichotic listening. *Journal of Experimental Psychology: Human Perception and Performance,* 1982, *8,* 315–324.

Dennis, I. Component problems in dichotic listening. *Quarterly Journal of Experimental Psychology,* 1977, *29,* 437–450.

Deutsch, J.A., & Deutsch, D. Attention: Some theoretical considerations. *Psychological Review,* 1963, *70,* 80–90.

Dixon, N.F. *Subliminal perception. The nature of a controversy.* New York: McGraw Hill, 1971.

Dixon, N.F. *Preconscious processing.* Chichester: Wiley, 1981.

Dixon, P. Algorithms and selective attention. *Memory & Cognition,* 1981, *9,* 177–184.

Duncan, J. The locus of interference in the perception of simultaneous stimuli. *Psychological Review,* 1980, *87,* 272–300.

Duncan, J. Directing attention in the visual field. *Perception & Psychophysics,* 1981, *30,* 90–93.

Dyer, F.N. The Stroop phenomenon and its use in the study of perceptual, cognitive and response processes. *Memory & Cognition,* 1973, *1,* 106–120.

Egan, D.E., & Schwartz, B. Chunking in recall of symbolic drawings. *Memory & Cognition,* 1979; *7,* 149–158.

Egeth, H.A. Attention and preattention. In G.H. Bower (Ed.), *The psychology of learning and motivation,* (Vol. 7). New York: Academic Press, 1977.

Egeth, H.A., Jonides, J., & Wall, S. Parallel processing of multielement displays. *Cognitive Psychology,* 1972, *3,* 674–698.

Eriksen, C.W., & Schultz, D.W. Information processing in visual search: A continuous flow conception and experimental results. *Perception & Psychophysics,* 1979, *25,* 249–263.

Fehrer, E., & Raab, D. Reaction time to stimuli masked by metacontrast. *Journal of Experimental Psychology,* 1962, *63,* 143–147.

Fischler, I. Associative facilitation without expectancy in a lexical decision task. *Journal of Experimental Psychology: Human Perception and Performance,* 1977, *3,* 18–26. (a)

Fischler, I. Semantic facilitation without association in a lexical decision task. *Memory & Cognition,* 1977, *5,* 335–339. (b)

Fischler, I. Research on context effects in word recognition: Ten years back and forth. *Cognition,* 1981, *10,* 89–95.

Fischler, I., & Bloom, P.A. Rapid processing of the meaning of sentences. *Memory & Cognition,* 1980, *8,* 216–225.

Fischler, I., & Goodman, G.O. Latency of associative activation in memory. *Journal of Experimental Psychology: Human Perception and Performance,* 1978, *4,* 455–470.

Fisher, D.L. Limited-channel models of automatic detection: Capacity and scanning in visual search. *Psychological Review,* 1982, *89,* 662–692.

Flowers, J.H., & Blair, B. Verbal interference with visual classification: Optimal processing and experimental design. *Bulletin of the Psychonomic Society,* 1976, *7,* 260–262.

Flowers, J.H., & Dutch, S. The use of visual and name codes in scanning and classifying colors. *Memory & Cognition,* 1976, *4,* 384–390.

Flowers, J.H., Polansky, M.L., & Kerl, S. Familiarity, redundancy, and the spatial control of visual attention. *Journal of Experimental Psychology: Human Perception and Performance,* 1981, *7,* 157–166.

Flowers, J.H., & Stoup, C.M. Selective attention between words, shapes and colors in speeded classification and vocalization tasks. *Memory & Cognition,* 1977, *5,* 299–307.

Forster, F.M., & Govier, E. Discrimination without awareness? *Quarterly Journal of Experimental Psychology,* 1978, *30,* 282–295.

Forster, K.I. Priming and the effect of sentence and lexical contexts on naming time: Evidence for autonomous lexical processing. *Quarterly Journal of Experimental Psychology,* 1981, *33A,* 465–495.

Foss, D.J. A discourse on semantic priming. *Cognitive Psychology,* 1982, *14,* 590–607.

Fowler, C.A., Wolford, G., Slade, R., & Tassinary, L. Lexical access with and without awareness. *Journal of Experimental Psychology: General,* 1981, *110,* 341–362.

Francolini, C.M., & Egeth, H.A. Perceptual selectivity is task dependent: The pop-out effect poops out. *Perception & Psychophysics,* 1979, *25,* 99–110.

Francolini, C.M., & Egeth, H.A. On the nonautomaticity of "automatic" activation: Evidence of selective seeing. *Perception & Psychophysics,* 1980, *27,* 331–342.

Friedman, A., & Polson, M. Hemispheres as independent resource systems: Limited-capacity processing and cerebral specialization. *Journal of Experimental Psychology: Human Perception and Performance,* 1981, *7,* 1031–1058.

Fromkin, V.A. (Ed.) *Speech errors as linguistic evidence.* The Hague, Paris: Mouton, 1973.

Fromkin, V.A. (Ed.) *Errors in linguistic performance. Slips of the tongue, ear, pen and hand.* New York: Academic Press, 1980.

Gatti, S.V., & Egeth, H.A. Failure of spatial selectivity in vision. *Bulletin of the Psychonomic Society,* 1978, *11,* 181–184.

Glaser, M.O., & Glaser, W.R. Time course analysis of the Stroop phenomenon. *Journal of Experimental Psychology: Human Perception and Performance,* 1982, *8,* 875–894.

Goodman, O., McClelland, J.L., & Gibbs, R.W. The role of syntactic context in word recognition. *Memory & Cognition,* 1981, *9,* 580–586.

Goolkasian, P. Retinal location and its effect on the processing of target and distractor information. *Journal of Experimental Psychology: Human Perception and Performance,* 1981, *7,* 1247–1257.

Gottsdanker, R. The attaining and maintaining of preparation. In P.M.A. Rabbitt & S. Dornič (Eds.), *Attention and performance V.* New York: Academic Press, 1975.

Govier, E., & Pitts, M. The contextual disambiguation of a polysemous word in an unattended message. *British Journal of Psychology,* 1982, *73,* 537–545.

de Groot, A.M., Thomassen, A.J., & Hudson, P.T. Associative facilitation of word recognition as measured from a neutral prime. *Memory & Cognition,* 1982, *10,* 358–370.

Haber, R.N. Perception and thought: An information processing analysis. In J.F. Voss (Ed.), *Approaches to thought.* Pittsburgh: University of Pittsburgh Press, 1967.

Hatano, G., Miyake, Y., & Binks, M.G. Performance of expert abakus operators. *Cognition,* 1977, *5,* 57–71.

Hirst, W., Spelke, E.S., Reaves, C.C., Caharack, G., & Neisser, U. Dividing attention without alternation or automaticity. *Journal of Experimental Psychology: General,* 1980, *109,* 98–117.

Hoffman, J.E. Interaction between global and local levels of a form. *Journal of Experimental Psychology: Human Perception and Performance,* 1980, *6,* 222–234.

Hoffman, J.E., & Nelson, B. Spatial selectivity in visual search. *Perception & Psychophysics,* 1981, *30,* 282–290.

Inhoff, A.W. Parafoveal word perception: A further case against semantic preprocessing. *Journal of Experimental Psychology: Human Perception and Performance,* 1982, *8,* 137–145.

Inhoff, A.W., & Rayner, K. Parafoveal word perception: A case against semantic processing. *Perception & Psychophysics,* 1980, *27,* 457–464.

Iwasaki, S. The limits of attention and its expansion. *Japanese Psychological Research,* 1978, *20,* 133–142.

Jensen, A.R., & Rohwer, W.D. The Stroop color-word test: A review. *Acta Psychologica,* 1966, *25,* 36–93.

Johnston, W.A., & Dark, V.J. In defense of intraperceptual theories of attention. *Journal of Experimental Psychology: Human Perception and Performance,* 1982, *8,* 407–421.

Johnston, W.A., & Heinz, S.P. Depth of nontarget processing in an attention task. *Journal of Experimental Psychology,* 1979, *5,* 168–175.

Johnston, W.A., & Wilson, J. Perceptual processing of nontargets in an attention task. *Memory & Cognition,* 1980, *8,* 372–377.

Jonides, J. Voluntary versus automatic control over the mind's eye's movement. In J. Long & A. Baddeley (Eds.), *Attention and performance IX.* Hillsdale, NJ: Erlbaum, 1981.

Jonides, J., & Irwin, D.E. Capturing attention. *Cognition,* 1981, *10,* 145–150.

Kahneman, D., & Henik, A. Perceptual organization and attention. In M. Kubovy & J.R. Pomerantz (Eds.), *Perceptual organization.* Hillsdale, NJ: Erlbaum, 1981.

Keele, S.W. Attention demands of memory retrieval. *Journal of Experimental Psychology,* 1972, *93,* 245–248.

Keele, S.W. *Attention and human performance.* Pacific Palisades: Goodyear, 1973.

Keren, G., O'Hara, W.P., & Skelton, J.M. Levels of noise processing and attentional control. *Journal of Experimental Psychology: Human Perception and Performance,* 1977, *3,* 653–664.

Kerr, B. Processing demands during mental operations. *Memory & Cognition,* 1973, *1,* 401–412.

Kleiman, G.M. Sentence frame contexts and lexical decisions: Sentence-acceptability and word-relatedness effects. *Memory & Cognition,* 1980, *8,* 336–344.

Klintman, H. Cognitive interaction in the visual field: A reaction time study. *Lund University Psychological Research Bulletin,* 1971, *11,* 5.

Koriat, A. Semantic facilitation in lexical decision as a function of prime-target association. *Memory & Cognition,* 1981, *9,* 587–598.

LaBerge, D. Automatic information processing: A review. In J. Long & A. Baddeley (Eds.), *Attention and performance IX.* Hillsdale, NJ: Erlbaum, 1981.

LaBerge, D., & Samuels, S.J. Toward a theory of automatic information processing in reading. *Cognitive Psychology,* 1974, *6,* 293–323.

Leonard, J.A. Tactual choice reactions. *Quarterly Journal of Experimental Psychology,* 1959, *11,* 76–83.

Levy, B.A. Interactive processing during reading. In A.M. Lesgold & C.A. Perfetti (Eds.), *Interactive processes in reading.* Hillsdale, NJ: Erlbaum, 1981.

Lewis, J.L Semantic processing of unattended messages using dichotic listening. *Journal of Experimental Psychology,* 1970, *85,* 225–228.

Logan, G.D. Attention in character-classification tasks: Evidence for the automaticity of component stages. *Journal of Experimental Psychology: General,* 1978, *107,* 32–63.

Logan, G.D. On the use of concurrent memory load to measure attention and automaticity. *Journal of Experimental Psychology: Human Perception and Performance,* 1979, *5,* 189–207.

Logan, G.D. Attention and automaticity in Stroop and priming tasks: Theory and data. *Cognitive Psychology,* 1980, *12,* 523–553.

Logan, G.D. Attention, automaticity and the ability to stop a speeded choice response. In J. Long & A. Baddeley (Eds.), *Attention and performance IX.* Hillsdale, NJ: Erlbaum, 1981.

Logan, G.D. On the ability to inhibit complex movements: A stop-signal study of typewriting. *Journal of Experimental Psychology: Human Perception and Performance,* 1982, *8,* 778–792.

Logan, G.D., & Zbrodoff, N.J. Constraints on strategy construction in a speeded discrimination task. *Journal of Experimental Psychology: Human Percpetion and Performance,* 1982, *8,* 502–520.

Lowe, D.G., & Mitterer, J.O. Selective and divided attention in a Stroop task. *Canadian Journal of Psychology,* 1982, *36,* 684–700.

Lupker, S.J., & Katz, A.N. Input, decision and response factors in picture-word interference. *Journal of Experimental Psychology: Human Learning and Memory,* 1981, *7,* 269–282.

Lupker, S.J., & Katz, A.N. Can automatic picture processing influence word judgments? *Journal of Experimental Psychology: Learning, Memory and Cognition,* 1982, *8,* 418–434.

Lynn, R. *Attention, arousal and the orienting reaction.* New York: Pergamon Press, 1966.

MacKay, D.G. Aspects of the theory of comprehension, memory and attention. *Quarterly Journal of Experimental Psychology,* 1973, *25,* 22–40.

MacKay, D.G. The problem of flexibility, fluency, and speed-accuracy trade-off in skilled behavior. *Psychological Review,* 1982, *89,* 483–506.

Marcel, A.J. Unconscious reading: Experiments on people who do not know they are reading. *Visible Language,* 1978, *12,* 392–404.

Marcel, A.J. Conscious and preconscious recognition of polysemous words: Locating selective effects of prior verbal context. In R.S. Nickerson (Ed.), *Attention and performance VII.* Hillsdale, NJ: Erlbaum, 1980.

Marcel, A.J., & Patterson, K.E. Word recognition and production: Reciprocity in clinical and normal studies. In J. Requin (Ed.), *Attention and performance VII.* Hillsdale, NJ: Erlbaum, 1978.

Martin, M. Speech recoding in silent reading. *Memory & Cognition,* 1978, *6,* 108–114.

McCauley, Ch., Parmelee, C.M., Sperber, R.D., & Carr, Th.H. Early extraction of meaning from pictures and its relation to conscious identification. *Journal of Experimental Psychology: Human Perception and Performance,* 1980, *6,* 265–276.

McLeod, P.D. A dual-task response modality effect: Support for multiprocessor models of attention. *Quarterly Journal of Experimental Psychology,* 1977, *29,* 651–667.

Mohnkopf, W. Zur Automatisierung willkürlicher Bewegungen. *Zeitschrift für Psychologie,* 1933, *130,* 235–299.

Moray, N. Attention in dichotic listening: Affective cues and the influence of instructions. *Quarterly Journal of Experimental Psychology,* 1959, *11,* 56–60.

Morton, J. The effects of context on the visual duration threshold for words. *British Journal of Psychology,* 1964, *55,* 165–180.

Morton, J. Categories of interference: Verbal mediation and conflict in card sorting. *British Journal of Psychology,* 1969, *60,* 329–346.

Morton, J. A functional model of memory. In D.A. Norman (Ed.), *Models of human memory.* New York: Academic Press, 1970.

Mowbray, G.H., & Rhoades, M.V. On the reduction of choice reaction times with practice. *Quarterly Journal of Experimental Psychology,* 1959, *11,* 16–23.

Münsterberg, H. *Beiträge zur Experimentellen Psychologie, Heft 1.* Freiburg: Mohr, 1889.

Myers, J.L., & Lorch, R.F., jr. Interference and facilitation effects of primes upon verification processes. *Memory & Cognition,* 1980, *8,* 405–414.

Navon, D. Forest before trees: the precedence of global features in visual perception. *Cognitive Psychology,* 1977, *9,* 353–383.

Navon, D., & Gopher, D. On the economy of the human processing system. *Psychological Review,* 1979, *86,* 214–255.

Navon, D., & Gopher, D. task difficulty, resources and dual-task performance. In R.S. Nickerson (Ed.), *Attention and performance VII.* Hillsdale, NJ: Erlbaum, 1980.
Neely, J.H. Semantic priming and retrieval from lexical memory: Roles of inhibitionless spreading activation and limited-capacity attention. *Journal of Experimental Psychology: General,* 1977, *106,* 226–254.
Neisser, U. Decision-time without reaction-time: Experiments in visual scanning. *American Journal of Psychology,* 1963, *76,* 376–385.
Neisser, U. *Cognitive psychology.* New York: Appleton-Century-Crofts, 1967.
Neisser, U. *Cognition and reality.* San Francisco: Freeman, 1976.
Neisser, U., & Becklen, R. Selective looking: Attending to visually specified events. *Cognitive Psychology,* 1975, *7,* 480–494.
Neisser, U., Hirst, W., & Spelke, E.S. Limited capacity theories and the notion of automaticity: Reply to Lucas and Bub. *Journal of Experimental Psychology: General,* 1981, *110,* 499–500.
Neumann, O. Aufmerksamkeit. In J. Ritter (Ed.), *Historisches Wörterbuch der Philosophie,* (Vol. 1). Basel, Stuttgart: Schwabe, 1971.
Neumann, O. Empfindung, In J. Ritter (Ed.), *Historisches Wörterbuch der Philosophie,* (Vol. 2). Basel, Stuttgart: Schwabe, 1972.
Neumann, O. Visuelle Aufmerksamkeit und der Mechanismus des Metakontrasts. In L. Eckensberger (Ed.), *Bericht über den 31. Kongreß der Deutschen Gesellschaft für Psychologie.* Göttingen: Hogrefe, 1979.
Newstead, S.E., & Dennis, I. Lexical and grammatical processing of unshadowed messages: A re-examination of the MacKay effect. *Quarterly Journal of Experimental Psychology,* 1979, *31,* 477–488.
Noble, M.E., Sanders, A.F., & Trumbo, D.A. Concurrence costs in double stimulation tasks. *Acta Psychologica,* 1981, *49,* 141–158.
Norman, D.A. Toward a theory of memory and attention. *Psychological Review,* 1968, *75,* 522–536.
Ostry, D., Moray, N., & Marks, G. Attention, practice and semantic targets. *Journal of Experimental Psychology: Human Perception and Performance,* 1976, *2,* 326–336.
Paap, K.R., & Ogden, W.C. Letter encoding is an obligatory but capacity-demanding operation. *Journal of Experimental Psychology: Human Perception and Performance,* 1981, *7,* 518–527.
Poltrock, S.E., Lansman, M., & Hunt, E. Automatic and controlled attention processes in auditory target detection. *Journal of Experimental Psychology: Human Perception and Performance,* 1982, *8,* 37–45.
Posner, M.I. *Chronometric explorations of mind. The Third Paul M. Fitts Lectures.* Hillsdale, NJ: Erlbaum, 1978.
Posner, M.I. Orienting of attention. *Quarterly Journal of Experimental Psychology,* 1980, *32,* 3–25.
Posner, M.I. Cumulative development of attentional theory. *American Psychologist,* 1982, *37,* 168–179.
Posner, M.I., & Boies, S.J. Components of attention. *Psychological Review,* 1971, *78,* 391–408.
Posner, M.I., & Cohen, Y. Attention and the control of movements. In G.E. Stelmach & J. Requin (Eds.), *Tutorials in motor behavior.* Amsterdam: North-Holland, 1980.
Posner, M.I., & Rogers, M.G. Chronometric analysis of abstraction and recognition. In W.K. Estes (Ed.), *Handbook of learning and congitive processes,* (Vol. 5). Hillsdale, NJ: Erlbaum, 1978.
Posner, M.I., & Snyder, C.R.R. Attention and cognitive control. In R. Solso (Ed.), *Information processing and cognition: The Loyola symposium.* Potomac, MD: Erlbaum, 1975. (a)
Posner, M.I., & Snyder, C.R.R. Facilitation and inhibition in the processing of signals. In P.M.A. Rabbitt & S. Dornič (Eds.), *Attention and performance V.* New York: Academic Press, 1975. (b)

Posner, M.I., Snyder, C.R.R., & Davidson, B.J. Attention and the detection of signals. *Journal of Experimental Psychology: General*, 1980, *109*, 160–174.

Proctor, R.W., & Proctor, J.D. Secondary task modality, expectancy, and measurement of attention capacity. *Journal of Experimental Psychology: Human Perception and Performance*, 1979, *5*, 610–624.

Rappard, H.V. A monistic interpretation of Wundt's psychology. *Psychological Research*, 1980, *42*, 123–234.

Rappard, H.V., Sanders, C., & De Swart, J.H. Wilhelm Wundt and the cognitive shift. *Acta Psychologica*, 1980, *46*, 235–255.

Reason, J. Actions not as planned: The price of automatization. In G. Underwood & R. Stevens (Eds.), *Aspects of consciousness*. New York: Academic Press, 1979.

Regan, J.E. Automaticity and learning: Effects of familiarity on naming letters. *Journal of Experimental Psychology: Human Perception and Performance*, 1981, *7*, 180–195.

Reitman, J. Skilled perception in Go: Deducing memory structures from inter-response times. *Cognitive Psychology*, 1976, *8*, 336–356.

Requin, J. Toward a psychobiology of preparation for action. In G.E. Stelmach & J. Requin (Eds.), *Tutorials in motor behavior*. Amsterdam: North-Holland, 1980.

Rosenbaum, D.A. Human movement initiation: Specification of arm, direction and extent. *Journal of Experimental Psychology: General*, 1980, *4*, 444–474.

Rumelhart, D.E., Norman, D.A. Simulating a skilled typist: A study of skilled cognitive-motor performance. *Cognitive Science*, 1982, *6*, 1–36.

Salthouse, T.A., & Somberg, B.L. Skilled performance: Effects of adult age and experience on elementary processes. *Journal of Experimental Psychology: General*, 1982, *111*, 176–207.

Sanders, A.F. Some remarks on mental load. In N. Moray (Ed.), *Mental workload: Its theory and measurement*. New York: Plenum Press, 1979.

Sanders, A.F. Towards a model of stress and human performance. *Acta Psychologica*, 1983, *53*, 61–97.

Schaeffer, B. Skill integration during cognitive development. In A. Kennedy & A. Wilkes (Eds.), *Studies in long-term memory*. New York: Wiley, 1975.

Schneider, W., & Fisk, A.D. Concurrent automatic and controlled visual search: Can processing occur without resource cost? *Journal of Experimental Psychology: Learning, Memory, and Cognition*, 1982, *8*, 261–278. (a)

Schneider, W., & Fisk, A.D. Degree of consistent training: Improvements in search performance and automatic process development. *Perception & Psychophysics*, 1982, *31*, 160–168. (b)

Schneider, W., & Shiffrin, R.W. Controlled and automatic human information processing: I. Decision, search, and attention. *Psychological Review*, 1977, *84*, 1–66.

Schuberth, R.E., Spoehr, K.T., & Lane, D.M. Effects of stimulus and contextual information on the lexial decision process. *Memory & Cognition*, 1981, *9*, 68–77.

Schulz, T. Stroop-Interferenz: Theorien und Daten. Ein Beitrat zur Anwendung der Kognitiven Psychologie. *Psychologische Beiträge*, 1978, *20*, 72–114.

Searleman, A. A review of right hemisphere linguistic capabilities. *Psychological Bulletin*, 1977, *84*, 503–528.

Seymour, P.H.K. Conceptual encoding and the locus of the Stroop effect. *Quarterly Journal of Experimental Psychology*, 1977, *29*, 245–265.

Shaffer, L.H. Multiple attention in continuous verbal tasks. In P.M.A. Rabbitt & S. Dornič (Eds.), *Attention and performance V*. New York: Academic Press, 1975.

Shaffer, L.H. Timing in the motor programming of typing. *Quarterly Journal of Experimental Psychology*, 1978, *30*, 333–345.

Shaffer, W.O., & LaBerge, D. Automatic semantic processing of unattended words. *Journal of Verbal Learning and Verbal Behavior*, 1979, *18*, 413–426.

Shallice, T. The dominant action system: An information-processing approach to consciousness. In K.S. Pope & J.L. Singer (Eds.), *The stream of consciousness*. New York: Plenum Press, 1978.

Shiffrin, R.M. The locus and role of attention in memory systems. In P.M.A. Rabbitt & S. Dornič (Eds.), *Attention and performance V.* New York: Academic Press, 1975.

Shiffrin, R.M., Dumais, S.T., & Schneider, W. Characteristics of automatism. In J. Long & A. Baddeley (Eds.), *Attention and performance IX.* Hillsdale, NJ: Erlbaum, 1981.

Shiffrin, R.M., & Schneider, W. Controlled and automatic information processing: II. Perceptual learning, automatic attending, and a general theory. *Psychological Review,* 1977, *84,* 127–190.

Siebeck, H. *Geschichte der Psychologie.* Reprint, Amsterdam: Schippers, 1961 (originally published 1879).

Sokolov, E.N. *Perception and the conditioned reflex.* Oxford: Pergamon Press, 1963.

Solomons, L., & Stein, G. Normal motor automatism. *Psychological Review,* 1896, *3,* 492–512.

Spelke, E., Hirst, W., & Neisser, U. Skills of divided attention. *Cognition,* 1976, *4,* 215–230.

Sperber, R.D., McCauley, L., Ragain, R., & Weil, C. Semantic priming effects on picture and word processing. *Memory & Cognition,* 1979, *7,* 339–345.

Sperling, G. Successive approximations to a model for short term memory. *Acta Psychologica,* 1967, *27,* 285–292.

Stanovich, K.E., & West, R.F. Mechanisms of sentence context effects in reading: Automatic activation and conscious attention. *Memory & Cognition,* 1979, *7,* 77–85.

Stanovich, K.E., & West, R.F. The effect of sentence context on ongoing word recognition: Test of a two-process theory. *Journal of Experimental Psychology: Human Perception and Performance,* 1981, *7,* 658–672.

Sternberg, S. The discovery of processing stages: Extension of Donders' method. *Acta Psychologica,* 1969, *30,* 276–315.

Stirling, N. Stroop interference: An input and an output phenomenon. *Quarterly Journal of Experimental Psychology,* 1979, *31,* 121–132.

Stroop, J.R. Studies of interference in serial verbal reactions. *Journal of Experimental Psychology,* 1935, *18,* 643–662.

Tecce, J.J., & Happ, S.J. Effects of shock-arousal on a card-sorting test of color-word interference. *Perceptual and Motor Skills,* 1964, *19,* 905–906.

Treisman, A.M. Contextual cues in selective listening. *Quarterly Journal of Experimental Psychology,* 1960, *12,* 242–248.

Treisman, A.M. Perceptual grouping and attention in visual search for features and objects. *Journal of Experimental Psychology: Human Perception and Performance,* 1982, *8,* 194–214.

Treisman, A.M., & Gelade, G. A feature-integration theory of attention. *Cognitive Psychology,* 1980, *12,* 97–136.

Treisman, A.M., Squire, R., & Green, J. Semantic processing in dichotic listening? A replication. *Memory & Cognition,* 1974, *2,* 641–646.

Trevarthen, C.B. Modes of perceiving and modes of acting. In H.I. Pick & E. Saltzman (Eds.), *Modes of perceiving and processing information.* Hillsdale, NJ: Erlbaum, 1978.

Tweedy, J.R., & Lapinski, R.H. Facilitating word recogniton: Evidence for strategic and automatic factors. *Quarterly Journal of Experimental Psychology,* 1981, *33A,* 51–59.

Tweedy, J.R., Lapinski, R.H., & Schvaneveldt, R.W. Semantic context effects on word recognition: Influence of varying the proportion of terms presented in an appropriate context. *Memory & Cognition,* 1977, *5,* 84–98.

Underwood, G. Moray vs. the rest: The effects of extended shadowing practice. *Quarterly Journal of Experimental Psychology,* 1974, *26,* 368–372.

Underwood, G. Contextual facilitation from attended and unattended messages. *Journal of Verbal Learning and Verbal Behavior,* 1977, *16,* 99–106.

Underwood, G. (Ed.) *Stategies of information processing.* New York: Academic Press, 1978.

Venables, P.H. Cognitive disorder. In J.K. Wing (Ed.), *Schizophrenia. Towards a new synthesis.* London: Academic Press, 1978.

Ward, L.M. Determinants of attention to local and global features of visual forms. *Journal of Experimental Psychology: Human Perception and Performance,* 1982, *8,* 562–581.

Wardlaw, K.A., & Kroll, N.E.A. Autonomous responses to shock-associated words: A failure to replicate. *Journal of Experimental Psychology: Human Perception and Performance,* 1976, *2,* 357–360.

Warren, R.E. Time and the spread of activation. *Journal of Experimental Psychology: Human Learning and Memory,* 1977, *3,* 458–466.

Waugh, N., & Norman, D.A. Primary memory. *Psychological Review,* 1965, *72,* 89–104.

Weiskrantz, C., Warrington, E.K., Sanders, M.D., & Marshall, J. Visual capacity in the hemianopic field following a restricted occipital ablation. *Brain,* 1974, *97,* 709–728.

West, R.F., & Stanovich, K.E. Source of inhibition in experiments on the effect of sentence context on word recognition. *Journal of Experimental Psychology: Learning, Memory, and Cognition,* 1982, *8,* 385–399.

Wickens, C.D. The structure of attentional resources. In R.S. Nickerson (Ed.), *Attention and performance VII.* Hillsdale, NJ: Erlbaum, 1980.

Wolford, G., & Morrison, F. Processing of unattended visual information. *Memory & Cognition,* 1980, *8,* 521–527.

v. Wright, J.M., Anderson, K., & Stenman, U. Generalization of conditioned CSRs in dichotic listening. In P.M.A. Rabbitt & S. Dornič (Eds.), *Attention and performance V.* New York: Academic Press, 1975.

Wundt, W. *Grundzüge der Physiologischen Psychologie,* (5th edn., Vol. 3). Leipzig: Engelmann, 1903.

Summary. This chapter compares two concepts of automaticity. The first concept, which is rooted in cognitive research and has recently been popularized by the "two process" approach (e.g., Posner & Snyder, Schneider & Shiffrin), views automaticity essentially as a matter of *direction of processing:* Automatic processes occur in a passive, bottom-up fashion, irrespective of intentions and free from interference. The second view, which dates back to Wundt and has been developed mainly in motor research, regards automaticity as a matter of *level of control:* Automatization is the acquisition of skills that enable actions or parts of actions to be controlled at a level not associated with conscious awareness. The main part of the chapter contains a critical review of recent data relevant to these concepts. It is concluded that most findings are better accommodated by the "level of control" view than by the "direction of processing" view. In the final section, a theory of automaticity is outlined which is based on this view. It conceptualizes automaticity as a *mode of parameter specification:* A process is automatic, if all its parameters are specified by a skill (a procedure stored in long-term memory) in conjunction with environmental information. If these two sources of constraint cannot specify all parameters, further constraints must be provided by attentional mechanisms.

18 Motor Learning as a Process of Structural Constriction and Displacement

HERBERT HEUER

Contents

1 Introduction

The purpose of this paper is to argue for a particular view of motor learning which is derived from two different sets of data. The first set stems from studies of interference between motor tasks on the one hand and tasks of a more cognitive nature on the other hand, which are simultaneously performed at different stages of training of the motor task. As is evident from the early work of Mohnkopf (1933) and Bahrick and Shelly (1958), the original aim of these studies was to trace the time course of automation. The second set of data stems from correlational studies of individual differences. Originally, these studies were concerned with the prediction of training success in motor tasks, and for this purpose motor performance at different levels of skill was correlated with a variety of other test scores.

Despite the apparent unrelatedness of the experimental approaches, their main results may be linked by two concepts which I will call structural constriction and structural displacement. I shall summarize briefly the general results from both approaches and discuss the linking concepts. This is followed by a more concrete example and some comments on implications.

Cognition and Motor Processes
Ed. by W. Prinz and A.F. Sanders

2 Experimental Results

In most studies the *correlations* between predictor tests and motor performance at different levels of skill have been subjected to factor analysis. In view of the serious criticisms of such analyses (e.g. Bechtoldt, 1970), we shall only consider the correlations. The main issue is whether and in what direction the correlations may change as a function of the amount of training of the motor task. For each predictor test we computed the rank correlation between performance correlations on one hand and stage of practice on the other. These correlations indicate the trend of the performance correlations: when the performance correlations increase with practice, the rank correlation is positive, while it is negative when performance correlations decline with practice.

Figure 1 displays the distribution of the rank correlations from three different studies (Fleishman & Hempel, 1954, 1955; Fleishman, 1955). In these studies three different motor tasks were used, the complex coordination test (CCT), discriminative reaction time (DRT), and rotary pursuit (RP). The predictor tests were partly identical. Most of the trends are negative, as are those reported by Reynolds (1952), Adams (1957), and others. Of course, the predominance of negative trends could depend on the sample of the tests; but these tests covered a relatively broad area ranging from word knowledge to simple reaction time. Therefore, it seems reasonable to conclude that during motor learning correlations with most other performance scores decline.

Figure 1 does not show only a majority of negative trends, but also a minority of distinct positive trends. For some tests the correlation with motor performance increases during learning. Whatever factors are measured by these tests, they appear to become progressively more important rather than less important for motor performance.

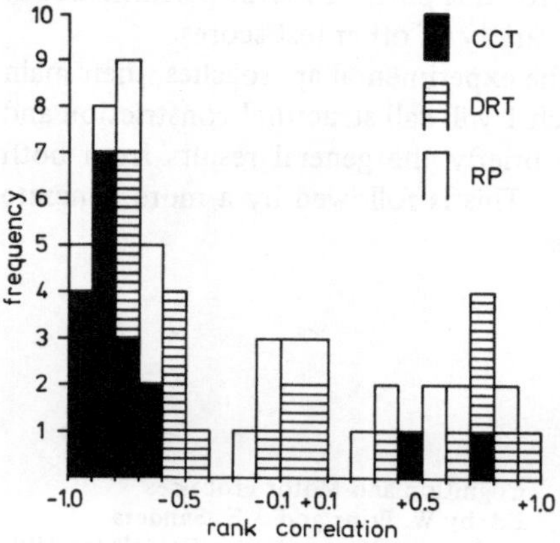

Figure 1. Trends in the changes of correlations between motor performance and different tests in the course of learning three different motor tasks: complex coordination test (CCT; Fleishman & Hempel, 1954), discriminative reaction time (DRT; Fleishman & Hempel, 1955), rotary pursuit (RP; Fleishman, 1960)

The general results of *dual-task studies* are similar to those of correlational studies. In a majority of the experiments dual-task interference declines as the motor task is practiced (e.g. Mohnkopf, 1933; Bahrick & Shelly, 1958; Truijens, Trumbo, & Wagenaar, 1976. Heuer & Merz, Note 1; Damos & Wickens, 1980). This is consistent with the familiar notion of automation in the course of motor learning. However, there are two types of exceptions.

The first type is found in experiments in which the learning curves for either tracking performance alone or in combination with a second task (e.g. serial anticipation of numbers or generation of a random sequence of numbers) do not converge (Noble, Trumbo, & Fowler, 1967; McLeod, 1973). That is, the amount of dual-task interference observed in tracking performance remains roughly the same in the course of learning. In the experiment of McLeod (1973) there were also control conditions in which the secondary tasks were performed alone. Again, the learning curves for either the secondary tasks alone or in combination with tracking did not converge; there was even a small trend towards divergence. Therefore, the non-decreasing interference in the tracking data cannot be attributed to a strategy that aims at a constant performance decrement in tracking and at improving performance in the secondary task (see Navon & Gopher, 1979).

Exceptions of the second type occur in experiments on tracking with predictable and unpredictable tracks. Performance improves more with predictable tracks, and after some practice is is well above that with unpredictable ones. Since dual--task interference generally declines as a tracking task is practiced, one would expect the performance decrement to be smaller with the predictable than with the unpredictable track when a secondary task is introduced late in training. Yet, the opposite is observed (Trumbo, Noble, & Quigley, 1968; Pew, 1974). Although there is more improvement in performance with predictable tracks, performance in such tasks is more disturbed by the secondary task than performance in tasks with unpredictable tracks. These results do not imply that dual-task interference may actually increase in the course of training of a motor task, but they show that more practice benefits are not necessarily associated with less dual-task interference.

Thus, the basic results from correlational and dual-task studies may be summarized as follows: When a motor task is trained, correlations with performance in most other tasks decline, and interference with most other tasks performed simultaneously declines as well. However, there are exceptions in that correlations with performance in some other tasks increase rather than decrease. Similarly, the amount of interference sometimes remains about the same.

3 Structural Constriction and Displacement

There may be various ways to arrive at a common interpretation for the convergent results of the correlational and dual-task studies. The interpretation I would like to suggest is based on two concepts – structural constriction and displacement –

which are admittedly somewhat global and maybe embedded in different theoretical frameworks. The framework chosen here concerns multiple resource theory (Navon & Gopher, 1979). We shall briefly comment on other possible frameworks in a later section, but the basic ideas will remain the same.

Consider the hypothetical example of Figure 2. In the left half a hypothetical resource demand composition is shown of some motor task early in training. The demands of the task on a certain resource pool indicate the average relative amount of resources of that type invested during performance. Interference will be largest with those secondary tasks that have high demands on the same resource types, that is on types 1 and 2. When a secondary task has low demands on types 1 and 2 but high demands on types 3 and 4, interference will be either small or absent.

Now, assume that there are interindividual differences in the efficiency of the various types of resources. These differences will have similar effects on tasks with similar resource demand compositions. In other words, performance in such tasks will correlate. In contrast, performance in tasks with dissimilar resource demand compositions will show lower correlations as long as the efficiency of the various resource types varies independently, at least to some degree.

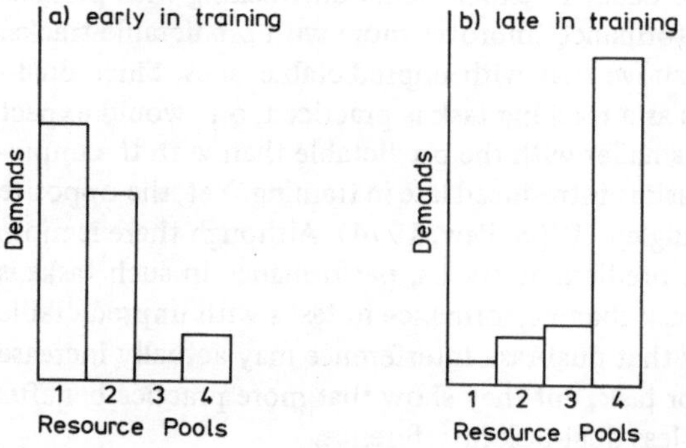

Figure 2a, b. An illustrative example of structural constriction and displacement

In the right half of Figure 2 the hypothetical resource demand composition is illustrated of the same motor task after intensive practice. The changes are twofold: The fist change I call *structural constriction.* In the course of learning there is a decrease of the number of resource pools substantially involved in motor performance. The second change I call *structural displacement.* Late in training other resource pools are important for motor performance than early in training.

The concepts of structural constriction and structural displacement refer only to the change in the resource demand composition and not to the origins of this change. Basically, two such origins could exist. The first could be a change in strategy. After practice subjects may change the way in which they perform the task. This can increase demands on resources which were hardly used in the initial stages of training. The second origin is an increase in the efficiency of resources which is likely to reduce the mean demands by hitting data limits. This increase in efficiency may differ for the various resource types and thus also contribute to the change described as constriction and displacement.

The observable consequences of structural constriction and displacement are the following:

1. Structural constriction is likely to produce a decreasing amount of dual-task interference with the majority of secondary tasks. Early in training (Figure 2) interference will be large with all tasks that have high demands on resource types 1 and 2, but late in training interference will be large only with those tasks that make high demands on resource type 4. A decrease of dual-task interference will be observed even more frequently when the different resource types are involved in different numbers of tasks, and when the demands on broader resource types are more likely to decline in the course of motor learning.

A similar argument holds with respect to correlations. At the onset of training, performance in the motor task will correlate with all other tasks which also have high demands on resource types 1 and 2 (and this correlation will be larger as the demands on types 3 and 4 are low). All these correlations will decrease, and these will be more than the correlations which increase. The latter are correlations with those tasks which have high demands on resource type 4 (and the increase will be larger when the demands on types 1–3 are low). In conclusion, structural constriction is the concept used to explain the majority of decreasing correlations and the majority of findings of decreasing dual-task interference.

2. Structural displacement is the concept used to explain the minority of increasing correlations and findings of unchanged dual-task interference. Correlations will increase with tasks that have high demands on resource type 4, and dual-task interference is likely to increase with such tasks as well. A problem is that examples of increasing dual-task interference appear to be almost non-existent. We know only of one such example reported by Bornemann (1942) which is statistically somewhat unreliable. Thus it seems that an actual increase of the demands on certain resource types is a rare event.

4 An Illustrative Example

To provide a more concrete example of the hypothesized changes consider the discriminative reaction time task used as a criterion task by Fleishman and Hempel (1955) and by Adams (1957). Figure 3 shows the somewhat complicated rule for the stimulus-response relationship. The rule was: Activate that switch that is in the same relative position to the other switches as is the red light relative to the green one. This rule is obviously spatial.

For this task one might distinguish three stages in the reaction process, namely stimulus identification, response selection, and response programming, although,

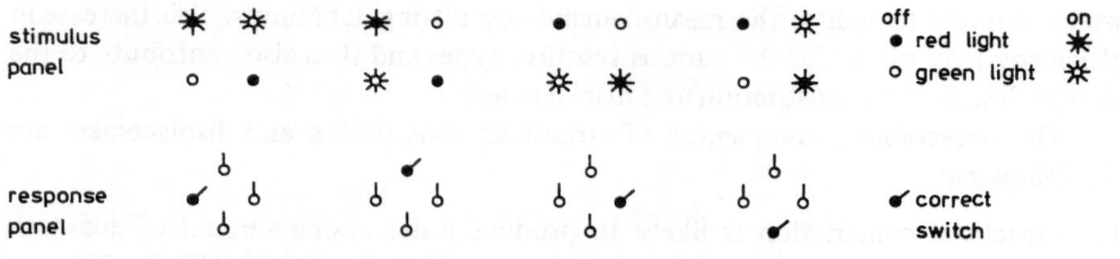

Figure 3. Stimulus-response relationship in the discriminative reaction time task of Fleishman and Hempel (1955) and Adams (1957)

of course, there could be more detailed subdivisions (see Sanders, 1980). Following a suggestion of Wickens (1980) stages can be considered as resource pools. It should be stressed that a one-to-one correspondence between stages and resource pools is unlikely (see Gopher & Sanders, this volume), but here the assumption serves an expositive purpose and is an approximation to a situation where different stages have unequal demands on different resource types.

Figure 4 depicts a hypothetical resource demand composition for the task early and late in training. The main change is the decrease in the demands of the response selection stage on the corresponding type of resources. This produces a general decrease in the demands and a different ranking of the three resource types with respect to their relative importance.

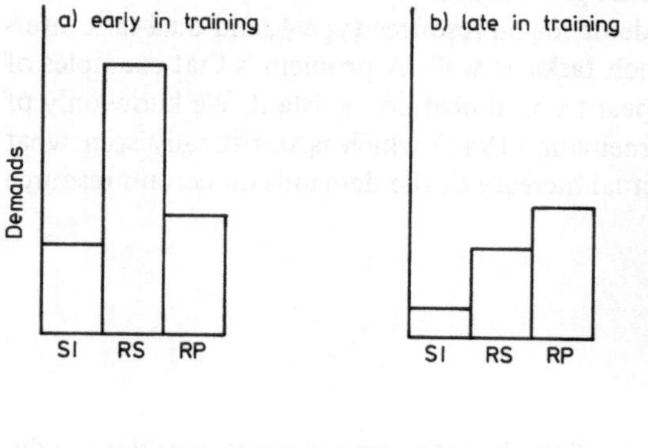

Figure 4a, b. Hypothetical demands of different stages of the discriminative reaction time task. (*SI:* stimulus identification; *RS:* response selection; *RP:* response programming.) For each stage a different resource type is assumed

For correlational data these hypothetical changes in resource demand composition imply that any test which has high demands on response selection resources should show decreasing correlations with the trained criterion task. Bearing in mind that the SR-relationship was spatially defined, tests involving spatial imagery could be of this type. On the other hand, tests with low demands on response selection resources and higher demands on response programming resources might show increasing correlations, since their resource demand composition is similar to that

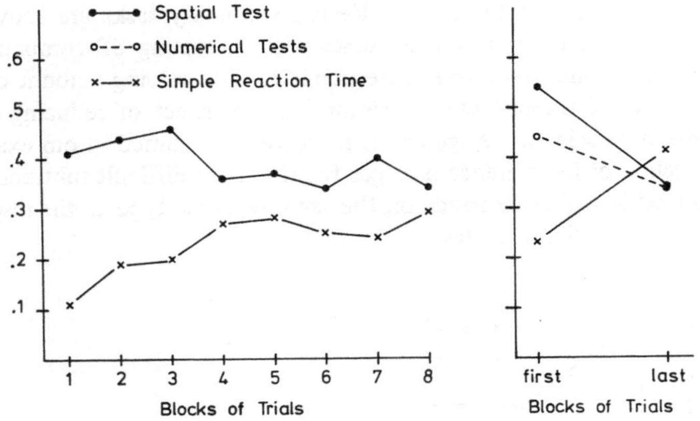

Figure 5. Correlations between discriminative reaction time and other tests in different stages of practice. *Left panel:* data from Fleishman and Hempel (1955). *Right panel:* data from Adams (1957). *Spatial test:* pattern comprehension. *Numerical tests:* arithmetic reasoning; numerical operations II

of the trained task. Simple reaction times are of this type. The corresponding results of Fleishman and Hempel (1955) and Adams (1957) are shown in Figure 5 in the left and right panels respectively. In addition, the decreasing correlation of a numerical test is shown, to which I shall return later.

What are the expectations for dual-task performance data? A task that demands resources only for response programming should show unchanged interference. A task with high demands on response selection resources should clearly show a decline of interference. In general, the decline of interference should be larger when the demands on response selection resources become larger relative to the demands on response programming resources.

Unfortunately, I have not found an interference study which parallels the correlational studies. But the hypothezised changes during learning might be mimicked by tasks that vary in their spatial SR-compatibility. Tasks with low compatibility have high demands on response selection resources and are similar to an untrained task, while tasks with high compatibility have low demands of response selection and might be considered as equivalent to a trained task. Tasks of this kind were used by Keele (1967). His secondary tasks were continuous mental subtractions of −1, −2, and −7. One might suppose that these variants affect the extent of demands on those resources which also supply the response selection stage of the reaction time task. At least this seems plausible in the light of the decreasing correlation between reaction time and numerical skills shown in the right panel of Figure 5. Thus we have tasks that mimic different phases of training and secondary tasks that vary in their relative demands on response selection resources.

The results for two of Keele's secondary tasks are shown in Figure 6. The amount of interference decreases with increasing SR-compatibility of the choice reaction time task. This corresponds to a decreasing amount of interference in the course of training when training has the effect of reducing the demands of the response selection stage on its resources as assumed in our example. Moreover, the decline of interference is larger for the more difficult subtraction task with its presumably higher demands on the same resource type as the response selection stage of the reaction time task.

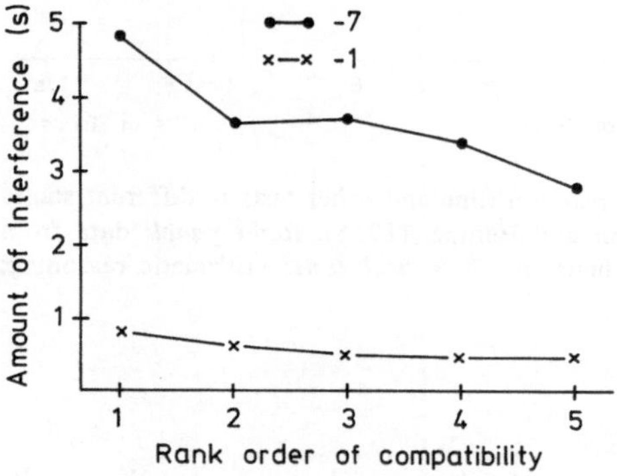

Figure 6. Amount of interference as a function of SR-compatibility for two difficulty levels of a mental substraction task (from Keele, 1967)

5 Other Theoretical Frameworks for Structural Constriction and Displacement

Although I have introduced the notions of structural constriction and displacement in the framework of multiple resources and processing stages – the relationship between these is more fully discussed by Gopher and Sanders (this volume) – there is no necessity to do so. Other theoretical frameworks also allow inclusion of these concepts as a description of changes that occur during motor learning. For example, consider the model of functional cerebral space (Kinsbourne & Hicks, 1978). In this framework structural constriction means that progressively less cerebral space is involved in performance during motor learning. The notion of structural displacement points to the possibility of shifts in respect to the maximally involved brain areas. A comparison of this framework with that of resource pools and processing stages makes it clear that structural constriction and displacement may refer to purely conceptual elements as well as to (at least in principle) anatomically definable ones.

Another general concept in terms of which structural constriction and displacement may be formulated is the hierarchical framework as outlined by Pew (this volume). In a distributed processing system the relative importance of the different processing units for task performance changes during training. The number of substantially involved processing units decreases. Again, on the final level of skill other processing units are most important than in early practice.

The traditional concept of abilities is a further general framework for structural constriction and displacement. Fleishman (1966) concludes that in the course of motor learning the number of abilities involved in performance decreases while there is also a shift in relative importance of different abilities. This can be described as constriction and displacement, but these notions are not intimately associated with an ability framework.

In conclusion, the notions of structural constriction and displacement are not confined to a certain framework, but they can be included in any general conception which defines elements on which constriction and displacement can take place. Whatever these elements may be, in the course of motor learning their involvement in performance appears to change in certain ways.

6 Some Implications

We believe that there is some heuristic value in considering sensorimotor learning in part as a process of structural constriction and displacement. Which questions arise from this view? The most important set of questions might well be concerned with the nature of structural displacements, and a general approach is by way of dual-task experiments. A motor task of interest has to be combined with at least two different secondary tasks at different stages of practice. These secondary tasks have to be selected such that they differ in particular processing demands. The changing pattern of interference between the motor task and both secondary tasks will in principle allow for conclusions about changing processing demands of the motor task in the course of practice, that is, about structural displacements.

The correlational data may be used in two ways. First, the available data might give some hints concerning the type of secondary tasks that could be most profitably used in the search for structural displacement. Second, correlations could be used as a convergent operation to validate the conclusions based upon dual-task interference data. As an example, correlational data suggest that, in pursuit tracking, imagery is involved in early stages of practice, but not in later stages (Fleishman & Rich, 1963). Interference data show also that imagery is involved early in practice (Baddeley, Grant, Wight, & Thomson 1975; Baddeley & Lieberman, 1980). This pattern thus poses the question whether interference between tracking and imagery decreases in the course of training which in turn might be compared with the interference pattern observed for a task demanding kinesthetic discriminations. For this type of task Fleishman and Rich found an increasing correlation in the course of motor learning.

Questions about the nature of structural displacements are, of course, not very different from those studied earlier in the context of the ability concept. But now the conceptual framework is different, the experimental techniques are different and not restricted to correlations only, and most likely the secondary tasks will differ from the tasks used in traditional correlational analyses. Thus, although psychology may be again moving in circles here, they could well be arranged in a spiral. If this spiral has a vertical orientation, it cannot be excluded that the movement is in an upward direction.

Reference Note

1. Heuer, H., & Merz, F. Einfluß von psychischen Belastungen auf das Erlernen und die Struktur motorischer Handlungen. *Berichte aus dem Fachbereich Psychologie der Philipps-Universität Marburg/Lahn,* Nr. 67, 1979.

References

Adams, J.A. The relationship between certain measures of ability and the acquisition of a psychomotor criterion response. *Journal of General Psychology,* 1957, *56,* 121–134.

Baddeley, A.D., Grant, S., Wight, E., & Thomson, N. Imagery and visual working memory. In P.M.A. Rabbitt & S. Dornic (Eds.), *Attention and performance V.* New York: Academic Press, 1975.

Baddeley, A.D., & Lieberman K. Spatial working memory. In R.S. Nickerson (Ed.), *Attention and performance VIII.* Hillsdale, N.J.: Erlbaum, 1980.

Bahrick, H.P., & Shelly, C. Time sharing as an index of automatization. *Journal of Experimental Psychology,* 1958, *56,* 288–293.

Bechtholdt, H.P. Motor abilities in studies of motor learning. In L.E. Smith (Ed.), *Psychology of motor learning.* Chicago: The Athletic Institute, 1970.

Bornemann, E. Untersuchungen über den Grad der geistigen Beanspruchung. II. Teil. Praktische Ergebnisse. *Arbeitsphysiologie,* 1942, *12,* 173–191.

Damos, D.L., & Wickens, C.D. The identification and transfer of timesharing skills. *Acta Psychologica,* 1980, *46,* 15–39.

Fleishman, E.A. Abilities at different stages of practice in rotary pursuit performance. *Journal of Experimental Psychology,* 1960, *60,* 162–171.

Fleishman, E.A. Human abilities and the acquisition of skill. In E.A. Bilodeau (Ed.), *Acquisition of skill.* New York: Academic Press, 1966.

Fleishman, E.A., & Hempel, W.E. Changes in factor structure of a complex psychomotor test as a function of practice. *Psychometrika,* 1954, *19,* 239–252.

Fleishman, E.A., & Hempel, W.E. The relation between abilities and improvement with practice in a visual discrimination reaction task. *Journal of Experimental Psychology,* 1955, *49,* 301–312.

Fleishman, E.A., & Rich, S. Role of kinesthetic and spatial-visual abilities in perceptual motor learning. *Journal of Experimental Psychology,* 1963, *66,* 6–11.

Keele, S.W. Compatibility and time-sharing in serial reaction time. *Journal of Experimental Psychology,* 1967, *75,* 529–539.

Kinsbourne, M., & Hicks, R.E. Functional cerebral space: A model of overflow, transfer and interference effects in human performance: A tutorial review. In J. Requin (Ed.), *Attention and performance VII.* Hillsdale, N.J.: Erlbaum, 1978.

McLeod, P.D. Interference of "attend to and learn" tasks with tracking. *Journal of Experimental Psychology,* 1973, *99,* 330–333.

Mohnkopf, W. Zur Automatisierung willkürlicher Bewegungen (Zugleich ein Beitrag zur Lehre von der Enge des Bewußtseins). *Zeitschrift für Psychologie,* 1933, *130,* 235–299.

Navon, D., & Gopher, D. On the economy of the human processing system. *Psychological Review,* 1979, *86,* 214–255.

Noble, A., Trumbo, D.A., & Fowler, F. Further evidence on secondary task interference in tracking. *Journal of Experimental Psychology,* 1967, *73,* 146–149.

Pew, R.W. Levels of analysis in motor control. *Brain Research,* 1974, *71,* 393–400.

Reynolds, B. The effect of learning on the predictability of psychomotor performance. *Journal of Experimental Psychology,* 1952, *44,* 189–198.

Sanders, A.F. Stage analysis of reaction processes. In G.E. Stelmach & J. Requin (Eds.), *Tutorials in motor behavior.* Amsterdam: North-Holland, 1980.

Truijens, C.L., Trumbo, D.A., & Wagenaar, W.A. Amphetamine and barbiturate effects on two tasks performed singly and in combination. *Acta Psychologica,* 1976, *40,* 239–244.

Trumbo, D.A., Noble, M., & Quigley, J. Sequential probabilities and the performance of serial tasks. *Journal of Experimental Psychology,* 1968, *76,* 364–372.

Wickens, C.D. The structure of attentional resources. In R.S. Nickerson (Ed.), *Attention and performance VIII.* Hillsdale, N.J.: Erlbaum, 1980.

Summary. Correlations between motor task performance and a variety of other test scores tend to decline when the motor task is practiced, but sometimes they increase. Similarly, dual-task interference between a motor task and a variety of other tasks mostly decreases in the course of training, but sometimes not. It is argued that both sets of results can be interpreted in terms of structural constriction and displacement. These notions can be used in different theoretical frameworks, e.g. those of multiple resources or functional cerebral space. In general terms they denote two changes that occur during motor learning. First, the number of structures (resource pools etc.) substantially involved in performance decreases, and second, those structures maximally involved may be different at the onset and end of training. An implication of the common interpretation for two different sets of data is that correlational and interference studies may be used as convergent operations to trace the nature of structural changes in the course of motor learning.

V Interactions Between Cognition and Action in Development

19 Cognition and Action in Development: A Tutorial Discussion

HERBERT L. PICK

Contents

1 Introduction

The relation between cognition and action in development is two-sided. On the one hand, action has profound implications for cognitive development. (The meaning of cognition is intended to include perception as well as the more usual cognitive processes.) On the other hand, cognition including perception is intimately involved in the development of action or skilled behavior. Indeed, current dissatisfaction with consideration of motor performance and perception as separate aspects of behavior has led a number of investigators to seek approaches which provide for tighter integration of these two concepts. One systematic analysis that will serve as a point of departure for the current discussion has been provided by Reed (1982). He argues for a concept of *action system* which is a functional or goal directed organization of behavior including both sensory and motor components. In fact, he argues quite cogently that these components cannot really be separated. Action systems in animals are defined by a variety of functions. These include ingestion of food, communication, play, information acquisition, object manipulation, locomotion, etc.

The plan of the present discussion is to try to illustrate both sides of the cognition-action relation with examples from current developmental research. Much of the most interesting current research is with infants and it will figure heavily in the present discussion. The state of the art in development of action is very uneven.

Cognition and Motor Processes
Ed. by W. Prinz and A.F. Sanders

Significant progress has been made but there are large gaps to be filled. Some of the research will also illustrate persisting theoretical and meta-theoretical issues in development of action. In the concluding section an attempt will be made both to highlight areas where empirical work is needed and to suggest issues that require theoretical development.

2 Relation of Action to Perceptual Cognitive Development

Action has been related to various aspects of cognition, including the acquisition of knowledge, its organization, and memory and retrieval. For example, in the case of organization of knowledge, intermediate implicit (or explicit) motor responses have often been hypothesized as mediating links between stimuli and responses and as links to account for generalization of responses. Motor routines have often been used as aids in memory tasks. In development nowhere is the relation of action to perception and cognition clearer than with acquisition of information.

There are a number of primitive motor skills and coordinations whose function seems to be the promotion of information acquisition. Two examples of these will be considered: early eye movement coordination and motor behaviors resulting in cross-modal processing of information.

2.1 Eye Movements

It would not be surprising if movements of the eyes were easier for babies to control than movements of many organs. The mass of the eye is quite small, requiring little muscular strength for movement. The resting position of the eyes with respect to the body is relatively constant providing a stable reference position for initiating eye movements. How are babies' eye movements organized to promote information acquisition? The most systematic study of eye movements of neonates has been reported by Haith (1980). He examined the nature of the eye movements of newborns by photographic methods when they were exposed to uniform homogeneous visual fields in the dark, in the light, and when presented vertical or horizontal edges, bars, and angles. With rather elegant infrared technology he was able to photograph eye movements in the dark and compare them with eye movements in the homogeneous lighted fields. There were rather marked differences in these two conditions. In the dark, the babies' eyes were open wider, the movements were more regular (i.e., they were rated as less jerky and astigmatic, showing more normal saccades and drifts), and the movements were, on the whole, smaller than in the uniform light field. These differences seemed to be qualitative in nature rather than endpoints of a quantitative continuum since changes of overall brightness when the field was lighted did not make any difference in the type of eye movemens. Haith suggested that the scanning in the dark might be mediated by an endogeneous motor organization whose functional rules were something like: (1) if awake and alert and the illumination is low – open eyes, and (2) if in darkness – search. The jerky, nystagmic, large eye movements in the homogeneous light fields might reflect the system shifting to a mode of control guided by external stimuli; however, the uniform nature of the stimulus field does not provide the right kind of structure for such control.

This possibility led Haith to perform further experiments adding structure to the visual field in the form of edges, bars, and angles. Generally, the eye movements were similar for both edges and bars. Fixations were clustered close to the edge or bar for both horizontally and vertically oriented stimuli in comparison with a no-stimulus control. Eye movements in the presence of vertical or horizontal stimuli were larger than in their absence and tended to result in edge crossing. If, as with adults, these saccadic eye movements cannot be controlled during movement, the observation of edge crossing implies anticipation of the result of an eye movement and perhaps a form of planning. Angles elicited fixation no more often than the simpler bar and edge stimuli. All together, these results suggested an additional rule: If in an illuminated environment and an edge is found, stay near the edge and attempt to cross it. Further research and analysis of the literature suggested to Haith the additional rule: When near an edge, reduce the dispersion of fixations perpendicular to the edge as contour density increases. Clearly the systematic nature of the neonates' eye movements would have the effect of ensuring that the baby focuses on informative parts of the visual field.

Whether the function of this systematic behavior is actually to gain information or something else, such as ensuring a high rate of cortical activity, is not known. A step forward answering this question might be provided by adding more meaningful structure to the stimulus field in this kind of research.

Let us consider a second instance of eye movement coordination that would seem to function to make information more available to the infant – accommodative convergence. Typically, when viewing objects binocularly humans change the convergence of their eyes as the distance of an object changes. An important function of this is the avoidance of double images or diplopia. However, it has been noted that adults converge their eyes even when viewing approaching objects monocularly. It seems that there is a link between the accommodation system and the vergence system. As an object approaches and the eye accommodates to maintain a clear image, the eyes also converge. Recently Aslin and Jackson (1979) have shown that this association between accommodation and convergence exists in 2-month-old babies. They presented an approaching target to infants monocularly and measured the amount of convergence of the two eyes. The percentage of convergence of the two eyes for a target distance of 15 cm in relation to a distance of 40 cm was about 2.9, which was about 70% of the adult value. By 6 months of age it had increased to 3.6%, which was about 86% of the adult value. This again would seem to be an eye movement mechanism which functions to provide the visual system with good information. It is interesting to note that this improvement in accommodative convergence parallels the improvement in accommodation per se between 2 and 6 months of age as observed by a number of investigators, e.g., Braddock and Atkinson (1979). See also Banks (1980) for a general review. The early presence of this rather exquisite coordination suggests a system genetically organized in form but one which is posibly finely tuned with experience; the system must be flexible enough to adjust the amount of convergence for developmental changes in interpupillary distance.

2.2 Cross-Modal Information Processing

The second example of primitive motor coordination whose function seems to promote information acquisition is provided by early instances of cross-modal information processing, e.g., the auditory-visual coordination indicated by eye movements directed toward sound sources. Wertheimer (1961) first reported this with the example of a newborn looking in the direction of clicking sound. Butterworth and Castillo (1976) studied this coordiation more systematically and found that eye movements would be directed toward or away from a sound source depending on its intensity. This result not only demonstrates auditory-visual localization but also an approach-avoidance adjustment of the motor behavior depending on the stimulus intensity. Of course, the well known rooting response is an example of a similar positive tactual and orienting response and Riser, Yonas, and Wilkner (1976) demonstrated an analogous appropriate avoidance of noxious odors by newborns.

As argued in the case of eye movements, one might look for examples of information acquiring motor behavior in skills where the infant is relatively proficient, where there is a relatively light constant muscular load, and where there is considerable constraint on the movements involved. Oral (especially tongue) movements provide another arena with some of these properties. In addition, the oral cavity provides a holding and fixing mechanism for objects inserted into it and the location of the tongue almost insures exposure to objects in the mouth. Meltzoff and Borton (1979) demonstrated a tactual-visual cross-modal transfer in neonates who were presented with a stimulus orally in the form of a pacifier inserted into their mouth. The baby was allowed to suck on one of two pacifiers for 90 sec and then was shown two shapes – one that matched the shape of the pacifier previously in the mouth and one that did not. The babies looked reliably longer at the tactually familiar shape than at the novel one. (The effective stimulus dimension in this situation appears to be surface texture or roughness. The test stimuli had the same overall shapes but differed in that one had a smooth surface while the other had a number of knob-like protuberances scattered over its surface.)

In another tactual-visual cross modal study Allan, Coranado, Herrera, and Rocha (Note 1) exposed 3-month-old infants to one of two pacifier-stimuli *similar* to those of Meltzoff and Borton. By means of a pressure transducer embedded in the pacifier they were able to record the actual sucking movements of the babies. After a 90 sec exposure to the oral stimulus the baby was again given a choice of visual stimuli – one that was the same as the oral stimulus and another that differed in surface texture. As a group, the babies showed only a slight but marginally significant looking preference for the familiar visual stimulus.

However, when the babies were divided into two groups according to whether their sucking behavior had diminished during the oral exposure period or not, a strong relation was found with visual preference. Babies whose sucking rate had significantly diminished, i.e., who had habituated to the oral stimulus, showed a looking preference for the familiar stimulus, while those infants who had maintained a relatively constant sucking rate all during oral exposure displayed no preference. This result is consistent with an interpretation that oral manipulation of the pacifier by the habituating babies was controlled by the novelty of the stimulus, i.e., the information available from the pacifier. When the novelty had worn off, when there was no more information available, they ceased sucking. Then when the visual choice was presented they related the visual stimuli to the oral stimulus; they recognized and were attracted to look at the familiar one. The non-habituating babies did not extract sufficient information from the oral stimulus either because they were slow information processors or their sucking was being maintained for its affective rather than its informative consequences.

The differential stimulus dimension of surface texture, which was critical in both these oral-visual experiments, is one that does not require precise guidance or perception of the exploratory tongue movements. This would ease the problem of information acquisition by the infants. Another cross-modal information acquisition experiment similar to these in design was reported by Walker and Gibson (Note 2). In this study the critical stimulus dimension was rigidity-elasticity rather than surface texture. A rigid or elastic cylinder was inserted into the mouths of 1-month-old infants. After 60 sec exposure, the babies were shown a pair of moving pictures – one depicting a cylinder undergoing a rigid transformation, the other depicting it undergoing an elastic transformation. The babies showed a significant preference for the novel display, i.e., the one undergoing the novel transformation [1] It is interesting to note that, again, this stimulus dimension is one that does not require precise spatial guidance or perception and control of movements; hence, it is less surprising that the young infants are able to use their tongue movements to acquire relevant information.

The paradigm of the cross-modal studies does not require finely controlled motor behavior for information acquisition. Nor do the eye movement studies described show any very sophisticated scanning strategies. Yet obviously, as the

1 Reliable preference for either the novel stimulus as here or the familiar stimulus as above are equally good indices for discrimination by the babies. It is not yet clear why in some habituation studies infants show a test preference for the novel stimulus, and in others for the familiar stimulus. The difference may be related to the degree of novelty of the display in the new modality. If the display as a whole is very different, detection of a familiar stimulus would lead to preferential looking. On the other hand, if the display as a whole is relatively familiar the novel stimulus might elicit the preference. Verifying such a hypothesis as this would require somehow assessing (or manipulating) the relative novelty of the display in the new modality

child matures it is able to use more varied and sophisticated motor skills in the acquisition of information. We might consider briefly a somewhat older child's acquisition and use of information for cross-modal matching of stimulus objects.

There is extensive literature on the development of cross-modal information processing in children. In the better types of studies inter-modal tasks are compared with intra-modal tasks. For example, visual-visual and tactual-tactual matching tasks are compared between themselves and with visual-tactual and tactual-visual tasks. It is quite typical in these studies for visual-visual tasks to be the easiest. Sometimes the tactual-tactual task is next easier or it is as easy as the other two types of tasks. If the tactual-tactual task is easier than the other two, the implication is that within-modality tasks are easier than between-modality tasks. At other times the order of difficulty after visual-visual is visual-tactual, tactual-visual, and then tactual-tactual. This order would seem to imply that visual tasks are easier than tactual. Of course, some combination of both these possibilities may be true, which might account for the pattern of visual-visual being easiest, and the other combinations equally more difficult. A more analytic hypothesis was proposed some time ago by Goodnow (1971). She suggested very simply that it is not modality that is primarily important or even whether one or two sense modalities are involved, but rather, the degree to which the information acquiring behavior in the exposure to the standard and comparison stimuli ensures that the same information is acquired. An example of this from my own research (Pick & Pick, 1966) which was replicated by Goodnow (1969) involves visual-visual or tactual-tactual comparisons of stimuli distinguished by curvature of lines or by orientation. The results for children of kindergarten age and younger were that stimuli differing in orientation (rotation, inversion, or left right reversal) were easier to distinguish by touch than by vision and stimuli differing in the curvature of outline were easier to distinguish visually than tactually. The relative ease in detection of orientation differences by hand seemed to be due to the simultaneous exploration of the two stimuli such that congruent exploratory movements did not result in identical patterns of stimulation. In contrast, the susceptibility of the visual system to mirror image confusions especially in children is well known. On the other hand, visual perception of curvature may well be mediated by fairly sensitive feature detectors while tactually it must be registered sequentially by movements of an exploratory organ and the small gradual changes of direction of a slightly curved line are difficult to detect and would possibly vary according to the limb movements involved in exploring the object.

As children develop more sophisticated motor skills a considerable amount of the variance in comparison of the efficiency of within- and between-modality information can be explained. These skills could incorporate conscious strategies and we thus would have cognitive effects on action aimed at gathering information relative to further cognitive development. This regression emphasizes the tightly knit relation between motor and cognitive aspects of the action system concerned with information acquisition.

2.3 Relation of Perception and Cognition to Development of Action

As just implied, the increasing cognitive sophistication of children as a function of age provides a potential basis for more sophistication in motor skill performance. While the correlation between these two aspects of behavior is obvious, the mechanism(s) by which cognition affects performance is (are) not clear, if indeed it can truly be said that there is a general causal link between cognitive development and motor performance. The nature of this relationship can be addressed at many levels. One issue is whether improvement in motor performance is based on greater perceptual sensitivity to critical aspects of stimulation, particularly those generated by and informative of one's own behavior. Examples of investigations of such sensitivity include studies by Lee and Aronson (1973) and Butterworth and Hicks (1976) of children's sensitivity to dynamic visual stimulation specifying posture and studies by Lasky (1977), McDonnell (1975), and Hay (1979; this volume) of sensitivity to visual feedback from children's own reaching movements. Another issue is whether the nature of mental representation of the external (and internal) environment changes in a way which affects motor performance. This has been studied most systematically by Mounoud (1982) in the context of children lifting weights and by a variety of authors in the context of spatial locomotion, e.g., Acredolo (1981) and Lockman (Note 3). Questions of the nature of mental representation begin to impinge on issues of voluntary or intentional acts as opposed to reflexive, automatic, or unconscious acts and are reflected in current growing interest in developmental psychology with setting of goals, planning of action, and meta-cognition or self awareness of one's own cognitive processes (Flavell & Wellman, 1977; Wellman & Somerville, in press). Researchers such as Connolly (1980) and Newell and Barclay (1982) have pointed out the importance of this level of analysis for motor performance but little substantive progress has been made.

The present discussion will focus on development of behavior involved in making the world accessible, namely reaching and locomotion.

2.4 Development of Reaching

The earliest behavior displayed by young children in accessing the world is reaching. There has been a lively controversy in the literature about how early infants begin to reach. Bower and his colleagues (Bower, Broughton, & Moor, 1970a, b) claimed that directed reaching was observable in neonates. Not only did they report that babies made hand movements in the direction of targets but they also showed considerable disruption of their behavior if they failed to make contact with an object such as occurred when they were presented with a stereosopically generated *virtual* object instead of a real palpable object. Attempts to replicate, at least in principle, the results of Bower et al. by Dodwell, Muir, and DiFranco (1976) and by Ruff and Halton (1978) were rather unsuccessful. These studies employed some control conditions that were missing in the original studies by Bower et al. and perhaps used stricter and more specific criteria for defining a reach. They did not obtain anywhere near the degree of reaching reported by Bower et al. More recently, however, two additional careful studies of neonates' reaching behavior yielded positive results (Rader & Stern, 1982; von Hofsten, 1982). Both of these studies employ strict criteria for defining reaches and controls to distinguish between reaching in the presence and absence of objects. (For example, Rader and Stern define a reach as consisting of three components – lateral extension of the arm, arcing toward midline, and flexion of arm toward upper half of body.)

It seems that infants do indeed exhibit a primitive form of reaching at birth. The primitive reaching that has been observed in the neonatal period, except for Bower's report, seems to be an all-or-nothing phenomenon, i.e., it is either present or not; the accuracy of the reach may or may not be high. There is little evidence for systematic correction of reaches. This may be a function of the poor sensitivity of our methods – the most sophisticated signal detectability methods may be needed just to show that reaching exists (McDonnel, 1979), or it may be that at this early age reaching, in fact, is elicited but not guided. There is evidence in the animal literature for a dissociation between visual eliciting and visual guidance of behavior. Hein (1972, 1974) for example, showed that differential rearing of kittens could selectively retard visually-elicited paw extension as opposed to visually guided placement. A smililar separation of function is also implied by the work of Hay (1979; this volume) who has demonstrated an age progression in the degree to which children's reaching is visually guided as opposed to simply elicited in ballistic form.

There is growing evidence that the precocious reaching of infants described above diminishes and perhaps even disappears. Rader and Stern found the incidence of reaching to decrease between 8 and 16 days of age. In another study, von Hofsten (Note 4) studied prereaching behavior in infants between 1 and 16 weeks of age. Prereaching was defined as early reaching before grasping occurred. He found that such prereaching behaviors diminished to almost zero about 7 weeks of age and their form changed about that time as well; before and after this stage, the hand was open during the arm extension while at 7 weeks practically all extensions were with closed fist. As the prereaching movements decline, assuming in fact that they do, visually guided reaching develops. Similar results were also reported by Trevarthen (1974). The most exciting recent and systematic evidence for this has been provided by von Hofsten (1979, 1980, 1982). In contrast to most previous studies, von Hofsten presented infants with moving objects as targets rather than stationary objects. Superficially, this might seem to pose the infant a more difficult problem. However, in fact, infants seem to display more sophisticated reaching under these conditions. The situation that he uses involves a small, brightly colored object moving laterally in front of the infant at various selected velocities. Infants were studied from about 12 to 36 weeks of age at approximately 3 week intervals. (A time motion analysis of the movements was used which permitted dividing each movement into elements. In this analysis position of the hand was plotted every 100 msec. Then instantaneous velocity was calculated hy measuring changes in velocity at each recording point.

Movement elements were defined by one acceleration and one deceleration phase which, in fact, are also equivalent to successive minima in the velocity curve.)

Various forms of goal directed behavior were identified including "reaches" in which the arm was stretched out toward the object accompanied by grasping or contact or, if contact was missed, the arresting of movement of hand in midair as if trying to make contact. Other goal directed behaviors included following the moving object, forward swiping, etc. The number of reaches out of possible attempts increased monotonically with age. There were more reaches at slower object speeds especially for the younger babies, suggesting that babies do not reach unless they know they could be successful. However, the reaches that did occur were more precise as the object velocities increased. That is, the distance actually moved from start to contact was closer to the minimum possible distance, the speed of movement was greater for the faster moving objects and the number of movement elements was least for the faster objects. The work of von Hofsten suggests that infant reaches are elicited by moving targets and that the direction of reach is a vector sum of the current position of the target and its direction (and speed) of motion. This results in the initial reach being made close to where the target will be at moment of contact and not in the direction of the target at moment of initiating reach. In addition, the reach is under visual guidance with corrections being made during reach. Such corrections occur more often for slower target speeds and for younger babies. The less frequent visual guidance of the older babies may be due to the fact that they don't need to correct their reaches as much. That is, their initial movements are more accurate and do not have to be corrected. Alternatively, it may be that the reaches of younger babies are organized differently from older babies, i.e., perhaps they are more inclined to use visual guidance while older babies use ballistic movements. We know from the work of McDonnell and Abraham (1979) that the older babies will correct if their reaches go awry. These investigators showed that by 5.5 months babies reaching for a target viewed through a wedge prism, which displaces its apparent position laterally, corrected the reach as their hand came into view. Similarly, at about this age, Lasky (1977) showed that babies' reaching was disrupted by the unusual visual feedback of not seeing their own hand as they reached under a mirror toward the virtual image of a target. Earlier, infants did not seem to require this feedback. The work of Lauretta Hay (this volume) shows something of an opposite shift at a later age. It would be interesting to try to integrate our understanding of these opposing trends.

The elicitation and guidance of reaches, even though sometimes difficult to distinguish from the noise of random movements in young infants, can be directly observed while the *planning* of reaching, on the other hand, must be inferred. It becomes most apparent when something interferes and a goal object cannot be reached in a direct way. This is the situation when detour behavior is called for by a barrier placed in the way of an attractive object. Lockman (Note 3) studied the development of simple reaching detours in a group of infants from 8 to 12 months of age. The goal object was lifted over a barrier in front of the child. The barrier was high enough so that the babies could only retrieve the object by reaching around one or another of the ends of the barrier. Not until an average of about 40 weeks did the infants solve this problem. Below this age they attacked the barrier directly, either trying to follow the object over the top of the barrier or going straight through it. Interestingly, the same problem with a transparent barrier was not solved until 3 weeks later. Apparently the object actually in view was so strong on elicitor of direct reaching that the more planful detour behavior did not occur till a somewhat later age.

The use of more sophisticated strategies in planning reaching-like movements continues at least through the early teens. Smith and Greene (1963) reported an abrupt shift in ability of children between the ages of 9 and 13 years to adjust to inversion and reversals of the visual field accomplished by closed circuit TV means. The younger of these children were completely unable to perform a writing task under such transformation. The older children, although somewhat bothered by the transformation, could accomplish the task rather effectively. Although their data appear to show a discontinuity in this kind of perceptual-motor performance, it seemed that perhaps discontinuity was more apparent than real, and had to do with the radical nature of the optical transformation of which their young subjects were unaware. Smothergill, Martin, and Pick (1971) examined the effect of various degrees of distortion on children of different ages. Children of 7, 9, and 11 years of age and adults were exposed to various rotations of the optical field ranging from 0° to 180° in 30° steps. The task was to move a stylus along a track without touching the sides. Performance deteriorates for all age groups with greater optical distortion up to 120°. For the younger children performance remains low at the greatest distortions but for the older children, at the greater distortions, there is marked improvement and this improvement occurs at distortions less than 180° for the oldest subjects. It is as if the older subjects could adopt a strategy of reversing all directions at 180° and even with slightly less distortion for the oldest and most sophisticated subjects.

2.5 Development of Locomotory Skills

Locomotory skills serve along with reaching and grasping to make the world more accessible. The earliest locomotory behavior is crawling. This behavior typically occurs after the age of 6 months and seems to be elicited by almost any kind of attractive object – very often by social objects such as the mother. Advantage is taken of this fact in the now classic visual cliff research in which the mother calls to the infant to come to her across the deep and shallow sides of a visual cliff (Walk & Gibson, 1961). One interpretation of the typically observed avoidance of the deep side of the cliff is that guidance of crawling behavior is dependent on the presence of a visually specified, rigid surface. At what age does this controlling context become effective? Is the avoidance dependent on the ability to make locomotory movements per se, or is it dependent on knowing when such movements are appropriate and inappropriate? A hint is provided by Campos' work examining the heart rate reaction of 5- and 9-month-old infants passively placed down on the deep and shallow sides of a visual cliff (Schwartz, Campos, & Baisel, 1975). The precrawling 5-month-olds showed a heart rate deceleration typical of orientation and interest to both sides of the visual cliff. The deceleration was greater to the deep than shallow sides, indicating discrimination of the difference but there is no indication of appreciation of the depths or danger of the deep side. The 9-month-olds, on the other hand, showed a heart rate acceleration to the deep side typically indicating fear, while they continued to show the orienting deceleration to the shallow side. Does this heart rate index actually reflect an infant's capacity to use spatially appropriate behavior? Rader (Note 5) has recently investigated whether precrawling infants, given an ability to make locomotory movements, would "know" when to use these appropriately. She tested 2- to 5-month-old infants using crawligators on the visual cliff. These are devices consisting essentially of formed boards on wheels on which young infants can lay and propel themselves by use of arm movements. Using such devices, all infants that she tested crossed the deep side of the cliff as readily as the shallow. The presence of a visually specified surface did not seem to be a necessary feature for them to engage in this precocious locomotory behavior. One might imagine various attentional explanations for this lack of sensitivity to the visual information or its absence. However, Rader also has provided converging evidence with somewhat older infants (Rader, Bausano, & Richards, 1980). Here similarly, she provided recently crawling infants with a means of "walking" earlier than they might ordinarily be capable of it – by means of walkers. These are devides on wheels in which babies can sit with their legs touching the ground. Normal rhythmic moving of the legs propels the baby but the baby is supported in an upright posture by the walker. Babies can walk with these devices about the same time they start to crawl. Young infants, who avoided the visual cliff when crawling, moved happily across it in the walker. Rader and her colleagues suggested that there is a disassociation between fear of the deep side of the visual cliff and avoidance. A crawling infant may avoid the depth initially not because of fear, but because there is no visually specified surface as context for the behavior. According to their view the motor program for crawling will only be released in presence of a visual surface. The motor program for the leg movements in walking does not have this requirement of a visually specified surface and hence babies will walk across the deep side. Some infants do crawl across the deep side of the visual cliff, and Rader determined that these infants tended to be those who started crawling at the youngest age rather than those who had most experience crawling. She suggested the intriguing hypothesis that crawling starts at early ages under tactual control and only at later ages does it come under visual control. For precocious crawlers, tactual control of crawling is maintained rather longer, hence they are more likely to crawl over the deep side of the visual cliff which does provide a tactual surface[2].

As with reaching behavior planning in locomotory behavior again can perhaps be inferred in detour problems. Lockman (Note 3) investigated babies' ability to

2 Recent evidence by Rader (Note 6) makes this hypothesis less likely

make crawling detours using the same general method and in the same babies as he had studied reaching detours. The average age for solving a simple detour problem by crawling was 45 weeks for an opaque barrier and 3 weeks later for an transparent barrier – the same pattern but at a slightly older age than with the reaching detour. It is noteworthy that there is a slight delay between the reaching and locomotory solutions since the problems are formally identical. It may be the case that the baby is freer for the cognitive demands of the problem at an earlier age when reaching since that is a skill developed earlier and more practised, whereas crawling itself requires attentional capacity which distracts from its appropriate use in this situation.

3 Conclusion: Theoretical and Meta-Theoretical Issues at the Interface of Cognition and Action in Development

There are a number of theoretical and meta-theoretical issues embedded in the previous discussion of experimental evidence. First, there is the early appearance and apparent disappearance of forms of behavior and then their reappearance again perhaps in a more sophisticated form. This is a phenomenon that has often been noted in the past but is still not completely understood (Bower, 1974; Trevarthen, 1974, 1982). One explanation has been that the early form of such behavior is reflex-like and this disappears as infants come to be able to inhibit the behavior that is previously reflexively triggered. When behavior can be inhibited young organisms are in a better position to select and organize components of skills and sequence them appropriately, etc. This idea, of course, raises the whole issue of voluntary behavior. There has been considerable interest but no firm answers on criteria for deciding that behavior is voluntary and/or intentional. The issue can arise at various levels of performance of behavior. We may be willing to say that a child *decided* to execute some act like reaching but the guidance of the act might not be under voluntary control. (A common observation is that one can decide to run up the stairs but if one thinks about it while doing it he/she is likely to misstep.)

We have seen that in a number of activites one may be able to specify how a behavior is elicited. However, for many mature actions along with an eliciting stimulus or situation, the action requires what might be called planning such as in the examples of detour behavior and adjustment to rotation of the visual field. A further complicating factor may be the specificity of the motor behavior itself. Thus, in the case of the detour problem, it was seen that the appropriate behavior occurred with reaching but not with crawling although the child was quite able to crawl. Again in the case of the visual cliff behavior, avoidance of the deep side was specific to crawling but did not occur when the child was given the possibility of walking. The eliciting of certain forms of behavior may be relatively finely tuned to situational factors. For example, both Field (1976) and Gordon and Yonas (1976) have reported evidence that infants will not reach for objects which are too far away to be attained.

With respect to the guidance of behavior we have seen that it can be dissociated from elicitation of behavior and that young infants are capable of visually guiding their reaching as evidenced by correction of reaches going amiss due to prism or mirror distorted vision. However, we do not know the details of the course of development of such guidance. Does the more frequent correction of younger babies during reaching, as observed by von Hofsten, reflect more dependence on visual feedback? Then how does one interpret a shift from ballistic to guided movement found by Hay for older children? In a rather different domain we know that auditory feedback is not necessary for speech production by mature speakers of a language (Pick, Siegel, & Garber, 1982). People can speak quite well even though deprived of auditory feedback by adventitious deafness or experimental masking noise. On the other hand, adults are more responsive to changes in intensity of auditory feedback than children. Adults will decrease their vocal intensity to a greater degree than children when auditory feedback is amplified. Furthermore, while adults can voluntarily accentuate or enhance their use of auditory feedback in this case they have considerably greater difficulty inhibiting their response to amplification of auditory feedback.

The last observation leads naturally to the more general question of how explicit knowledge about movement can be used in production of movement. It is almost axiomatic that with many motor skills telling someone explicitly what to do does not help. But, of course, that is too general a statement. *When* does telling someone what to do help? How should they be told? How is the task of using such conscious information different for people of different ages and levels of experience? The answer to such questions has both theoretical and educational importance.

In conclusion, one might ask why study the relation of cognition and action in development. There is an obvious practical answer: It is an important developmental problem. By understanding the development of cognition and action we can help children learn more easily the important motor skills they need to acquire and we can detect and help remedy cases where normal development is absent. There is an equally important theoretical reason. Understanding the development of processes is one way to approach investigation of mature functioning. See, for example, Bernstein's (1967) analysis of the development of walking movements. So far the contribution of developmental studies in relation to this latter reason has not been very great. However, I would like to describe one recent study which exemplifies how a developmental study might help inform us about mature functioning and, in fact, serve as a bridge between the theoretical and practical reasong for studying development.

Linguists have developed a number of distinctive feature systems for describing the phonemic contrasts used in language, e.g., Ladefoged (1971) and Chomsky and Halle (1968). While these descriptions are quite elegant in systematically organizing all the phonemic distinctions of particular languages the psychological reality of such systems is not completely clear. A case study by Broen (in press) exemplifies one approach to investigating the psychological meaningfulness of feature systems. She reported the analysis of four children under treatment for speech articulation problems. Basically her strategy was to train a child to make a particular phonemic contrast in one context and to note the degree to which the contrast was achieved in other untaught contexts. Positive results from such generalization tests would support the psychological reality of feature systems. For example, one of her children did not make a velar-alveolar distinction in producing stop consonants and articulated /d/ for both /d/ and /g/, and /t/ for both /t/ and /k/. The child was taught the /t/ versus /k/ distinction in final position and correct production of that particular velar-alveolar distinction generalized to /d/ versus /g/ in both initial and final positions. In general, the pattern of generalization of trained to untrained phonemic distinctions she found for the four children argues for the psychological meaningfulness of such feature systems. Furthermore, this type of study has implications for how to use speech therapy with children more effectively. Where should one train to achieve greatest generality? The accumulation of such data would also provide a means for deciding whether one feature system was a better description than another of how speakers of a given language actually organized their phonology and whether this organization changes as a function of age.

The study of both normal and abnormal development can, in this way, provide clues about the organization of specific domains of motor skills. By studying the development of skills the structure and organization of mature performance may be more easily detectable – through such patterns of generalization or by more accentuated invariant patterns in timing and intensity of movement.

Acknowledgment. Preparation of this manuscript was supported by the Center for Research in Human Learning in part from a grant from the National Institute of Child Health and Human Development, Grant No. HD 01136.

Reference Notes

1. Allan, T.W., Coranado, S., Herrera, J., & Rocha, P. *Object discrimination in oral-oral and oral-visual sensory context.* Paper presented at the Meetings of the International Society of Infant Development, Austin, Texas, April, 1982.
2. Walker, A.S., & Gibson, E.J. *Intermodal perception of substance.* Paper presented at the Meetings of the International Society of Infant Development, Austin, Texas, April, 1982.
3. Lockman, J.J. *The development of detour knowledge during infancy.* Unpublished Ph. D. Dissertation, University of Minnesota, 1980.
4. von Hofsten, C. *Developmental changes in the organization of pre-reaching movements.* In preparation, 1983.
5. Rader, N. *Visual cliff behavior in pre-crawling infants.* Unpublished manuscript. (Available from N. Rader, Psychology Dept., University of California, Los Angeles.)
6. Rader, N. Personal communication, September 1982.

References

Acredolo, L.P. Small- and large-scale spatial concepts in infancy and childhood. In L.S. Liben, A.H. Patterson, & N. Newcombe (eds.), *Spatial representation and behavior across the life span.* New York: Academic Press, 1981.

Aslin, R.N., & Jackson, R.W. Accommodative-convergence in young infants: Development of a synergistic sensory-motor system. *Canadian Journal of Psychology,* 1979, *33,* 222–231.

Banks, M.S. Infant refraction and accommodation. *International Ophthalmology Clinics,* 1980, *20,* 205–232.

Bernstein, N. *The co-ordination and regulation of movements.* Oxford: Pergamon Press, 1967.

Bower, T.G.R., Broughton, J.M., & Moore, M.K. Demonstration of intention in the reaching behavior of neonate humans. *Nature,* 1980, *228,* 669–681. (a)

Bower, T.G.R., Broughton, J.M., & Moore, M.K. The coordination of visual and tactual input in infants. *Perception and Psychology,* 1970, *8,* 51–53. (b)

Bower, T.G.R. *Development in infancy.* San Francisco: Freeman, 1974.

Braddock, O., Atkinson, J., Howland, H.C., & French, J. A photorefractive study of infant accommodation. *Vision Research,* 1979, *19,* 319.

Broen, P.A. Patterns of misarticulation and patterns of articulation change. In N.J. Lass (ed.), *Speech and language: Advances in basic research and practice* (Vol. 8). In press.

Butterworth, G., & Castillo, M. Coordination of auditory and visual space in newborn infants. *Perception,* 1976, *5,* 155–161.

Butterworth, G., & Hicks. L. Visual proprioception and postural stability in infancy: A developmental study. *Perception,* 1976, *6,* 255–262.

Chomsky, N., & Halle, M. *Sound patterns of English.* New York: Harper and Row, 1968.

Connolly, K. Development of competence in motor skills, In C.H. Naudeau, W.R. Halliwell, K.M. Newell, & G.C. Roberts (Eds.), *Psychology of motor behavior and sport – 1979.* Champaign, IL: Human Kinetics, 1980.

Dodwell, P.C., Muir, D., & DiFranco, D. Responses of infants to visually presented objects. *Science,* 1976, *194,* 209–211.

Field, J. Relation of young infants' reaching behaviors to stimulus distance and solidity. *Development Psychology,* 1976, *12,* 444–448.

Flavell, J.H., & Wellman, H.M. Metamemory. In R.V. Kail & J.W. Hagen (Eds.), *Developmental perspectives on memory and cognition.* Hillsdale, N.J.: Erlbaum, 1977.

Goodnow, J.J. Eye and hand: Differential sampling of form and orientation properties. *Neuropsychologia,* 1969, *7,* 365–373.

Goodnow, J.J. The role of modalities in perceptual and cognitive development. In J.P. Hill (Ed.), *Minnesota Symposium on Child Psychology* (Vol. 5). Minneapolis: University of Minnesota Press, 1971.

Gordon, F.R., & Yonas, A. Sensitivity of binocular depth information in infants. *Journal of Experimental Child Psychology,* 1976, *22,* 413–422.

Haith, M.M. *Rules that babies look by.* Hillsdale, N.J.: Erlbaum, 1980.

Hay, L. Spatial-temporal analysis of movements in children: Motor programs versus feedback in the development of reaching. *Journal of Motor Behavior,* 1979, *11,* 189–200.

Hein, A. Acquiring components of visually guided behavior. In A.D. Pick (Ed.), *Minnesota Symposia on Child Psychology* (Vol. 6). Minneapolis: University of Minnesota Press, 1972.

Hein, A. Prerequisite for development of visually guided reaching in the kitten. *Brain Research,* 1974, *71,* 259–263.

von Hofsten, C. Development of visually directed reaching: The approach phase. *Journal of Human Movement Studies,* 1979, *5,* 160–178.

von Hofsten, C. Predictive reaching for moving objects by human infants. *Journal of Experimental Child Psychology,* 1980, *30,* 369–382.

Lasky, R.E. The effect of visual feedback of the hand on reaching and retrieval behavior of young infants. *Child Development,* 1977, *48,* 112–117.

Ladefoged, P. *Preliminaries to linguistic phonetics.* Chicago, IL: University of Chicago Press, 1971.

Lee, D.N., & Aronson, E. Visual proprioceptive control of standing in human infants. *Perception and Psychophysics,* 1973, *15,* 529–532.

McDonnell, P.M. The development of visually guided reaching. *Perception and Psychophysics,* 1975, *18,* 181–185.

McDonnell, P.M. Patterns of eye-hand coordination in the first year of life. *Canadian Journal of Psychology,* 1979, *33,* 253–267.

McDonnell, P.M., & Abrahm, W.C. Adaptation to displacing prisms in human infants. *Perception,* 1979, *8,* 175–185.

Meltzoff, A.N., & Borton, R.W. Intermodal matching by human neonates. *Nature,* 1979, *282,* 403–404

Mounoud, P. Development of sensorimotor organization in young children: Grasping and lifting objects. In G.E. Gorsman (Ed.), *Action and thought: From sensorimotor schemes to symbolic operations.* New York: Academic Press, 1982.

Newell, K.M., & Barclay, C.R. Developing knowledge about action. In J.A.S. Kelso & J.E. Clark (Eds.), *The development of movement control and coordiantion.* New York: Wiley, 1982.

Pick, A.D., & Pick, H.L., Nr. A developmental study of tactual discrimination in blind and sighted children and adults. *Psychonomic Science,* 1966, *6,* 367–368.

Pick. H.L., Jr., Siegel, G.M., & Garber, S.R. Development of speech production as a perceptual-motor task. In J.A.S. Kelso & J.E. Clark (Eds.), *Development of movement control and coordination.* New York: Wiley, 1982.

Rader, N., Bausano, M., & Richards, J.E. On the nature of visual-cliff-avoidance response in human infants. *Child Development,* 1980, *51,* 61–68.

Rader, N., & Stern, J.D. Visually elicited reaching in neonates. *Child Development,* 1982, *53,* 1004–1007.

Reed, E.S. An outline of a theory of action systems. *Journal of Motor Behavior,* 1982, *14,* 98–134.

Rieser, J., Yonas, A., & Wilkner, K. Radial localization of odors by human newborns. *Child Development,* 1976, *47,* 856–859.

Ruff, H., & Halton, A. Is there directed reaching in the human neonate? *Developmental Psychology,* 1978, *14,* 425–426.

Schwartz, A., Campos J., & Baisel, E. The visual cliff: Cardiac and behavioral correlates on the deep and shallow sides at five and nine months of age. *Journal of Experimental Child Psychology,* 1975, *15,* 86–99.

Smith, K.U., & Green, P. A critical period in maturation of performance with space-displaced vision. *Perceptual and Motor Skills,* 1963, *17,* 627–639.

Smothergill, D.W., Martin, R., & Pick, H.L., Jr. Perceptual-motor performance under rotation of the central field. *Journal of Experimental Psychology,* 1971, *87,* 64–70.

Trevarthen, C. The psychobiology of speech development. In E.H. Lenneberg (Ed.), Language and brain: Developmental aspects. *Neurosciences Research Program Bulletin,* 1974, *12,* 570–585.

Trevarthen, C. Basic patterns of phylogenetic change in infancy. In T. Bever (Ed.), *Dips in learning.* Hillsdale, N.J.: Erlbaum, 1982.

Walk, R.D., & Gibson, E.J. A comparative and analytic study of visual depth perception. *Psychological Monographs,* 1961, *75* (15, Whole No. 519).

Wellman, H.M., & Somerville, S.C. The development of human search ability. In M.E. Lamb & A.L. Brown (Eds.), *Advances in developmental psychology* (Vol. 2). Hillsdale, N.J.: Erlbaum. (in press)

Wertheimer, M. Psychomotor coordination of auditory-visual space at birth. *Science,* 1961, *134,* 162.

Summary. One important role action plays in cognitive development involves making available the information necessary for developing mental processes. This function may be seen already in infancy where eye movements seem elegantly adjusted to provide a young organism with informative stimulation. Eye scanning movements and the relation between accommodative and convergent movements are good examples of this. Also early in infancy exploratory movements of the tongue and later manipulatory movements of the hand are well adapted to provide information about the substance and identity of objects. Even at an early age the use of this information is not specific to the modality of input, i.e., there exists from a very early age cross modal sensory processes. Such processing in the service of information acquisition continues to become more sophisticated at later ages.

On the other hand, the reciprocal influence of perception and cognition on the development of action is equally apparent. This side of the relation between cognition and action can be exemplified in both the development of reaching and the development of locomotion. In the development of reaching perception already plays an important role in infancy both in how reaches are elicited and in their visual guidance. Again, these components increase considerably in sophistication with increasing age. Similarly, perception plays an important role in eliciting and guiding early locomotory (crawling) behavior in babies and increasingly sophisticated cognitive strategies are apparent with later development.

Important theoretical issues in the relation between cognition and action include the nature of voluntary behavior. When and how does this develop and exactly how does planning affect action? The final section is concerned with these questions.

20 Biodynamic Structures, Cognitive Correlates of Motive Sets and the Development of Motives in Infants

COLWYN TREVARTHEN

Contents

1 Introduction – Bernstein's Theory of Motor Coordination

Bernstein (1967) proposed that movements are created and maintained by the "biodynamic structures" of body and brain. He compared genesis of movement to morphogenesis of tissues and organs. In both processes the elements differentiate in time relative to each other within a coherent whole. He described human locomotion in unprecedented detail, showing how the form of movement was a resultant of cyclical forces in muscles and reactive forces arising peripherally – from gravitation, inertia of body parts and pressures and resistances of surrounding media. His analyses of walking and running demonstrated how the central excitatory program complements and exploits or channels the peripheral forces generated by limb oscillations with astonishing precision and regularity. They clarified the relationship between the command of neuromuscular excitation and the responsive sensitivity of the central program to information feedback from receptors.

This brilliant treatment of locomotion as a natural biochemical problem led Bernstein to consider kinds of work in which objects are displaced while gripped by the hands (e.g., hammering) or struck by the fingers (e.g., typing). He argued that the perception of any object must be founded on an accurate identification of its

Cognition and Motor Processes
Ed. by W. Prinz and A.F. Sanders

potentialities for use. Perception of an instrument must anticipate the ways in which its attachment to the body will change the mechanics of limb action. In other words, prediction of what Gibson has called an object's perceptual "affordances" for action (Gibson, 1979) is necessary if the brain is to mobilise appropriate motor programs and set in motion an effective sequence of body-object interaction. In the same vein, when considering the physiognomy of written letters, Bernstein likened the topological form of a letter, by which it is recognised in spite of distortions or variations in calligraphic style or orientation, to the equation for the movement that writes it readily and uniformly in different postural contexts, on different surfaces and with different instruments or materials.

Bernstein's theory of how movements are controlled provides a natural basis for developmental studies. He stressed that "biodynamic structures live and develop". By minutely measuring changes of force curves in walking and running from infancy to adulthood, and the involution of control in senescence, he extracted evidence for changes in the competence of the central nervous system for assimilating sensory data on the progress of the interactions between neuromotor output (muscle contractions) and peripheral kinetic forces. Toddlers walk with a central patterning of cyclic limb displacements that lacks precision and refinement in its uptake of proprioceptive feedback. The development of more efficient walking and the emergence of running as a distinct variant of locomotion involve acquisition of more comprehensive predictions of how events will evolve peripherally, quicker response to nullify accidental perturbations and more forceful propulsion to lift the body from the substrate so that it is in "flight" for a larger proportion of time. Bernstein compared the development of more efficient walking and of flight in running to the stages in learning a gymnastic skill such as skiing.

Recent microanalysis of the behaviour of infants, discussed below, confirm the merit of Bernstein's descriptive method. Studies of both spontaneous and reactive movements of neonates give abundant evidence of innate motor programs that are preadapted to coordinate standing, locomotion, orientation of head, eyes and limbs round the body axis and prehension and manipulation of objects (Trevarthen, Murray, & Hubley, 1981; Trevarthen, 1983c). They show how development systematically adds more and more efficient acceptance of proprioceptive, exproprioceptive and exteroceptive guidance, increasing the effectiveness of movements.

Development of perception and of cognition in infants and of their motivation to engage in and solve problems of movement may be fruitfully explored with a focus on this evolution in biodynamic strategies of the infant's brain. Infants' "concepts" of space, of events and of objects are built by enrichment of pre-existing motives to engage in profitable movements. The nervous system contributes a vital structural competence of taking up experience which undergoes systematic growth and elaboration. This development, while dependent on input from a facilitating and informing environment, is at no stage wholly plastic or passive. Temporal and spatial characteristics of infant movements indicate that their biodynamic structures are self-originating and self-maintaining. The rate and direction of their development follows a highly predictable course in all infants, regardless of variations in the environment, as long as the circumstances give "adequate" support to infant motives and they do not degenerate into pathological variants. Piaget (1953), and his fertile predecessor Baldwin (1894) who ennunciated the principle of circular reactions by which motor acts recreate experiences and develop by "imitating" the afferent effects they create, both failed to give adequate recognition to the generative and cognitive potential of programs for movement arising fron inside the brain. These programs, it must be admitted, are the product of neuromorphogenetic processes of embryo and fetus of which little definite is known.

2 Where Preparation for Movement Changes the Significance of Stimuli

2.1 Observations Contributing to Sperry's Motor Theory of Perception

My ideas on the importance of motor coordinating processes for perception and cognition are strongly influencedby the work of Roger Sperry with whom I worked as a research student in the early 1960s. Sperry's experiments in the 1940s and 1950s on the regrowth after transection of sensory, motor and central circuits in the brains of fish and amphibia proved that connections were made to the same target locations as before surgery, even if, because of surgical rearrangements, the sensory-motor coordinations so formed were maladaptive for the animal (Sperry, 1944, 1963).

In the course of this work Sperry presented two observations which illustrate how adjustments of body motility may change the signal value of stimuli. He described how the escape movement of the leg of a spinal frog, with brain disconnected or removed, reverses when the posture of the leg is changed. A noxious stimulation of the toe of a retracted foot causes it to extend, but a foot which is extended pulls back when it is hurt by the same stimulus. Therefore, even the spinal cord has the capacity to reset reflex circuits automatically according to limb position and thus change the "sign value" of a stimulus. By rearranging afferent nerves to the spinal cord or the connections of motor nerves to muscles, Sperry showed that these motor adjustments expressed the anatomy of nerve pathways. The circuits could not adapt to an artificial rearrangement of this wiring. He inferred they were innately produced. Now motor physiologists are used to the idea that the complex locomotor automatisms of spinal mammals can be exchanged, the cycles of limb extension and retraction responding by a change of program to proprioceptive feedback from the limbs when the animal is stepping to keep up with a moving belt on which its paws are supported (Grillner, 1975). As the motion of the belt is accelerated progressively, a spinal cat or dog will "shift gears" at particular rates of slip to change from walking to trotting to galloping.

Sperry's second observation concerned the orientation of the whole body with respect to the visual frame of reference (Sperry, 1950). He was carrying out tests of regeneration of eye-to-brain connections in fish which involved inversion or transplantation of one eye and covering or removal of the other. Cut retinotectal connections from the experimental eye regrew after a short time and the fish regained visual guidance of locomotion. However, the responses of the fish were abnormal in direction, in agreement with the hypothesis that each retinal ganglion cell had found contacts with the same tectal region as before operation. When the retina was placed in the head with naso-temporal axes reversed, maladaptive turning movements were made. The operated fish could rest stationary in the tank, but as soon as they started to swim they would twist with accelerating speed and spin around until exhausted. Sperry suggested that the forced circling was due, not to reflex reaction to kinesthetic afference, but to an automatic effort by the brain of the fish to generate a motor output that would match or balance a normal ("expected") direction of flow in the retinal image. Naso-temporal reversal of the retina caused every forward or turning movement to give a signal in the brain like that to be expected from a turning movement opposite to the one actually made. Compensatory acceleration would increase the mismatch. Sperry named his hypothetical adjustment in sensory receiving processes a "corollary discharge of motor patterns" and he used this concept to explain how the brain normally achieves perception of a stable world during locomotion and distinguishes unintended drift of the body or evidence of objects in independent motion. Sperry's theory is the same as the "efference copy" idea independently arrived at by von Holst and Mittelstaedt about the same time to explain the abnormal movements of a preying mantis when its head was twisted round and cemented in an upside down relationship to the body (von Holst & Mittelstaedt, 1950).

2.2 Theories of How Perception and Action Inform Each Other

The notion of efference copy or corollary discharge influences thinking about possible neural mechanisms of perceptual constancy, and about processes involved in detection of the stable, invariant, or permanent attributes of space and objects by moving subjects (von Holst, 1954). In particular, it formed the basis of Held's theory of the adaptation of locomotion, reaching and awareness to prismatic displacement of the visual field (Held, 1961), and Teuber's concepts of perceptual distortions following brain lesions outside sensory receiving areas (Teuber, 1961). It led Schneider, Held, Ingle and me to the Two Visions Theory which emphasises the importance of relating kinds of visual information processing and their anatomical pathways to the motor functions which are being informed (Schneider, 1967; Held, 1968; Ingle, 1967; Trevarthen, 1968).

In a theoretical paper published in 1952, Sperry insisted that the anatomical organization of the brain mechanisms for perception or for recognition of objects or symbols can best be approached by tracing the location of the various motor coordinating structures in the brain. This paper, entitled *Neurology and the Mind-Brain Problem,* criticises the classic psychophysical stimulus-to-brain approach to perception which takes no account of how movements may be coordinated in central circuits. It also takes issue with anticonnectionist approaches of Lashley and the Gestalt psychologists who considered form perception to arise in transitory interference effects, or magnetic or electrical fields radiating through the cortical grey matter. Since all the output of the brain, but for a specialized glandular component, takes the form of movements, patterns of psychic experience must be translatable into motor and premotor patterns. Sperry states that once we accept that this relationship is *necessary* for the neural correlates of psychic experience we can exclude "numerous forms of brain code which otherwise might seem reasonable but which fail to meet this criterion". Current advances in neuro-physiological study of sensory detector systems and in theory of processing of input do not change the argument which led Sperry to conclude that in some sense a "preparation to respond" *is* the perception itself. Neuro-anatomical evidence shows that the systems coordinating movement and perception are wired into somatotopic arrays set up in a bilaterally symmetrical field with antero-posterior and dorso-ventral polarization that originates from early steps in neuromorphogenesis in the embryo, when organization of the C.N.S. and the body have close inductive relations. I have attempted elsewhere to begin to work out how morphogenetic processes in brains might generate a hierarchy of representations of space for action of the body, as well as general categories for discimination of events, objects, forms and symbols relative to body action (Trevarthen, 1973, 1974a, 1979a).

There is a fundamental difference between a nativistic motor theory of perception, such as Sperry's, and empiricist theories such as Hebb's concept of neuronal assembly (Hebb, 1949) which provided the theoretical background for research which called itself "physiological psychology" in the 1960s. Both Sperry and Hebb were concerned to explain phenomena such as perceptual constancy, equivalence and completion where psychic content does not fit stimuli in any simple way. However, Hebb emphasised that the intermediate integrative or coordinating principles could be built up in development by the brain keeping traces of the sensory effects of movements made to track stimuli. Research on the development of visual perception in infants and kittens does not support Hebb's thesis. Sperry's proposal that certain elements and categories of perceptual processing must be formulated prior to experience and that they must congenitally match the cerebral and corporal systems of movement and coordinated actions appears more correct, even though there is abundant evidence for plastic adaptations of perceptive systems to stimulus patterns in the early postnatal period for their refinement and calibration (Trevarthen, 1979a).

Gibson's theory of perception represents a completely different argument which attempts to show that "direct" sensitivity to the natural regularities or invariance in stimulation under any displacement of receptors could enable the subject to dispence with internal feedback from the motor signal (Gibson, 1966). He argued that there is sufficient evidence in the stimulus transformations received by an animal when it moves to identify every kind of self-produced movement, and that there are other unconfusable kinds of evidence in stimuli for defining motion independent of the subject. However, the selective power of awareness and its object-constructing abilities cannot be fully explained by this theory of the ecology of stimuli. Eventually Gibson had to invoke the subject's motor purposes in his Theory of Affordances (Gibson, 1979) to explain how the perceiving organism chooses ways of moving to exploit environmental information, and chooses information that will complete plans for movement.

That is where we are now. There is a upsurge of interest in "coordinative structures" and "perceptuo-motor schemata". The influence of attention to goals, preparatory sets, prior categorizations and cognitive strategies on how perceptual information is taken up, processed, retained or retrieved is widely admitted. Nevertheless, a majority of cognitive psychologists still prefer to rationalize their experimental data in terms of progressive processing of receptor information by almost inert subjects, just as the psychophysicists of last century sought to do.

2.3 Split-Brain Studies of Eye-Hand Coordination and Hemispheric Cognitive Modes

Observations I made when testing adult split-brain rhesus macaques or baboons and human commissurotomy patients have led me to conclude that consciousness is strongly affected by back-flow from impulses to action at the level of the cerebral hemispheres.

1. In one experiment, I trained monkeys with optic chiasm, corpus callosum and anterior commissure sectioned to learn with polarizing filters in front of their eyes and polarized stimuli. In each trial the monkey could see conflicting overlapping patterns on two response screens with each of his two eyes and two hemispheres (Trevarthen, 1962). I obtained double independent learning. The split-brain subjects learned both of the conflicting discriminations in the two disconnected, equal-sized, mirror-symmetric forebrain visual systems. For example, the left eye-hemisphere system learned to push + and not ○ to obtain a reward while the right half of the visual system learned the opposite, ○ was correct and + incorrect. This double learning proved the mechanism of visual retention in these subjects to be completely divided in two equal parts that could be functionally balanced so as to learn at the same rate.

After the above training was complete, the polarizing filters were removed and both halves of the brain were given the same stimulation. Now, consider the case of a trial where there was + on the left screen and ○ on the right screen. According to previous learning, the right half brain should expect a reward from a push to the ○ on the right screen and the left brain should see the + on the left as the one to push. In fact the monkey acted, with little hesitation, as if perceiving with only the eye opposite the hand chosen for response. When the subject's manual preference was changed by blocking the preferred hand, the monkey, after a couple of hesitant attempts, changed over to the other hand. Then the choices of visual stimulus became appropriate to the previous experience of the second half of the brain.

Either spontaneous or forced use of one hand appeared to activate awareness in the contralateral hemisphere, that is the hemisphere with strongest sensory and motor connections to the hand that was moving. Information received by the second eye, on the same side as the hand being used, seemed at that time to be ignored or not perceived, just as if the monkey were experiencing resolution of a perceptual rivalry between the conflicting visual images in the left and right hemispheres by suppression of the image in one side. The conflict and its resolution appeared to be mediated through motor adjustment.

2. With a split-brain baboon I filmed a tug-of-war between the two hands as one hand sought to grasp food held in the other hand while the animal was not looking at his hands (Trevarthen, 1965). Here the impulse to grasp the food with one hand did not appear to have been communicated to the other hand, so it reacted as if another animal was trying to remove what it held. Frame-by-frame analysis showed that the conflict was resolved within 2 or 3 sec when the animal, startled at the commotion, looked down at its hands. Then the withdrawing hand gave up the food to its partner. Unity of motor coordination had been re-established.

3. A comissurotomy patient, NG, was trying to aim a pen held in the left hand to the centre of a polygon cut out of white card and placed on a black table in the right visual field, about 45° to the right of the point where she kept her gaze fixated. She could not respond and described the shape as vanishing immediately the movement began (Trevarthen, 1974b, c). Here the patient's testimony was that the image of the object, initially seen, was blotted out of awareness in the left hemisphere the instant a movement, initiated by the right hemisphere, had started. Sensory guidance of the reaching movement to its intended target was no longer possible, and the movement was aborted. While the subject kept in readiness to respond, intently fixating a point 40 cm in front of her on the table, the white card about 20 cm long was invisible, even when moved, until the instant it crossed the vertical meridian. Appearance of the corner of the card past the midline in the left visual field triggered a forceful and rapid response with the waiting left hand. The preparatory motor set in the right hemisphere had received perceptual confirmation of its target.

4. Finally, I cite the shifting visual awareness observed when Jerre Levy and I presented chimeric stimuli and changed the type of judgement or cognitive strategy required of the commissurotomy subject (Levy, Trevarthen, & Sperry, 1972; Levy & Trevarthen, 1977). The bilateral chimeric stimuli, which we gave tachistoscopically for 150 msec, showed different half-objects, patterns or words to left and right of the vertical meridian of the subject's visual field. When commissurotomy patients were asked to say what they saw, they reported only the right side stimulus that had been seen by the left hemisphere. This hemisphere retains essentially normal powers of speech in these subjects while the right hemisphere is essentially mute. When, however, the patients were asked to point with either hand to a picture matching the stimulus in form or appearance, the left half of the chimera, seen by the right half brain, was chosen. In numerous tests we found we could switch the

swareness or retention from one side of the divided brain to the other by varying the task or cognitive set of the subject. Verbal report activated the left hemisphere, visual matching and pointing activated the right hemisphere. Here the factor influencing the outcome of a competition in visual awareness or in retrieval from a short term memory, was a mode of acting or reasoning elicited by verbal instruction.

Human beings are unique in having different kinds of cognitive process in the two cerebral hemispheres which regulate different kinds of performance. Early tests by Sperry and Gazzaniga (1967) had shown that verbal output was controlled almost entirely from the disconnected left hemisphere of commissurotomy patients. Levy (1974) and Nebes (1974) demonstrated, by comparing tactual awareness of left and right hands, that the right hemisphere disconnected from its partner has a superior ability to detect shapes. These findings are in agreement with a century of evidence on the effects of unilateral cortical lesions (Trevarthen, 1983b). Communication by speaking, writing, hand signs, or by any highly practised formal manual skill (such as typing or virtuoso piano playing) tends, in the majority of human beings, to involve the left hemisphere more than the right. This hemisphere also seems more in control of logical propositional sequencing, and even simple arbitrary temporal ordering of acts seems to be best understood by it. Kimura has proposed that there is a specialized motor sequencer related to manual signing in the left cortex (Kimura, 1979). The right side of the brain is superior at formulating how the body moves in an organized space or territory, for initial adjustment to the spatial configuration of objects and for recognising the configuration of shapes that are hard to break up conceptually into elements, including recognition of individuals from their faces. The right hemisphere also appears to be best equipped for perceiving emotions and it may dominate in regulation of interpersonal contacts by emotion. These high level mental operations which tend to have unequal representation in the hemispheres are all slow to mature and highly sensitive to training or education, but the anatomical basis for their left-right segregation would appear to be defined in fetal stages.

The most plausible theory of the evolution of hemispheric asymmetry of function is that it took root in the adaptation of cerebral circuits of the left hemisphere to direct indicative and propositional social signalling and protolinguistic communication by signing with the right hand. However, human cerebral asymmetry of function reflects the development of a number of parallel cognitive and expressive systems most of which have no obvious or direct relationships to control of one-handed movement. A majority of persons are right handed for signing movements and for manipulative skills of cultural value, but preference for the left hand for these activities, which occurs spontaneously in a sizeable fraction of normal humans is not usually associated with right hemisphere dominance for language. There appears to be great individual variation in slow-to-mature anatomical arrangements within the cerebral cortex, reflecting the freedom with which morphogenetic processes may generate the connective relations that determine cognitive learning in infancy and childhood. How endowment or genetic predispositions and education or human environment interact in this process remains a major question for brain science.

Levy and I found that cognitive sets of commissurotomy patients did not automatically favour the hemisphere best adapted to respond to the task presented (Levy & Trevarthen, 1976). Perceptual processing and decision making could be channelled into the relatively incompetent hemisphere. We called the directive process of this kind, setting hemisphericity of response, "metacontrol". It may involve brain stem attentional mechanisms which certainly contribute to the unity of voluntary action and awareness which commissurotomy patients retain in spite of the interruption of all the commissural nerve axons between neocortical and associated structures of the two hemispheres.

Sperry and I found that the visual system of the midbrain could contribute to an immediate awareness of the ambient spatial array in a commissurotomy patient (Trevarthen & Sperry, 1973). In our tests of unity of visual experience in these subjects we observed numerous instances of oscillating or fluctuating awareness, one or other half of the visual field assuming transitory dominance in cosciousness depending on the type of action to be made, or the occurrence of animated events in the field when the subject was more or less inert. In the test described above where a commissurotomy patient reported erasure or blotting out of the percept of a piece of white card, even when the card was in motion, the condition of stationary readiness to make a movement with one hand evidently prevented any bilateral unification of visual awareness for the target through the brain stem. This accords with the hypothesis that focal awareness associated with precision hand movements is essentially neocortical, while whole body displacement and large scale orientations are more likely to engage visuomotor systems below the hemispheres (Trevarthen, 1968).

The experiments with split brain monkeys and commissurotomy patients described above give us glimpses of shifts in consciousness associated with changes in motor initiative. All this may not deny that object identification, pattern recognition and message scanning processes are the result of progressive filtering of information in inward flow systems that extract features, count their relative representations and encode laws of association to define new categories of features. It does draw attention to the backstream of central integrations that pass judgment on the afferent data or change steps in the process of information uptake. If, as it would appear, these internal discharges can edit at the level of primary visual awareness, then they must be able to influence associative learning and the acquisition of new feature-extraction rules. They can be expected to play a directive role not only in moment-by-moment awareness of adults, but also in the development of consciousness or of learning to perceive throughout childhood (Trevarthen, 1979d).

3 First Stages in Postnatal Development of Biodynamic Structures of Humans

3.1 Action and Perception in Infancy

Research with infants has produced evidence that cognitive activity couples the processing of perceptual information to coordinated movements even in newborn infants (Trevarthen, Murray, & Hubley, 1981). Though profoundly immature and physiologically dependent, the newborn human can move in a coordinated way to select experiences. The evidence for this was scarce in pre-1960 research because theories of development had assumed that infants must begin life without mental representations or motives. They were thought to have no perceptual categories except those built into the receptor fields of reflexes. Observers were impressed with the motoric weakness of infants and the unpredictability of their reactions to imposed stimuli; they missed the small movements which constantly animate the receptors, changing their orientations and channelling experience. As long as infant subjects were required to be passive responders, as they were in the traditional experimental tests, confirmation was obtained for infantile incompetence, and also for a slow overcoming of shortcomings in awareness and the inferences they make as they enter childhood.

A change of tactics that allows young subjects to control situations in more normal ways has, in the past 20 years, changed the visibility of their spontaneous mental processes. With operant methods and variants of Piaget's "clinical" techniques even undergraduates are now able to demonstrate object awareness and reasoning in subjects too young to speak. Babies who cannot walk or reach for objects are found to use representations of time, space and object distinctness to direct their awareness, to control events and to regulate their expressive behaviour (e.g., Bower, 1974; Papousek, 1969).

It is true that newborns are often sleepy or withdrawn and individuals differ greatly in their altertness. But when a baby alerts to, fixates and tracks a nearby object or event, it ignores other stimulus changes that are just as large physically. Attention is most likely to be caught by effects who are changing or displacing, especially if the changes are animated by persons who are showing responsiveness or "contingency reactivity" to the baby's movements. A newborn may also show preferential fixation for static contrasts, textures or figures. The orienting movements frequently involve the whole body in a total pattern of displacement and they make adjustments adapted to overcome variation in the force of gravitation with the body axis in different positions. Since 1967 I have been making detailed recordings of how young infants move. I have found many regularities indicative of cerebral control at a high level (Trevarthen 1974d, e, 1982b, 1983c; Trevarthen, Hubley, & Sheeran, 1975).

A newborn can aim a well-formed reach-to-grasp movement towards an object that has been located visually, or to a sound. The movement has a cyclical form and regulated tempo. But this immature stereotypical "pre-reaching" is rarely projected far enough to touch an object. Contact follows if the object happens to be located where the reaching movement peaks and the object is sometimes grasped if it touches the hand at the right phase of the movement. Hand positioning and opening of the fingers are probably not adapted to visual information about object size and orientation. In spite of these limitations this rudimentary reaching is organised enough to show that an infant is equipped at birth to perceive the directions of displacement, radial location, distance and size of objects.

Selective awareness of objects is expressed in the choice of direction of fixation of head and eyes. These orientations require elaborate coupling of contractions in neck nuscles with activity of the extrinsic oculomotor muscles. Both saccadic (stepping) and smooth rotations of the eyes are seen from the moment of birth, the latter being triggered by relative displacement of the whole visual field or as compensatory back-rotations accompanying head turning. Infants under 15 weeks of age appear to be unable to track a displacing object with smooth eye rotation. The eyes jump after the object by a chain of saccades. The saccades of newborns are conjugate (yoking the two eyes) and have a velocity profile remarkably close to the oculomotor saccades of adults. Statistical analysis of the distribution of saccades in time shows that they are spaced; that is, they are triggered by an internal neural pacemaker or relaxation oscillator which may be driven at a maximum rate of three per second, which is the same as the peak tempo of saccades exhibited by an adult who is exploring a moderately large field (Trevarthen, 1974d, e; Trevarthen, Hubley, & Sheeran, 1975).

Pre-reaching movements and head turning also manifest pulsing or surging at 3 to 4 per sec (Trevarthen, Murray, & Hubley, 1981). The surges of arm displacement are frequently coupled in synchrony with steps of the eyes. Thus the infant's gaze is displaced in intimate association with extensions of the hands. The common time base for steps or orientation breaks up contact with environmental stimuli into packets. Contacts of different receptor members with the common object are automatically grouped and synchronized, creating predictable moments of focal attending.

Developments in looking and reaching behaviour in the first few months of an infant's life show that improvements in perception of objects outside the body are linked to growth of motor structures in the brain and to improved proprioceptive detection of the inertial forces that are created inside the body by displacements of trunk, head and limbs. The maturation of variable, guided arm extension and coordinated grasping of the hand under visual control in the first six months shows that antigravity support systems have a preset epoch of growth between 3 and 6 months when there is a dramatic improvement in the use of proximal and axial motor systems throughout the body of the baby (Trevarthen, 1982b, 1983c). From the start of normally elicited reaching, the movements to intercept displacing objects show evidence of anticipatory adjustment to the object's velocity (von Hofsten, 1979). Distal adjustments, particularly anticipatory adjustment of fingers to the visually perceived form of the object or to guide precision grip of fine details or small particles, start to mature significantly at a later time, in months 6–9. Thus different parts of the prehensile mechanisms of a baby attain functional efficiency at different ages. This would appear to be a manifestation of a general principle of development that achieves integration of voluntary behaviour by pitting different components of the perceptuomotor apparatus against each other inside the brain. Improvements in sensitivity of receptors and in sensory acuity appear to be linked to these motor developments (Trevarthen, Murray, & Hubley, 1981).

3.2 Laws of Perceptuo-Motor Control Related to Body Form and Implications for Development

Bernstein's observations of locomotion and tool use showed him that the distal parts of limbs were more complex to control and subject to strong centrifugal or whipping forces when proximal segments are in rapid movement (Bernstein, 1967; Trevarthen, 1983c). In stepping, jumping and running, and also in reaching to grasp or catch, a precise coordination of rotations in distal and proximal limb segments is therefore required to shield the distal components from injury when under high velocity or transportation by proximal rotations. These principles lead behaviour to be divided into epochs of two kinds: attention to detail, requiring stabilisation of proximal members, alternates with large scale movement during which there must be adjustments to protect the organs of focal attending (Trevarthen, 1978).

Distal and proximal systems have different functions in relation to the exploration of objects. Feeling movements of the hands have temporal and sequential features like those of the facial organs of exploratory awareness – the eyes, nostrils, lips and tongue. Each of these parts of the body may take up an instantaneous feedback of information about surfaces, forms, composition and constitution of objects, and they may exert enough force to grasp, displace or change objects. Eyes, ears and nose oriented by head rotations, collect information at a distance; they require no mechanical contact and have absolutely no effect on the object of interest, unless it is a sentient being, aware of being observed. Eye saccades may rotate the visual image on the retina while the rest of the body is stationary. Smooth pursuit eye movements may stabilize the image while the head is in motion, or track an object in motion. All exploratory or attentional movements may be ballistic and saccadic, in which case they express the tempo, serial ordering and spatial extension of scanning programs, some, at least, of which are built into the brain before birth.

Orienting and focalizing movements must be capable of precise coordination with movements of locomotion or prehension which require inertial forces of heavy parts of the body, or loading forces from heavy objects attached to the body, to be subdued or incorporated into the movement plan. Thus spatial and temporal parameters of exploratory perception need to fit the dynamic frequencies of whole body action. These requirements for coordination explain why infants have an integrated and preprogrammed readiness to orient the modalities of special sense within one egocentric behavioural and experiential frame that has one time base. Furthermore, conspicuous developments in both sensory acuity (Held, 1979) and object cognition (Bower, 1974) in infancy are closely attendant on maturational changes in the coordination and control of the proximal segments that affect orientation. This explains why focal attention to detail, which is necessary for efficient identification of objects remote from the body, develops in tandem with the infant's mastery of body support, turning and reaching.

Changes in the perceptuomotor coordination in development can now be related to developments in anatomically distinct central motor systems. Proximal and distal musculature are controlled by different parallel neural systems at all levels of the neuraxis (Kuypers, 1973). These systems have different developmental schedules.

In every area of activity, infants showed a predictable life history of motor control. Walking follows a similar pattern to reaching but matures later. The basic total pattern is seen in newborn stepping which is comparable to prereaching, then there is progressive mastery of control over the forces in an upright body that is to be propelled by placing the feet to obtain support while moving over irregular surfaces and around obstacles. Descriptions of motor development such as those of Gesell (Gesell, Thompson, & Armatruda, 1934) and McGraw (1963) contain a wealth of data on the significant changes. Studies of the effects of sensory defect, such as blindness, or of periods of enforced immobilization of limbs on the development of infants confirm that the capactiy to acquire precise sensory-motor coordination is based on an internally regulated time-table of growth in coordinative systems.

Psychologists have tended to think of motor development as "physical" and independent of cognitive growth which is "mental". They see cognition progressively internalizing learned representations of phenomena outside the body. But "non-linearities" of development in both performative and cognitive areas cannot be explained by the concept of associative learning and combination of reflex sensory-motor elements. Features of the developmental profile such as regressions, repetitions (decalages) and sudden emergences give evidence for morphogenetic processes working in cerebral structures in some degree of independence from experience, and for parallel systems of control competing within the body as they generate new levels of integrative functioning (Trevarthen, 1982b).

The growth strategy by which the infant's biodynamic structures "feed" on the environment and accommodate to changing body mechanics defines limits for the development of cognitive images, programs or "concepts" in the higher levels of the infant's brain. The "object concept" may have an autonomous vitality in the infant's mind and become larger as consciousness expands by reorganisations of representational schemes without intervention of testing movements, and in some respects perception of reality may anticipate manipulation and the performance of consumatory acts in its knowing the identity or "meaning" of things in the exterior world. Nevertheless, the development of perception must show effects of adjustments between cerebral motor components. These adjustments are not, as Piaget (1953) and Baldwin (1894) assumed, entirely the outcome of a need to repeat circular reactions that feed stimuli back from movements starting as reflex responses.

3.3 Communication in Infancy

Newborn infants make movements that are specifically fitted to communication with persons. This class of movements differs in important respects from those Bernstein considered, but they, too, must obey the general laws of coordination and regulation. They must be formulated to assimilate information from an appropriate perceptual field. Developmental psychologists have, since the 1930s, strongly resisted any concept of special innate potentialities for communication, but research of the past 15 years has opened up an entirely new perspective. Indeed, infants appear to have rudiments of all forms of communicative expression including some which are uniquely human.

Cooperative behaviour depends upon a system of signs and signal movements, plus an ability to perceive messages in these that trigger complementary social responses. All communication processes that lead to intelligent cooperation, and not just to synchronization of behaviour or congregation of individuals, depend on representational structures in one subject that can detect and distinguish signals by which to locate action of another subject and identify its causes. Human communication serves cooperative cultural life which is in all respects more complex than that of animals. However it originates in evolution, human communication in families and in society requires an awareness of purposive, self-expressive and cooperative motives in other persons, all inferred from the movements they make.

Interpersonal engagements in adult life are regulated by emotions which are expressed by much the same behaviours in all human societies. Facial, vocal, gestural and postural expressions excite a sympathetic or complementary emotional state in another person. Whereas locomotion and object prehension must manage a physical engagement between the body and the environment by a direct mechanical coupling, interpersonal communication manages a mental engagement mediated by signs of motive state (Trevarthen, 1979b, d, 1982a). Humans can perceive one another's interior state of awareness and readiness to act from certain movements of the face and vocal tract, probably all evolved from self-regulatory movements, which elaborately signal a person's central psychic processing. The obvious adaptive value of such a remarkable transfer of information between cerebral systems in different individuals is to promote a level of conscious cooperative action of great delicacy, power and efficiency.

Human communication also involves many non-affective kinds of expressive movement. Mutual awareness is guided and synchronized by an exchange of visual orientations, changes of body posture, reaching or manipulating movements and steps of locomotion. Complicated gesticulatory movements of the hands and articulated vocalizations have virtually unlimited potential for elaboration into conventional signs of a language. The communicative power of these movements derives from the ability of humans to perceive others as actors. In other words, human beings have a faculty for detecting human movements and for inferring the intentions and motivation behind them.

Experiments with event perception give glimpses of the special perception processes behind this awareness of human agency. For example, Michotte's classical studies of causality revealed a natural tendency of subjects to perceive certain categories of velocity relation or temporal contingency, between displacements of simple meaningless stimuli, as "animated" and "purposive" or "intelligently reactive" (Michotte, 1962). Some forms of interaction were perceived as expressive of emotions and they excited specific emotions in observers by "empathy" (Michotte, 1950). Johannson's studies of perception of human action reduced to filmed patterns of moving lights attached to limb segments show that these dynamic configurations are compellingly perceived as persons of identified gender and age, moving with appropriate vitality, grace and energy (Johansson, 1975). Of course, learned conventions of gesture and emotional expression by voice and facial movement guide adults in such inferences about the feelings and actions of others, and language adds immeasurably to precision of reference to both subjective and communicative states and objective reality. Nevertheless, the perception of human agency from its motoric roots is direct and compelling in the human mind and it would appear to be formulated prior to the learning of social conventions or language.

The evolution of more elaborate signalling systems that can transmit information about the specific contents of consciousness and about a subject's intentions to act in particular ways in relation to the world shared with other subjects leads to language. However, all factual, propositional or denotative communication rests upon a fundamental understanding of interpersonal relations and of the motivations in other persons with respect to each other. This intersubjective component of "acts of meaning" is present to a remarkable extent in the communicative actions of young infants.

In infancy there is obviously a special, and rather peculiar, need for cooperation. Infant and mother must engage in collaborative support of the infant's vulnerable life system and this caretaking aspect of infantile responses and signals is widely recognized. It is less widely understood that from birth infants also manifest readiness to engage in affectively controlled interaction of a kind that has no immediate relation to caretaking and that will lead to sharing experience of objects and of purposes behind actions (Trevarthen, 1977, 1979b, 1983a; Trevarthen & Hubley, 1978). The expression of affect is a necessary component of this prelinguistic communication, for it controls the delicate fitting together of cognitive and purposive processes without which language learning would not be possible.

Slowed down and subjected to microanalysis the interaction of a mother and her infant is revealed to be like a skilled ballet. Each watches and adjusts to the other until they can act together to express in unison and alternation with a joint control over their actions. As their relationship consolidates and develops, the mother may be observed to contribute her superior capacity for action to extend progressively the interaction of infant with herself and with the environment they share. This is an unconscious adaptation to the infant's expressions of interest and disinterest or pleasure and displeasure which is often described as teaching the infant how to communicate. It is more accurate to say that infant and mother share control of a specialized kind of communication rich in affective information. The infant repeatedly breaks away from the maternal guidance to try out new levels of autonomous control in interaction with the impersonal or physical world. Developments over the first year follow a clear pattern which is an expression of changing motivation in the infant, as well as of perceptuo-motor learning.

Infants exhibit movements that are pre-adapted to all forms of human interaction long before they speak. Expressivity in communication changes with age, not so much because emotions develop, but because the motives underlying communication with other persons are transformed as the infant becomes more aware of the world of objects and more involved in joint action on this world (Trevarthen, 1982a). The close companions of an infant, who constitute a vital human environment, must change their communications to fit these developments. Evidence for this conclusion comes from recent descriptive studies.

In early weeks an infant is responsive to the emotional tone of a person who approaches or picks the baby up. That a young infant is aware of faces and face movements as equivalent to their own expressive states is shown by tests of imitation. A newborn may imitate mouth opening or tongue protrusion and grimaces resembling expressions of joy, sadness or surprise (Maratos, Note 1; Meltzoff & Moore, 1977; Field, Woodson, Greenberg, & Cohen, 1982). Such automatic reflection of an exaggerated and isolated expressive movement (and newborns may also crudely imitate vocalizations and hand movements) seems to be a by-product of the cognitive apparatus on which natural reciprocal communication is based.

Normally newborns do not act to sustain face-to-face communication. They are often deliberately avoidant of this form of interaction. By two months, however, most infants are capable of sustained and highly responsive eye-to-eye orientation and reciprocal communication of affective facial, vocal and gestural signals with an attentive adult. Eye-to-eye contact has been thought due to some distinctive physical characteristics of the eye as a stimulus, but this is a tautological hypothesis. The infant's behaviour is compatible with the interpretation that visual awareness of eyes is in some sense an "imitation" or, rather, a "mirroring" by the baby of another person. A 2-month-old acts as if finding a mother's eyes looking, and her affectionate vocalizations, mouth openings, head movements, raised eye-brows and smiling to be both exciting and emotive. In response the baby expresses alternate concentrated wariness and smiling recognition (Oster, 1978), with periodic outbursts of intense though poorly organised gesticulatory movements. The signs of interest and smiling on the baby's part stimulate regular cycles of affectionate speech by the mother, "baby talk" that is marked by repetition and musical tempo and cadence, with head movements and caresses closely synchronized and parallel in expressive tone. A "protoconversational" interaction takes place in which the infant contributes a wide variety of communicative movements (Trevarthen, 1979b). If the mother becomes negative or unresponsive while interacting in this way, a 2-month-old shows immediate distress (Murray, Note 2; Trevarthen, 1983a; Trevarthen, Murray, & Hubley, 1981). Experiments show that the affective aspect of direct communication between a mother and infant is dependent on both of them being aware of what human feelings are appropriate to a positive interaction, and that it requires a particular timing in affectionate support from the partner (Murray, Note 2).

After 3 months, as the infant gains improved control of reaching and better support for the head, exploratory and performative behaviour that is increasingly sensitive to the presence of objects and occurrence of events in the nearby environment competes with and displaces orientation to the mother's face. The infant frequently shows a preference for looking away from the mother to impersonal surroundings (Sylvester-Bradley & Trevarthen, 1978; Trevarthen, 1983a). Communication is aided, however, by an increasing sense of play which a partner may exploit with systematic teasing (Trevarthen & Hubley, 1978; Trevarthen, 1983a). There is a growing interest in what mothers do with their hands, interest in hands appearing to increase as interest in the mother's face wanes.

Experimental studies with techniques that connect the infant to some kind of reactive servo device show that infants of this age (3–6 months) are strongly attracted to any events contingent on their efforts to move (Watson, 1977). Moreover, reactions to the success of failure of movements include signs of pleasure or displeasure (Papousek, 1969). This affective component of the response of an infant gives incentive and guidance to a human partner, and it leads to the presentation of games. With infants over 3–4 months of age, intimate play normally includes chanting of action rhymes and a "surprising" pattern of action, as in "Peek-aboo", "Round and round the garden like a teddy bear . . .", "Hoppe, hoppe Reiter", etc. The infants orient expectantly to an approaching assault or surprise and laugh after the climax. The dance-like rhythmical pattern of these action games indicates that their motivating force is derived from motor control. The beat is related to the intrinsic pulsing of exploratory and performatory movements described for neonates (Trevarthen, 1983c).

Developments in expression of affect and changes in reaction to approaches by unfamiliar persons indicate that around the middle of the first year infants become increasingly self-aware. A baby of this age pays increasing attention to his or her mirror image and also looks intently at other babies. Evidence has been presented that 6- to 9-month-olds have a preference to look at infants of the same sex (Lewis & Brooks, 1978). Thus it would appear that the child is becoming increasingly aware of the special features of the human environment that relate most directly to a self-image. None of the evidence would suggest that the infant is incapable of perceiving self distinct from the outside world or other persons before this age; indeed, adaptive interactions with persons or objects seen at 2 or 3 months would require such a cognitive differentiation. But the second half of the first year is a time of significant growth in awareness of social interactions and in definition of relationships. In a sense the child develops a much more critical dependence on familiar individuals and the particular form of communication that is possible with them.

By this age communication with familiars is naturally affectionate, subtle and skilled. It is founded on habits of play and an awareness of gestures and expressions in situations that are shared with the same persons every day. Moreover, the infant employs negative affect, anger, sadness or avoidance and withdrawal with increasing effect, to limit and direct interactions in ways that are adjusted to the familiarity of partners. By this means the infant asserts a new degree of independence and mastery in relation to familiars. An increased suspicion or fear of strangers which many infants exhibit at 7–9 months of age would appear to be a consequence of an increased awareness of dependence on those with whom a close relationship has formed. More negative reactions to mocking or teasing and more violent reaction to competition with siblings at this age may have a related explanation.

At about 40 weeks after birth infants change radically in their capacity to adjust their exploratory and performatory actions to interests and activities of others. By vocalization, gesture and shifts of gaze between persons and surroundings they increasingly express feelings, interest or purpose to others about events, objects and what others are doing. These "acts of meaning" support a vocal "proto-language" with many formal features of language but without verbalization (Halliday, 1975). Modulation of voice by the infant is appropriate to communicative intent and the speech of the adults who adress the baby, especially that of highly familiar adults who have an established communicative relationship to the child, is clearly perceived by the child to signal their intentions and interests.

A simple test of cooperative object use reveals that at this age a fundamental integrative process has occurred in the child's mind which permits sharing a task (Trevarthen & Hubley, 1978; Hubley & Trevarthen, 1979). If mother and infant are presented with a set of objects that may be combined to represent some construction or activity, such as a model truck and a set of small wooden dolls, the baby of 9–10 months increasingly looks at a mother and listens to her speech when she attempts to direct the manipulation. Prior to this time the infant's manipulation of the object was playful and self-absorbed. Now it begins to be deliberately shared with the mother who succeeds in partly directing a joint or cooperative manipulation (e.g., putting the dolls in the truck). Analysis of mothers' speech to infants in this situation reveals that the change in infants' awareness of joint action leads to a transformation in maternal speech (Trevarthen, 1983a, c). Questions, playful devices to attract attention and alerting statements decline and quiet commands increase. This change reveals again that communication of mothers with infants is constantly adapting to complement changes in infants' motivation.

In the second year, cooperative awareness of the familiar shared environment leads the child into imaginative exploitation of objects, places, occasions and personalities. Most research with this age group has been concerned either with Piaget's concept of the development, in the child's private world of isolated action, of "deferred imitation" and "representative causality", or with the emergence of speech. There is need for a deliberate investigation of the child's activities within the social and cultural context of the home and immediate community. By 18 months, children in all cultures are beginning to show a marked interest in customary instruments and tasks, in clothing and etiquette, in rituals and occasions. As they pay increasingly critical notice to what is said to them they begin to utter words to designate objects of common interest, persons with whom they interact and actions these persons are performing or wish to be performed.

The transition from protolanguage to language evidently follows upon a psychological transformation in the child's awareness of things that are defined by their customary uses or purposes. Thus infancy is brought to a close by the child voluntarily sharing the artificial world which constantly occupies the persons around him. I believe it will be impossible to understand this development without a sound theory of how infants become conscious of their own actions always in context of interaction with other persons. The motor coordinations responsible for this cognitive development are necessarily cooperative. If we are to understand them we will have to formulate a theory of how a voluntary and aware agent may construct a representation of the different volition and awareness of another similar agent.

4 Conclusions

Cognition is dependent on motor coordination; for the generation of primary experience, for the regulation and isolation of unambiguous afferent information, for controlled exploration to find new natural and dependable features of objects, for the coordination of modalities and for the discovery of the affordances of objects and situations for motor exploitation. In addition, human awareness, more than that of any other organism, must master the translation of messages conveyed by movements of one individual to another. Interpersonal or intersubjective awareness must originate in a process of recognition that matches received information about movements of others to the subject's own inherent motivating structure.

Coordinated voluntary action requires that the body be controlled by a neuronal system that represents all the parts of the body in relation to one another. Bernstein's theory of how movements are coordinated and regulated is important because it clarifies how integrated bodily activity requires adjustment of efferent signals to information about the body and its displacements, both from inside the body and from exteroceptive data about the surrounding world. We must somehow extend this conception of a neural representation of the body as a motor system with its envelopes of afferent input to the idea of a detector in each individual for other similar active bodies. Evidently the active and emotional self causes awareness of agency in others and responds to their feelings.

Observational and experimental research on infants shows that important ingredients of the coordinating machinery of the body and the perceptual apparatus which relates sensory input to body actions are innate. It also shows that development of both cognitive processes and skill in movement relies upon a systematic engagement of an active subject with an informative environment. Control of action by perception is certainly acquired in the sense that entirely new situations are assimilated and prior by correction and enlargement of prior impulses to act and prior perceptual categorizations. But basic categories are established by antenatal developments in the brain.

The studies of infants also shows that the cerebral structures for interpersonal communication are precociously developed, before schemas for manipulated objects, and that they enlist actions and awareness of adults to the aid of the mental development of the baby. It is likely that the greater part of the infant brain is devoted to interpretation of and response to the expressive meaning in actions of other persons. The development of structure and of the expression of structure in this component of the cerebral system for motor control is dependent on the formation of a stable and developing set of relationships with other persons (attachments) in which strategies for first affective and later practical cooperation are exercised and perfected (Trevarthen, 1980b, 1983a). The development of games played between mothers and infants illustrates clearly how the inherent motives of infants control the needed human environment, and how an infant's expressions of playfulness and curiosity stimulate an appropriate course of instruction. This instruction evidently stimulates the growth and development of the biodynamic structures of human cooperative awareness.

Once the infant has crossed the threshold into realization of genuine joint action in tasks in which objects are managed by the baby and another person making complementary moves, each recognising the purposes of the other, then the way is open for acquisition of conventions of object use (Hubley & Trevarthen, 1979; Trevarthen, 1979c, 1980b). This in turn leads naturally to a great expansion of signalling about awareness of the identity and potentialities of objects, events and personalities, a class of behaviours for which the infant has shown a latent readiness in preceding months.

Against this background, the emergence of language may be seen as an unfolding and enriching of biodynamic structures or motives specifically adapted for sharing with others mental control over the interface between the body and a world of useable objects. Shared concept and planned joint action on the world result in the creation of an endlessly elaborated artificial environment of culture in which human beings thrive as fish swim in the sea or birds fly in the air. By the end of infancy, as the child begins to acquire a vocabulary of the mother tongue, his or her involvement in the conventions of the culture is clear and pervasive.

The complex innate patterns of movement in infants and the way they develop in relation to each other express elaborate cerebral motive processes (Trevarthen, 1982a). These processes develop towards superior levels of integrative function by means of a competition for functional space in the brain networks (Trevarthen, 1982b). It is characteristic of human mental development that processes regulating interpersonal relations between the child and others have a primary role in the determination of how both motoric and cognitive skills develop (Trevarthen, 1983a).

Acknowledgements. Work with infants discussed in this chapter has been supported by the Medical Research Council and Social Science Research Council of the U.K., and the Spencer Foundation of Chicago. I am indebted to Janet Panther for help in preparing the manuscript.

Reference Notes

1. Maratos, I. *The origin and development of imitation in the first six months of life.* Ph. D. Thesis, University of Geneva, 1973.
2. Murray, L. *The sensitivities and expressive capacities of young infants in communication with their mothers.* Ph. D. Thesis, University of Edinburgh, 1980.

References

Baldwin, J.M. *Mental development in the child and the race.* New York: Macmillan, 1894.

Bernstein, N. *The coordination and regulation of movements.* Oxford: Pergamon, 1967.

Bower, T.G.R. *Development in infancy.* San Francisco: Freeman, 1974.

Field, T.M., Woodson, R., Greenberg, R., & Cohen, D. Discrimination and imitation of facial expressions by neonates. *Science,* 1982, *218,* 179–181.

Gesell, A.L., Thompson, H., & Armatruda, C.S. *Infant behaviour: Its genesis and growth.* New York: McGraw Hill, 1934.

Gibson, J.J. *The senses considered as perceptual systems.* Boston: Houghton-Mifflin, 1966.

Gibson, J.J. *The ecological approach to visual perception.* Boston: Houghton-Mifflin, 1979.

Grillner, S. Locomotion in vertebrates: central mechanisms and reflex interaction. *Physiological Reviews,* 1975, *55,* 247–303.

Halliday, M.A.K. *Learning how to mean: Exploration in the development of language.* London: Arnold, 1975.

Hebb, D.C. *The organisation of behaviour.* New York: Wiley, 1949.

Held, R. Exposure history as a factor in maintaining stability of perception and coordination. *Journal of Nervous and Mental Diseases,* 1961, *132,* 26–32.

Held, R. Dissociation of visual functions by deprivation and rearrangement. *Psychologische Forschung,* 1968, *31,* 338–348.

Held, R. Development of visual resolution. *Canadian Journal of Psychology/Review of Canadian Psychology,* 1979, *33,* 213–221.

von Hofsten, C. Development of visually directed reaching: The approach phase. *Journal of Human Movement Studies,* 1979, *5,* 160–178.

von Holst, E. Relations between the central nervous system and the peripheral organs. *British Journal of Animal Behaviour,* 1954, *2,* 89–94.

von Holst, E., & Mittelstaedt, H. Das Reafferenzprincip. *Naturwissenschaft,* 1950, *34,* 464–476.

Hubley, P., & Trevarthen, C. Sharing a task in infancy. In I. Uzgiris (Ed.), *Social interaction during infancy, new directions for child development* (Vol. 4, pp. 57–80). San Francisco: Jossey-Bass, 1979.

Ingle, D. Two visual mechanisms underlying the behaviour of fish. *Psychologische Forschung,* 1967, *31,* 44–51.

Johansson, G. Visual motion perception. *Scientific American,* 1975, *323,* (6), 76–88.

Kimura, D. Neuromotor mechanisms in the evolution of human communication. In H.D. Steklis & M.H. Raleigh (Eds.), *Neurobiology of social communication in primates.* New York: Academic Press, 1979.

Kuypers, H.G.J.M. The anatomical organisation of the descending pathways and their contributions to motor control, especially in primates. In T.E. Desmedt (Ed.), *New developments in E.E.G. and clinical neuropsychology* (Vol. 3, pp. 38–68). Basel: Karger, 1973.

Lewis, M., & Brooks, J. Self-knowledge and emotional development. In M. Lewis & L.A. Rosenblum (Eds.), *The development of affect* (pp. 205–225). New York: Wiley, 1978.

Levy, J. Cerebral asymmetries as manifested in split-brain man. In M. Kinsbourne & W.L. Smith (Eds.), *Hemispheric disconnection and cerebral function.* Springfield, IL: Thomas, 1974.

Levy, J., & Trevarthen, C. Metacontrol of hemispheric function in human split-brain patients. *Journal of Experimental Psychology: Human Perception and Performance,* 1976, *2,* 299–312.

Levy, J., & Trevarthen, C. Perceptual, semantic and phonetic aspects of elementary language processes in split-brain patients. *Brain,* 1977, *100,* 105–118.

Levy, J., Trevarthen, C., & Sperry, R.W. Perception of bilateral chimeric figures following hemispheric disconnection. *Brain,* 1972, *96,* 61–78.

McGraw, M.M. *The neuromuscular maturation of the human infant.* New York, London: Hafner, 1963.

Meltzoff, A.N., & Moore, M.H. Imitation of facial and manual gestures by human neonates. *Science,* 1977, *198,* 75–78.

Michotte, A. The emotions regarded as functional connections. In M.L. Reymert (Ed.), *Feelings and emotions.* New York: McGraw Hill, 1950.

Michotte, A. *Causalité, permenance et réalité phenoménales.* Louvain: Publications Universitaires, 1962.

Nebes, R.D. Dominance of the minor hemisphere in commissurotomized man for the perception of part-whole relationships. In M. Kinsbourne & W.L. Smith (Eds.), *Hemispheric disconnection and cerebral function.* Springfield, IL: Thomas, 1974.

Oster, H. Facial expressions and affect development. In M. Lewis & L.A. Rosenblum (Eds.), *Affect development.* New York: Wiley, 1978.

Papousek, H. Individual differences in learned responses in human infants. In R.J. Robinson (Ed.), *Brain and early behaviour.* London, New York: Academic Press, 1969.

Piaget, J. *The origins of intelligence in the child.* London: Routledge and Kegan Paul, 1953.

Schneider, G.E. Contrasting visuomotor functions of tectum and cortex in the golden hamster. *Psychologische Forschung,* 1967, *31,* 52–62.

Sperry, R.W. Optic nerve regeneration with return of vision in anurans. *Journal of Neurophysiology,* 1944, *7,* 57–70.

Sperry, R.W. Neural basis of the spontaneous optokinetic response produced by visual inversion. *Journal of Comparative and Physiological Psychology,* 1950, *43,* 482–489.

Sperry, R.W. Neurology and the mid-brain problem. *American Scientist,* 1952, *40,* 291–312.

Sperry, R.W. Chemoaffinity in the orderly growth of neural circuits. *Proceedings of the National Academy of Sciences,* 1963, *50,* 703–710.

Sperry, R.W., & Gazzaniga, M.S. Language following surgical disconnection of the commissures. In F.L. Darley (Ed.), *Brain mechanisms underlying speech and language.* New York: Grune & Stratton, 1967.

Sylvester-Bradley, B., & Trevarthen, C. "Baby-talk" as an adaptation to the infant's communication. In N. Waterson & K. Snow (Eds.), *Development of communication: Social and pragmatic factors in language acquisition.* London: Wiley, 1978.

Teuber, H.L. Perception. In T. Field, H.W. Magoun, & V.E. Hall (Eds.), *Handbook of physiology: Neurophysiology* (Vol. 3, pp 1595–1668). Washington: American Physiological Society, 1961.

Trevarthen, C. Double visual learning in split-brain monkeys. *Science,* 1962, *136,* 258–259.

Trevarthen, C. Functional interactions between the cerebral hemispheres of the split-brain monkey. In E.G. Ettlinger (Ed.), *Function of the corpus callosum.* Ciba Foundation Study Group, No. 20. London: Churchill, 1965.

Trevarthen, C. Two mechanisms of vision in primates. *Psychologische Forschung,* 1968, *31,* 299–337.

Trevarthen, C. Behavioral embryology. In E.C. Carterette & M.P. Friedman (Eds.), *Handbook of perception: Biology of perceptual systems* (Vol. 3, pp 89–117). New York: Academic Press, 1973.

Trevarthen, C. Cerebral embryology and the split-brain. In M. Kinsbourne & W.L. Smith (Eds.), *Hemispheric disconnection and cerebral function* (pp. 208–236). Springfield, IL: Thomas, 1974. (a)

Trevarthen, C. Analysis of cerebral activities that generate and regulate consciousness in commissurotomy patients. In S.J. Diamond & J.G. Beaumont (Eds.), *Hemisphere function in the human brain* (pp. 235–263). London: Paul Elek, 1974. (b)

Trevarthen, C. Functional relations of disconnected hemispheres with the brain stem and with each other: Monkey and man. In M. Kinsbourne & W.L. Smith (eds.), *Hemispheric disconnection and cerebral function* (pp 187–207). Springfield, IL: Thomas, 1974. (c)

Trevarthen, C. L'action das l'espace et la perception del'espace: Mecanisms cérébraux de base. In F. Bresson et al. (Eds.), *De l'espace corporel à l'espace écologique* (pp. 65–80). Paris: Presses Universitaires de France, 1974. (d)

Trevarthen, C. The psychobiology of speech development. In E.H. Lenneberg (Ed.), *Language and brain: Developmental aspects. Neursciences Research Program Bulletin* (Vol. 12, pp 570–585). Boston: Neurosciences Research Program, 1974. (e)

Trevarthen, C. Descriptive analyses of infant communication behaviour. In H.R. Schaffer (Ed.), *Studies in mother-infant interaction: The Loch Lomond Symposium* (pp 227–270). London: Academic Press, 1977.

Trevarthen, C. Modes of perceiving and modes of acting. In J.H. Pick (Ed.), *Psychological modes of perceiving and processing information* (pp. 99–136). Hillsdale, N.J.: Erlbaum, 1978.

Trevarthen, C. Neuroembryology and the development of perception. In F. Falkner & J.M. Tanner (Eds.), *Human growth: A comprehensive treatise* (Vol. 3, pp 3–96). New York: Plenum, 1979. (a)

Trevarthen, C. Communication and cooperation in early infancy. A description of primary intersubjectivity. In M. Bullowa (Ed.), *Before speech: The beginnings of human communication* (pp. 321–346). London: Cambridge University Press, 1979. (b)

Trevarthen, C. Insticts for human understanding and for cultural cooperation: their development in infancy. In M. von Cranach, K. Foppa, W. Lepenies, & D. Ploog (Eds.), *Human ethology* (pp. 530–571). Cambridge: Cambridge University Press, 1979. (c)

Trevarthen, C. The tasks of consciousness: how could the brain do them? *Brain and mind* (pp. 187–217). CIBA Foundation Symposium 69 (New Series). Amsterdam: Excerpta Medica, 1979. (d)

Trevarthen, C. Neurological development and the growth of psychological functions. In J. Sants (Ed.), *Developmental psychology and society* (pp. 46–95). London: Macmillan, 1980 (a)

Trevarthen, C. The foundations of intersubjectivity: Development of interpersonal and cooperative understanding in infants. In D. Olson (Ed.), *The social foundations of language and thought: Essay in honor of J.S. Bruner.* New York: Norton, 1980. (b)

Trevarthen, C. The primary motives for cooperative understanding. In G. Butterworth & P. Light (Eds.), *Social cognition: Studies of the development of understanding* (pp. 77–109). Brighton: Harvester, 1982. (a)

Trevarthen, C. Basic patterns of psychogenetic change in infancy. In T. Bever (Ed.), *Dips in learning.* Hillsdale, N.J.: Erlbaum, 1982. (b)

Trevarthen, C. Interpersonal abilities of infants and generators for transmission of language and culture. In A. Oliverio & M. Zapella (Eds.), *The behaviour of human infants.* London, New York: Plenum, 1983. (a)

Trevarthen, C. Hemispheric specialization. In I. Darian-Smith (Ed.), *Handbook of physiology: The nervous system* (Vol. 3). Washington: American Physiological Society, 1983. (b)

Trevarthen, C. How control of movements develops. In H.T.A. Whiting (Ed.), *Bernstein Reassessed.* Amsterdam: North Holland, 1983. (c)

Trevarthen, C., & Hubley, P. Secondary intersubjectivity: Confidence, confiding and acts of meaning in the first year. In A. Lock (Ed.), *Action, gesture and symbol.* New York: Academic Press, 1978.

Trevarthen, C., Hubley P., & Sherman, L. Les activités innées du nourrisson. *La Recherche*, 1975, *6*, 447–458.

Trevarthen, C., Murray, L., & Hubley, P. Psychology of infants. In J. Davis & J. Dobbing (Eds.), *Scientific foundations of clinical paediatrics* (2nd edn., pp. 211–274). London: Heinemann, 1981.

Trevarthen, C., & Sperry, R.W. Perceptual unity of the ambient visual field in human commissurotomy patients. *Brain*, 1973, *96*, 547–570.

Watson, J.S. Perception of contingency as a determinant of social responsiveness. In E.B. Thoman (Ed.), *Origins of the infant's social responsiveness.* Hillsdale, N.J.: Erlbaum, 1977.

Summary. Bernstein's concept of biodynamic structures of brain and body that actively coordinate movements in relation to peripheral factors clarifies the relationship between perception, cognition and motor control. It is also helpful in consideration of psychological development where it is necessary to trace how experience is incorporated into programs for movement.

Sperry's theory of inherent patterning of nerve circuits that regulate behaviour led him to a view of the relationship between movement and perception that is close to Bernstein's. Cases are discussed where the uptake of information in perception is powerfully influenced by preparations of the subject to move in a particular way. Studies of split-brain animals and of human commissurotomy patients provide illuminating examples.

Recent work on the development of motor coordinations in infants brings clear evidence for innate mechanisms that pattern movements before they become guided by perception, and that provide intrinsic constraints on experience. Ultimately, the origin of these mechanisms must be found in the embryology of cerebral systems.

In addition to movements pre-adapted to locomotion and prehension of objects, infants show elaborate communicative behaviour that expresses subtle aspects of motivation with respect to other human beings. The development of communication before language represents a particularly clear example of a biodynamic process achieving differentiation and functional efficiency through selective engagement with a reactive and supportive environment. By expressing emotions and responding to emotional expressions infants establish dynamic relationships with familiar caretakers. Development of other forms of signalling lead them into cooperative activities and shared understanding of objects, kinds of mental activity which are essential to the development of language.

21 Discontinuity in the Development of Motor Control in Children

L. HAY

Contents

1 Developmental Features of Goal-Directed Movements

The aim of this paper is to illustrate some aspects of discontinuity in the development of movement control in children, as exemplified by target aiming movements. Because this kind of movement is both simple and spatially oriented, it implicates both motor effector and receptor functions. The spatial-temporal parameters of this type of movement provide reliable evidence of the processes involved. In their fully developed form, the goal-directed movements comprise a programmed ballistic approach, generally followed by a visually guided "homing" onto the target. The ontogenetic priority of the ballistic movements over visual guidance is a well known fact in infants. They can perform successful ballistic reaching toward objects long before being able to guide their hand visually toward objects. The emergence of visual guidance, one of the most outstanding events in the development of movement control, supposes an increase in attentional capacity and leads to a higher flexibility in response. But the manner of controlling directed movements does not seem to be finally established even long after the emergence of visual guidance.

1.1 Spatial Accuracy

We shall consider the development of children's performance in a target aiming movement performed in the absence of vision of the hand (Hay, 1979). The task

Cognition and Motor Processes
Ed. by W. Prinz and A.F. Sanders

was to point with the finger as accurately as possible toward one of several targets which lit up randomly on a horizontal line in front of the subject. The subject pointed at the target with the right arm extended and moving horizontally from a starting point on the right side. Thus the target position was reached by adjusting the direction of the arm with a direct movement but without speed constrain. View of arm and hand was precluded by an opaque horizontal screen placed just under the targets. Children of 5, 7, 9 and 11 years of age were tested. Accuracy of movements as a function of age is presented in Figure 1, in terms of the mean constant error and in terms of the mean intraindividual variability. The lined areas represent the mean maximum range of error for each age group, that is the maximum intraindividual variability over each series of pointing movements. There is a systematic bias toward undershooting in all groups, but it is significantly higher at 7 and 9 years of age. An analysis of variance shows a significant effect of age ($P < 0.05$) and a significant quadratic component ($P < 0.01$). At 5 and 11, the mean accuracy is rather high and equal, but the intraindividual variability decreases considerably with age.

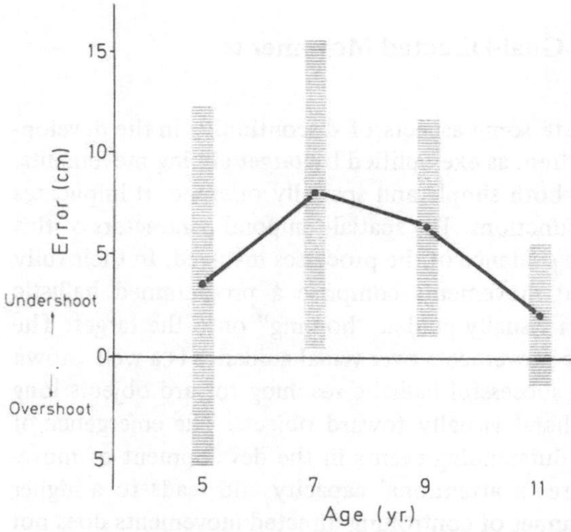

Figure 1. Accuracy of pointing without visual feedback as a function of age: mean constant error *(points)* and mean intraindividual range of errors *(lined areas).*

The non-monotonic development of this particular visuo-manual coordination seems to be related to the combined influence of two factors, as shown by other data. (1) Visuo-manual coordination is established by an arm movement, conversely to visual localization of the hand. In another experiment in which subjects were required to localize visually the final position of their unseen hand just after a movement, by stopping a visual index moving toward their finger tip, the mean constant spatial error did not change with age to a great extent, and particularly did not increase at 7, as it did in the present experiment. (2) The second factor is the lack of vision of the moving arm. When subjects could see their hand while adjusting their movement onto the target, the final accuracy did not change with age in a non-monotonic manner either, but it slightly increased, particularly between 5 and 7. So, regarding the present results, it seems that in a task involving an adjustment of movement onto a target, the lack of vision of the hand affects performance to various degrees according to age. The sudden increase in error at 7 suggests a change in the manner of movement control. It might be interpreted as an increased need for visual feedback control in order to achieve movements. On the other hand, the younger children's performance suggests a ballistic type of control which allows rather good performance on average although the intraindividual variability is rather high.

1.2 Timing

This non-monotonicity in age-related changes of constant error is also shown in an aspect of timing of the response. Figure 2 presents the distributions at the four ages of movement times for one of the movement amplitudes. The grey areas represent

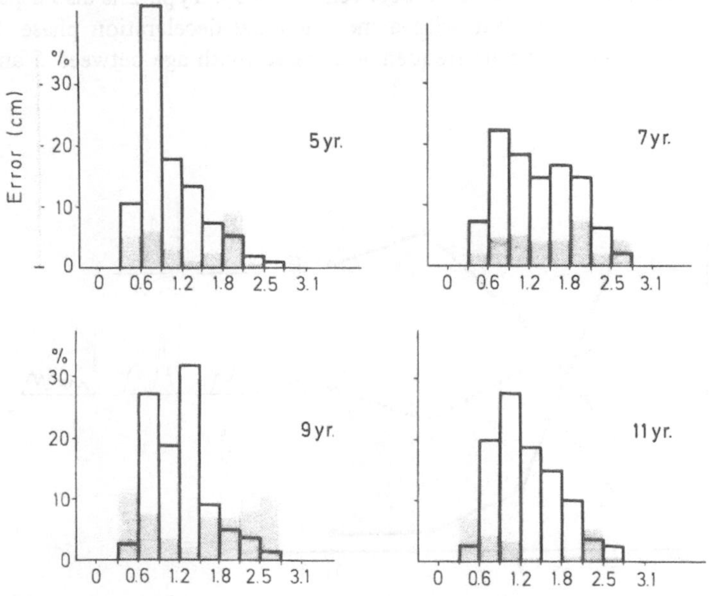

Figure 2. Movement time frequency (%) and related mean error of pointing *(grey areas)*, for each age group.

the mean error corresponding to each class of movement times. It must be emphasized that those movement times are not the optimal times but reflect spontaneous performance, the only requirement of the task being accuracy. On average, 5-year-old children trend to perform more rapid movements than older children in this task. But the main characteristic of the distribution of movement durations at this age is its sharp peak as opposed to the broad distribution of the 7-year-olds. However, the high reliability in movement time at age 5 does not correspond to optimal accuracy, as shown by a general negative correlation between accuracy and movement time frequency, suggesting a sort of stereotyped pattern. On the contrary, a more optimal strategy can be found at age 9 as shown by the positive correlation between accuracy and movement time frequency. This suggests a higher level of sensori-motor coordination.

1.3 Movement Patterns

If we consider the development of movement patterns, different types of patterns are encountered in children. The main types of velocity curves and their relative frequency in each age group are presented in Figure 3. Differences in frequency between groups were tested by means of the Chi-squared test. Type 1 is a ballistic-like pattern, with one maximum velocity and very sudden acceleration and deceleration. This pattern is very frequent at age 5, and rapidly decreases with age ($P < 0.001$ between 5 and 7; $P < 0.10$ between 7 and 9). Type 2 is also a pattern with one maximum velocity, but with a more gradual deceleration phase. This pattern is present in age 5, and its frequency increases with age between 5 and 9 ($P < 0.01$

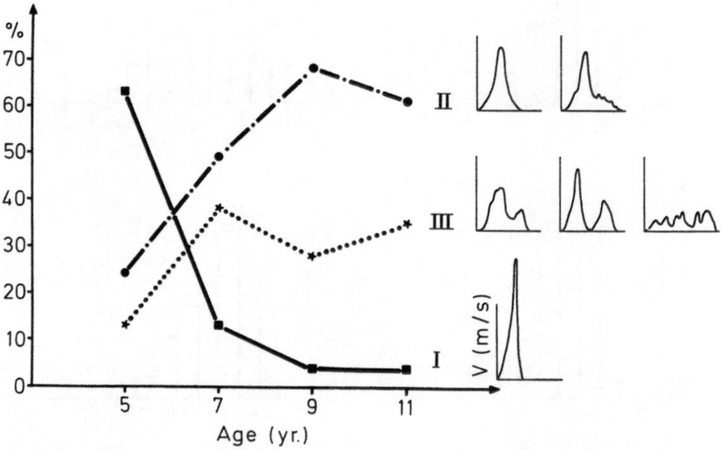

Figure 3. Percentage of each type of velocity patterns, as a function of age (288 movements per age).

between 5 and 7; $P < 0.10$ between 7 and 9). Type 3 comprises two sorts of patterns: (1) "step" movements, with several velocity maxima and several more or less pronounced accelerations and decelerations; (2) "ramp" movements with low and constant velocity and no acceleration. This pattern is rare at age 5; its frequency increases at 7 ($P < 0.001$), and remains at about the same level from 7 to 11.

It seems that there is a double developmental change in the patterns encountered during this age range, and in the type of movement control they express. First, the ballistic-like patterns of younger children are replaced by patterns involving a great amount of braking activity and feedback processing (proprioceptive feedback in this case, because visual feedback is not available), which occurs as continuous or discontinuous control (as shown by the two patterns of type 3). Second, feedback control becomes concentrated at the end of movements, which preserves the ballistic approach phase. (This second aspect is shown by the increase of type 2 movements which become dominant at 9 and 11 years.) So the feedback and feedforward components seem to be enhanced separately and successively during a phase of development, before being integrated into a smooth motor sequence, in which each of them takes part at the appropriate time, and so produces optimal efficiency.

2 Feedforward and Feedback Processes in Development

This phenomenon of instability in performance at age 7, manifested by various patterns (without a particular dominance of a pattern), by the variability in movement times, and by the spatial inaccuracy of movements, merits some comments. What underlies the instability and relative awkwardness of movements at this age? The cause does not seem to be related to any transitory disability in motor programming itself as suggested by results of another study of children from 6 to 11 years of age (Bard, Paillard, Bellec, & Fleury, 1982). They used another kind of aiming task which is simpler and chiefly involves programming activity. The task was to push a lever quickly in one of various directions indicated by luminous lines, without seeing the hand. The movements were just directional flinging of the hand and did not require any adjustment of movement in amplitude and velocity. Thus the task can be considered mainly a test of directional programming ability itself. The results show that directional error of movements decreases between 6 and 7 and does not change significantly afterwards, suggesting that programming ability, when operating separately, reaches its highest efficiency early in the age range under consideration here.

Thus the transitory awkwardness of movements seems to be related to the integration of feedback control and to the necessary modulation of velocity which accompanies use of feedback. These perturbations at age 7, and the subsequent development of control, could be compared with similar features of feedback processing during learning in adults. Studies have shown that in a visuo-motor task, there are privileged moments for integrating visual information in the system. These moments are rather frequent at the beginning of learning and their occurrence tends to perturb the motor sequence. For example in Pew's study (1966) in which subjects had to control an oscilloscope target by means of two keys, one producing target acceleration to the left and the other to the right, the effect of practice resulted in a decrease of feedback based responses. Learning then consists in choosing only the appropriate moments for visual control to occur, and in reducing its frequency of occurrence.

3 Visual Feedback Integration and Flexibility of Response

The adaptability of response under conditions of rich visual feedback can be illustrated by the sequential modification of movement trajectories performed under conditions of displaced vision. In one study (Hay, 1979) children had to perform a target aiming task, with displacement of the visual field including the target and hand trajectory. In this condition, the trajectory of the hand is bent, as it shifts from an incorrect initial direction (based on the virtual target position) to a correct direction (based on visual guidance of the hand in relation to the target). The extent of the visually corrected part of the trajectory was quantified by expressing the distance between the bend and the target, as a percentage of the total distance to be covered. The visually corrected part of the trajectory was averaged over the first three trials of the series, and submitted to an analysis of variance. The results showed that it tends to change with age ($P < 0.10$) with a significant cubic component ($P < 0.05$). It is longest at 7 and shortest at 5, as shown by Duncan's test at $P < 0.05$ threshold. In addition it should be noted that as the series of trials pro-

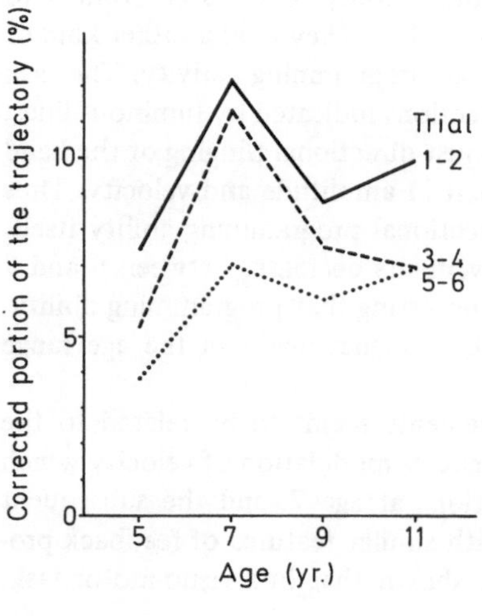

Figure 4. Visually corrected portion of pointing movement trajectories as a function of age and trial order in the series

gresses, the initial error in direction tends to decrease (according to an adaptative process), and the visually corrected part of trajectory also decreases in all groups. However, as shown in Figure 4, the greatest modification in response can be found in 7-year-old children, whose performance reaches the level of the older ones as early as the sixth trial, on average.

4 Non-Monotonicity in Early Development

Non-monotonic development in motor control processes has been noted frequently at earlier stages of development. In the case of reaching behaviour in infants, development looks like an alternation between periods when this action is performed as a whole such as in early neonate reaching, and other times when the different components are activated separately. For example, several authors (Amiel-Tison & Grenier, 1980; Bower, 1974; Trevarthen, Hubley, & Sheeran, 1975) report a type of very early reaching behaviour, visually elicited by objects, and even showing to some extent an anticipatory adaptation to the spatial characteristics of the objects. This early reaching, supposedly directed by an amodal sense of spatial direction, then generally disappears and is replaced by a top level reaching. But many components of later reaching style seem to be present in the young infant and to work separately. These include swiping with closed fist or grasping on the midline without approaching the object, while looking at the object. This particular feature of development was emphasized by Bruner and Koslowski (1972) who considered that the improvement of a skill such as reaching implies not only the differentiation and mastery of the constituent abilities (such as the visually elicited swiping, the visual guidance of the moving hand, the visually anticipated grasp patterns), but also implies their adequate serial ordering into an adapted and economical sequence. Furthermore breaking down an act into modular form is an essential preliminary to its inclusion in a more complete sequence.

This interpretation seems congruent with the description of development of motor control in children presented in this paper. Development includes an integration of both components of movement into the response. Another cycle of alternation between ballistic and feedback control has been found in a study of the regulation of grasp force relative to the weight of an object in children from 2 to 5 (Mounoud & Hauert, 1982). The spatial-temporal analysis of the movements used in supporting the object showed an alternate dominance of continuous and discontinuous patterns of movement during development.

A model involving systems theory might be adequate to account for such a developmental process and the related discontinuity in development. Using Bernstein's idea (Bernstein, 1967) that the achievement of control implies a reduction of or "mastery" over degrees of freedom in the action-system being regulated, Bruner asserted that mastery of visually guided activity in infancy and childhood implies a cycle of crude restriction of forms of movement, and of programmatic skill formation within the limits of that restriction. Skill moves to a next step only when restriction is altered. The child's progress depends on qualitative changes of skill, that is on formation of new streategies or programs of action, which involve an increase in degrees of freedom, and need subsequent consolidation and automatization (Bruner & Bruner, 1968).

The transitory disruptive character of a new control process may be exemplified by the saccadic approach to an object while glancing alternately from hand to object, as described by White, Castle, and Held (1964) during the period between 3 and 4.5 months. This period precedes the achievement of the top level reaching, with a smooth approach under visual control at 5 months. So visual control of the hand seems to have a disruptive influence on the motor sequence during the first stage of its integration to the control process of the on-going movement, suggesting an initial rivalry between both activities. Visual fixation of the hand has been classically considered a prerequisite for eye-hand coordination. But this rivalry and the existence of an early syncretic form of eye-hand coordination led some authors to deny this function of hand regard, which is rather considered as a factor of developing the control of the fine finger movements (Trevarthen, Hubley, & Sheeran, 1975).

5 Cognitive and Psychophysiological Concepts in Motor Development

In conclusion we would say that the results of the present studies of target aiming movements in children show a first phase in which the task is performed primarily on the basis of the subject's own internal organization. This is expressed in a learned movement pattern which can be modified in an adaptive way. Conversely, the subsequent relative awkwardness and flexibility of movement represents a second developmental phase when the subject tries to collect more information about the environment and about the ongoing action, in order to proceed with it. This receptivity to external information, in spite of its transitory perturbing influence, is a factor of subsequent improvement in performance, and also a factor of greater flexibility of responses. Paillard (1982) suggested a parallel between the flexibility allowed by feedback regulation and the assimilative capacities of the Piagetian sensorimotor schema. He identified the accommodative capacities of the schema with the feedforward process, because of the possible adaptive modification of the motor programs, i.e., the subject's internal organization. In one sense, the data presented here are an example of development in childhood which, to some extent, repeats the previous development in infancy. According to Paillard's view about Piagetian and psychophysiological concepts, this alternating dominance between the feedforward and feedback processes could be expressed in cognitive terms as an alternation between the accommodative (feedforward) and assimilative (feedback) capacities of the Piagetian schema.

References

Amiel-Tison, C., & Grenier, A. *Evaluation neurologique du nouveau-né et du nourrisson.* Paris: Masson, 1980.

Bard, C., Paillard, J., Bellec, J., & Fleury, M. Eye-head coordination for directional control of projecting aiming movement in children. *Psychology of motor behavior and sport.* North American Society for the Psychology of Sport and Physical Activity, University of Maryland, University Park, 1982.

Bernstein, N. *The coordination and regulation of movement.* New York: Pergamon Press, 1967.

Bower, T.G.R. *Development in infancy.* San Francisco: Freeman, 1974.

Bruner, J.S., & Bruner, B.M. On voluntary action and its hierarchical structure. *International Journal of Psychology,* 1968, *3,* 239–255.

Bruner, J.S., & Koslowski, B. Visually preadapted constituents of manipulatory action. *Perception,* 1972, *1,* 3–14.

Hay, L. Spatial-temporal analysis of movements in children: motor programs versus feedback in the development of reaching. *Journal Motor Behavior,* 1979, *11,* 189–200.

Mounoud, P., & Hauert, C.A. Development of sensori-motor organization in young children: grasping and lifting objects. In G. Forman (Ed.), *Action and thought: from sensori-motor schemes to symbolic operations.* New York: Academic Press, 1982.

Paillard, J. La psychophysiologie et l'oeuvre de Jean Piaget. *Archives de Psychologie,* 1982, *50,* 75–86.

Pew, R.W. Acquisition of hierarchical control over the temporal organization of a skill. *Journal of Experimental Psychology,* 1966, *71,* 764–771.

Trevarthen, C., Hubley, P., & Sheeran, L. Les activités innées du nourrison. *La Recherche,* 1975, *6,* 447–458.

White, B.L., Castle, P., & Held, R. Observations on the development of visually-directed reaching. *Child Development,* 1964, *35,* 349–364.

Summary. Some aspects of target aiming movements are described in children. The aiming task under consideration is a horizontal movement of the extended arm toward a target, without vision of the limb, with accuracy but no speed requirements. Age-related changes in spatial terminal accuracy, timing, and velocity patterns of movements, show a non-monotonic development between the ages of 5 and 11 years. These changes are interpreted as an alternate dominance of feedforward (5-year-olds) and feedback (7-year-olds) components of movement, followed in older children by integration of both processes into the motor sequence in which each of them takes part in an optimal manner. It is suggested that the transitory awkwardness of movements around 7 years of age is not related to a transitory disability in motor programming, but rather to the need of appropriate integration of feedback control and to the necessary modulation of velocity it involves. Flexibility of response allowed by rich visual feedback is exemplified by a study of goal-directed trajectories under prismatic vision, at the same ages. It is suggested that these data on children, as well as development of reaching style in infants, from neonate syncretic reaching to later "top-level" reaching, separated by a period of fragmented pre-reaching responses, are relevant to the notion of modularization of action as a developmental process. Another cycle of alternation between bal-

listic and feedback control is described in a study of the regulation of grasp force relative to the weight of an object between 2 and 5 years of age. Finally, a view drawing a parallel between psychophysiological feedforward and feedback concepts, and accommodative and assimilative capacities of the Piagetian schema, is mentioned.

Author Index

Page numbers in *italics* refer to the References

Subject Index